Patho
PHYSIOLOGY
for Medical and Allied Students

Patho
PHYSIOLOGY
for Medical and Allied Students

Mohan B Dikshit, MBBS, MD, Dip. Av. Med. FAMS

Notion Press

Old No. 38, New No. 6
McNichols Road, Chetpet
Chennai - 600 031

First Published by Notion Press 2016
Copyright © Mohan B Dikshit 2016
All Rights Reserved.

ISBN 978-1-945621-68-0

To the memory of my departed parents, Prof. BB Dikshit, my Father, renowned Medical scientist and educationist, teacher par excellence and above all a true leader of men, and Srimati Hirabai B Dikshit, my Mother, poetess, teacher and a tower of strength.

Table of Contents

Foreword . . . . *ix*

Preface . . . . *xi*

Acknowledgements . . . . *xiii*

Chapter - 1
Cellular Adaptations to Changed Environment 1

Chapter - 2
Fluid, Electrolyte and pH Disturbances 9

Chapter - 3
Some Aspects of Deranged Kidney Function 22

Chapter - 4
Physiological Disturbances in Cardiovascular Disease 40

Chapter - 5
Pathophysiology of Shock and Burns: Multi Organ Dysfunction 73

Chapter - 6
Physiological Disturbances in Respiratory Disease 89

Chapter - 7
Pathophysiology of Some Gastro-intestinal Disorders 109

Chapter - 8
Pathophysiology of Obesity and Weight Loss 134

Chapter - 9
Pathophysiology of Musculoskeletal Disorders 148

Chapter - 10
Circadian Rhythms and Pathophysiology of Stress 161

Chapter - 11
Pathophysiology of Common Blood Disorders 178

Chapter - 12
Some Aspects of the Pathophysiology of the Central Nervous System 200

Chapter - 13

Pathophysiology of Common Thermoregulatory Disorders............................221

Short Answer Questions and Their Solutions..237

Index..253

Foreword

It gives me great pleasure and I am indeed honoured to write this Foreword to Group Captain Professor Mohan Dikshit's, "Pathophysiology for Medical and Allied students." This book fills an important lacuna in the teaching of pre-clinical sciences while providing the right perspective to ensure that students of medical and other allied health disciplines appreciate that much that is learnt in the early years of their education in medical and health sciences schools has relevance to their attaining their goal as competent practitioners in their field. Mohan Dikshit given his experiences in Pune and later in Sultan Qaboos University in Muscat, Oman where I had the good fortune to evaluate the results of these novel approaches in practice, is most suited to introduce this approach through this book to Indian and other Asian teachers and students.

William Osler while emphasising that education is a life-long process had once said, "we can only instil principles, put the student in the right path, give him methods, teach him how to study, and early to discern between essentials and non-essentials." Addressing continuously the issue of what is 'essential' as distinct from the 'non-essential' is a challenge that all Preclinical teachers face on a daily basis. Problem Based Learning tries to move away from standard building block structures where a lot of content is shoved down the throats of students which they do not retain for long anyway. Problem-based approaches to learning should apply throughout the physician's career and the components have to be organized in sequential blocks with early exposure to patients and case management. And Mohan Dikshit's book makes just that point.

Teachers in India and Asia more generally often subscribe to the view that their students are generally passive when they relate to teachers and prefer to learn by rote and be assessed for their efforts in that direction in contrast to the more outspoken and irreverent conduct of students in the West. This is no longer true. The growth of the social media has changed that. Students of the present day discern and appreciate quite early that the problem solving model integrates the acquisition of knowledge and its application to solving problems thereby endowing them with skills that ensure lifelong learning. Mohan Dikshit's book on Pathophysiology is bound to be a success with this generation of students of medicine and allied health sciences.

Prakash Shetty, MD, PhD, FRCP
Former Professor of Physiology, St John's Medical College, Bangalore

Preface

Pathophysiology was not always a priority for pre-clinical teachers. In fact I remember I took a bit of flak from my senior professorial colleagues of the University of Pune when I had included, as a part of a question on maintenance of normal blood pressure, a note on hypertension and principles of its treatment (2 marks!). In today's world of 'advanced' medical education parlance, the question (I must confess) was poorly framed, but the intent was clear. It was time that we gave a "clinical" orientation to our pre-clinical teaching. As an IAF specialist in, I was fortunate to have in my classes various categories of students- Medical Corpsmen, Flight cadets, Fighter and Transport pilots and of course medical officers who undertook various training courses in aviation medicine and physiology including post graduate training. While talking to them it was obvious that they found physiology much more interesting if its phenomenon were linked with what they knew and saw as common illnesses and the symptoms and signs these produced. I continued to dabble in this direction of teaching pre-clinical students even after I left the Air Force to join a civilian medical college. But yet the term "Pathophysiology" as a part of the course curriculum remained absent.

My first opportunity to teach the subject as Pathophysiology came my way when I was invited as a visiting faculty by the College of Medicine and Health Sciences, Sultan Qaboos University (SQU) in Muscat, Oman. I was tasked with developing and delivering a course in pathophysiology to senior nursing students. It was soon apparent that the course was well accepted and the students felt that it helped them to understand clearly as to why the patients exhibited symptoms and signs as they did, and appreciate the principles of management prescribed for their patients. It was during this phase that the idea of writing a book on pathophysiology aimed at the Indian medical student community took seed in my mind. Later as we developed new courses at the SQU for pre-clinical medical students with pathophysiology as the back bone, the benefits became clearer and led to this book.

The book is not filled with details of every aspect of Physiology and its disruption. For example, the chapter on cardiovascular physiology remains confined to salient topics such as heart failure, ischemic heart disease and hypertension. In the respiratory chapter; airway disorders, diffusion and respiratory failure have been discussed. And so on. Each chapter has been introduced with some basic physiology to direct the student towards appreciating the consequences of the whys and hows of alteration of that physiology.

Why have only such topics been chosen? That is because these issues are directly correlated with the basic happenings in the organ systems which are read as a preclinical subject. For example, physiology of resistance to airflow is a part of regular physiology of respiration. What better then to have the student link this normal physiology to its change in a disease like airway obstruction? Such relationships between normal and disrupted physiology are relatively easy to understand. The aim is get the student to realize this so that he develops the confidence to apply it to any aspect of altered physiology which we call disease. This, I believe, is conceptualization of a process. I have not included chapters on Immune mechanisms and Endocrine glands. An attempt has been made to give a new dimension to endocrine physiology by introducing pathophysiology of stress and circadian rhythms. This is not a subject usually dealt with in standard textbooks of physiology, and in the current scenario I felt that the budding doctor must be made aware of this area in pathophysiology at the very inception of his/her training. Also introduced is a chapter on thermoregulatory disturbances. Again with the view that in a tropical country such as ours, heat illness in particular is widely prevalent, and the medical student should be aware of its implications in day to day life very early in his /her career.

A number of flow charts have been designed to aid the student to develop a thinking process and analyze the information given in the text. In this manner I hope the student learns to conceptualize the mechanism of disease presented as disturbed physiology. If the book is referred to while studying Physiology, this will become all the more clear to the student and assure him/her that basic science is useful.

In the text, a number of questions have been interposed. It is hoped that the student uses the information given in the text and in the flow charts to derive answers to these questions. My responses to those queries raised are given chapter wise in the Short Answer Question (SAQ) modules at the end of the book. I would welcome comments on the solutions I have provided from students who may think differently. It will be a privilege to get into such a dialogue with them on the internet.

I do hope that pre-clinical teachers have a look at this book. It may enthuse them to use pathophysiology more liberally while teaching. It may also enthuse some young mind to expand on the material in the book and make it much more comprehensive. I would also strongly encourage nursing staff/medics/students to read this book. Perhaps post graduate students may find it useful as a simple, quick reference source. The book may be noticed by administrators so that the idea that pathophysiology should be a part and parcel of pre-clinical teaching may germinate.

Mohan B Dikshit
Pune, India
justmohan@gmail.com; mohan_dikshit@hotmail.com

Acknowledgements

The initial chapters that I wrote were edited by my elder son Mohit. Though a software engineer, his mastery of English took care of whatever minor glitches he noticed in the write up. I wonder if I would have ever noticed them. My younger son Vivek helped by refining my preface, and material for the back cover. It reads so much better now. Krishna my wife always made solicitous enquiries about my slow yet definite progress.

Prof Brian D'Monte my friend of more than 55 years, and a senior colleague from AIIMS scrutinized every chapter. His learned inputs and constructive criticism went a long way in improving the manuscript. His appreciation of the idea of this book and his constant encouragement gave me added impetus. Prof CV Ramani my class mate from AIIMS now a learned neuro-psychiatrist in the US offered invaluable inputs.

And my school mate and dear friend Dr Sheikh Irfan! Every time we emailed each other or talked on the phone or met here in India, he would say "Mohan, your book? When do I get to see it?" Irfan, here it is.

And most importantly, the authors whose books and scientific works I referred, read and tried to imbibe in order to understand and develop my own perceptions which I then put down in the form of this book.

CHAPTER - 1

Cellular Adaptations to Changed Environment

Plan

1. General introduction
2. The normal cell
3. General responses ofthe cell to injury
    i. changes in size or number
    ii. changes in type an structure
    iii. reversible/irreversible injury
    iv. ROS in cellular injury
    v. reperfusion injury
4. Pathophysiology of aging
    i. definition of aging
    ii. general aspects
    iii. theories of aging

General Introduction

The smallest functional unit of the body is the **CELL.** Layers of these cells coalesce into **TISSUES** which in turn bind into **ORGANS**, and finally into the organ **SYSTEMS.** The cell is nurtured by the extracellular fluid (Claude Bernard's Milieu Interior-the internal environment), and all the organ systems work in unison to preserve the integrity of the internal environment even under the most trying circumstances, whether in the zero gravity of space, or in the hyperbaric surroundings of the deep sea. It is therefore a closed loop system: cell to organ system to the extra cellular fluid back to the cell. That is Physiology. When disease strikes, the integrity of the internal environment and the cells it bathes is disturbed. This alters normal Physiology to **pathophysiology** which manifests externally as symptoms and signs of disease. As physicians, we treat disease. That means we help to restore body physiology back to normal. In the process, cells may suffer temporary, or on occasions, permanent damage. Sometimes, body physiology may fail in its fight back to restore itself to normal in spite of whatever the physician tries. Under the circumstances, death is the terminal event.

The Normal Cell and Its Function

The reader should refer to a diagram of a normal cell in any standard text book of physiology. Newer staining techniques have revealed the presence of a new organelle, **Vaults**. These are cytoplasmic nucleo-proteins. They are shaped like the pores in the nuclear wall, and this has led to the speculation that these organelles may be passing through the nuclear pores, carrying messenger RNAs. They may also be involved in as yet an unknown way in cancer cells' resistance to drug treatment. New molecules secreted by various cells are constantly being identified. Each one probably helps us to understand better the functioning of the human body.

General Responses of the Cell to Changing Environment and Injury

Cellular morphology and function may change as a result of adaptation, injury, aging, or cancer. The range of response can vary from subtle changes in the cell which the inherent mechanisms can repair, to cell death and necrosis.

Cellular Adaptation to Injury

Adaptations may be the result of physiological or pathological stresses. In the latter, these changes may be successful only temporarily in the initial phases of the adverse situation. Severe and/ or prolonged adversity may overwhelm the adaptive process which may then progress to cell injury or death. Adaptation may result in a change in1. Size or number of cells or 2.Type and structure of the cells (**Fig 1**).These changes take time to get established, follow a predictable pattern, and are reversible in most instances when the stimulus is removed early enough in time.

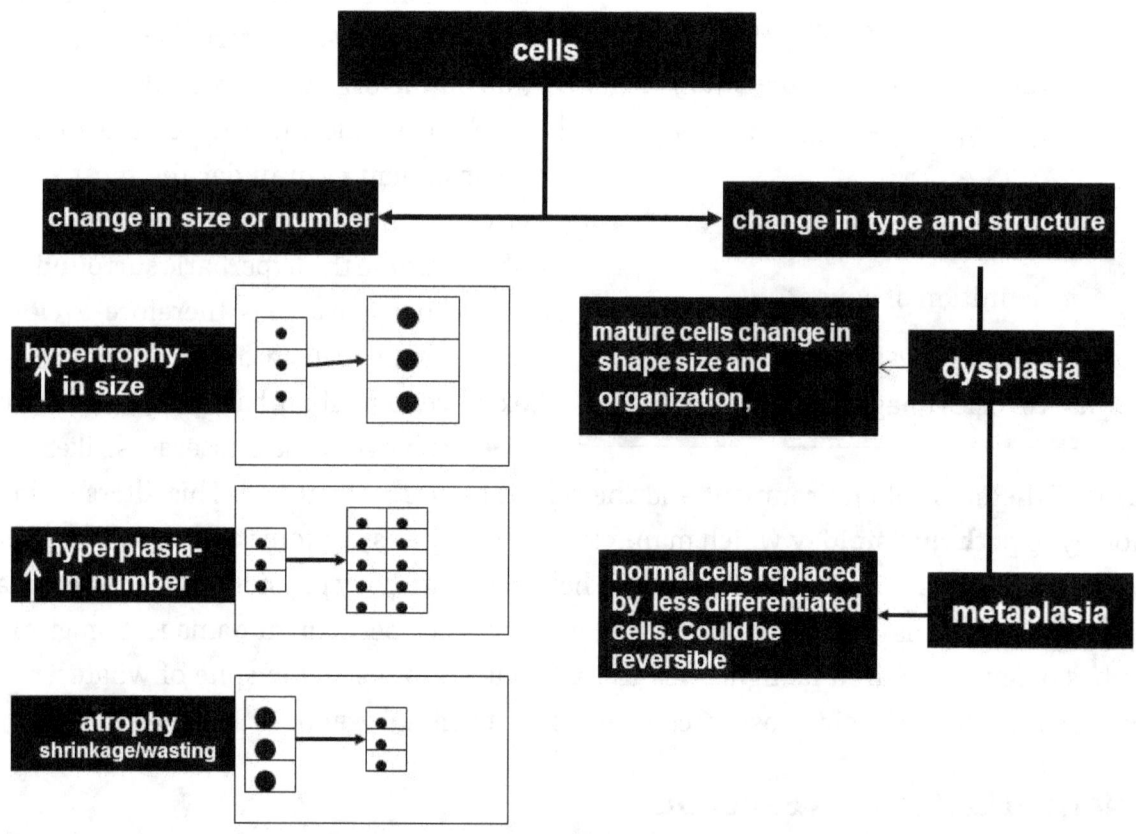

Fig. 1 Cellular adaptations to changing environment

Change in Size or Number

1. **Hypertrophy.** This involves an increase in size of the cells, and therefore the size of the organ involved. The intra-cellular changes include increased accumulation of protein in the mitochondria, endoplasmic reticulum, and the plasma membrane. The hypertrophy may occur as a result of tropic signals for example an increase in size of sex organs after puberty in response to high levels of sex hormones, or as result of increase in workload of the organ. The classical example of the latter is muscle hypertrophy on strength building exercise. If the excess work load is removed, the hypertrophy regresses. The heart may also have to work overtime in response to high blood pressure. This change however may be overwhelmed by the hypertension in its advanced stages, and lead to pathophysiology: a typical example of the initial beneficial adaptive change progressing on to cell damage. Hypertrophy occurs in cells which cannot divide. One of the exceptions to this rule is the pregnant uterus which undergoes an increase in size as well as number of its cells with advancing pregnancy.

2. **Hyperplasia.** This is a reversible increase in number of cells (not the size) because of increased rate of cell division. Hormones and chronic irritation may stimulate cells to develop this change. A typical example is liver regeneration after a part of it is removed. This is called compensatory hyperplasia. (*A person donates a part of his liver to his sibling. What will happen to this person's liver in 6 months time?*) Various factors likely to be involved in bringing about this response are the Hepatocyte Growth Factor, Transforming Growth Factor α, Tumour Necrosis Factor α, Epidermal Growth Factor, and Interleukin 6. During pregnancy, the uterine cells increase not only in size, but also in number (Hormonal hyperplasia). (*What makes the uterus regress to its near original size after pregnancy is over?*) Another illustrative example is the increase in red blood cell count which occurs on exposure to high altitude. Chronic trauma or irritation in the mouth such as an ill-fitting denture, may lead to mucosal membrane hyperplasia. Rarely hyperplasia may change into a cancer. Examples of this are bronchial epithelial hyperplasia in response to smoking, and endometriosis.

3. **Atrophy** is an acquired decrease in the size of cells, resulting in a shrinkage of the organ that this group of cells constitutes. The main changes at the cellular level are: i. shrinkage in size; ii. accumulation of lipofuscin, a pigment; iii. decrease in number of mitochondria, endoplasmic reticulum and myofilaments. There is a decrease in protein synthesis, or an increase in protein breakdown, or both. The possible pathway for protein breakdown is the ubiquitin –proteasome pathway which may be activated by glucocorticoids, and metabolic acidosis. Proteins are first conjugated with ubiquitin and then acted upon by proteosomal enzymes. Atrophy may occur as a physiological phenomenon. The thymus gland undergoes programmed atrophy as a child grows into an adult. Pathological atrophy is a result of 1. ischemia; 2. pressure; 3. disuse; 4. exhaustion or overuse (*Excess growth hormone may lead to a frank diabetes mellitus. Please explain*); 5. hormone deficiency;

6. loss of nerve supply. Whether the cause is physiological or pathological, the changes at the cellular level remain the same. Ischemia is by far the commonest triggering factor. Philosophically it might be argued that disuse atrophy is a means of economizing on body resources. For example, muscular atrophy occurs in a fractured limb which is being treated with a plaster cast.. This limb needs less nutrition during that phase, and hence the body physiology in its wisdom, produces autophagic vacuoles which auto digest part of the tissue not in use until the plaster is removed, and use of the limb is restarted. This is where the importance of physiotherapy becomes obvious. **(*Theoretically, can a tumor be treated by producing atrophy in it? If so how?*)**

If atrophy is a condition which develops in tissues which were normal at birth, **hypoplasia** means tissues/organ are underdeveloped and undersized at birth as a result of intra – uterine hypoxic injury, or nutritional deficiency. It may also happen if genetic signals required for the development of a tissue are turned off prematurely. Hypoplasia is a permanent change. If genetic signals are either defective or absent, **aplasia** of an organ may occur. ie the cells making up this organ do not develop at all, hence the organ does not develop. Occasionally the bone marrow may be come aplastic if it has been severely damaged by radiation, even though at birth, it was fully functional.

Change in Type and Structure of Cells (Fig. 1)

1. **Dysplasia** may happen in mature cells and involves a change in their shape, size, and organization. It is caused by abnormal differentiation and involves a problem of growth regulation, thus increasing its propensity towards a cancerous change Dysplastic changes may be reversible if the offending stimulus is removed early. Dysplasia often occurs in the cervical epithelium of the female reproductive tract, and bronchial epithelium of smokers. In both the situations the dysplasia may turn into cancer.

2. **Metaplasia.** In certain situations, less mature cells may be transformed into another type in order to cope with the changing environment. This happens because of cytokines and growth factors whose secretion may be induced by the new environment. For example, in chronic smokers, the pseudo-stratified columnar epithelium is replaced by stratified squamous epithelium as a protection against the chronic insult. When the insult is removed, the epithelium changes back to its original status. Metaplasia may be ordered by a reprogramming of stem cells of the epithelium. Occasionally a metaplasia may turn cancerous.

Cell injury Sometimes the environmental change may go beyond inducing just adaptations. They may injure the cells. The injury may be reversible (cells recover), or irreversible where in the cells die. The various outcomes are outlined in **Fig 2**.

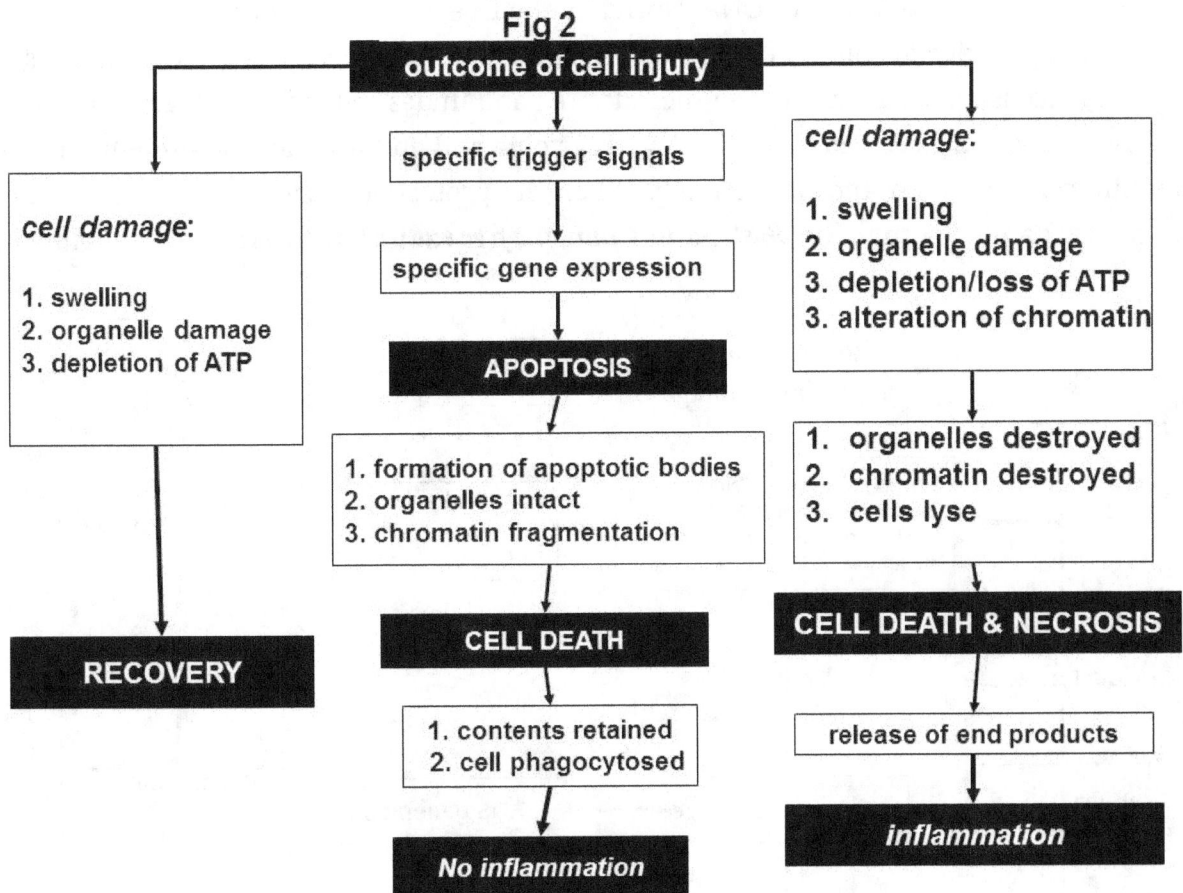

Fig 2

Two individuals exposed to the same stress may suffer different degrees of injury depending upon factors such as nutritional and immune status. The various injurious agents are hypoxia (usually as a result of ischemia), release of reactive oxygen species (ROS) (free oxygen radicals), chemicals, physical agents (trauma, heat, radiations), and infections. Ischemic hypoxia is by far the commonest cause of injury. The common themes in cellular injury are i. ATP depletion; ii. release of oxygen-derived free radicals; iii. changes in cell permeability; iv. Increase in cytosolic Ca⁺⁺. The possible pathophysiology involved is given in **Fig. 3** (***In a box in the flow chart, it is indicated that water retention takes place in the cell, resulting in its swelling. How does this happen?***)

Role of oxygen-derived free radicals in the pathophysiology of cell injury. An electrically uncharged atom, or a group of atoms with an unpaired electron constitute a "Free radical." Because of its free state, this particle seeks to bond with other substances such as lipids, proteins and carbohydrates. It is this chemical linkage that is injurious. Free radicals are naturally generated in the mitochondria, peroxisome, cytosol, and endoplasmic reticulum of the injured cells and activated phagocytes i. during normal metabolic processes; ii. electromagnetic radiations; iii. enzymatic metabolism of chemicals and drugs. The better recognized free radicals are superoxide, hydrogen peroxide (H_2O_2), hydroxyl radicals (OH-) and Nitric Oxide (NO), collectively called reactive oxygen species (ROS). These act as physiological modulators of some mitochondrial activity, but also cause damage by lipid peroxidation, protein fragmentation and alterations of DNA. Normally the free radicals are destroyed by naturally produced Superoxide dismutase (SoD), catalase and glutathione peroxidase (anti-oxidants) so that a balance is maintained between their desired and undesired effects. Other known anti-oxidants are Vitamins E, and C, amino acids- cysteine and glutathione, and proteins such as albumin, ceruloplasmin and transferring. When the protective enzymes are overwhelmed, free radical damage occurs. (***In what manner may the free radicals interfere with ionic pumps?***).

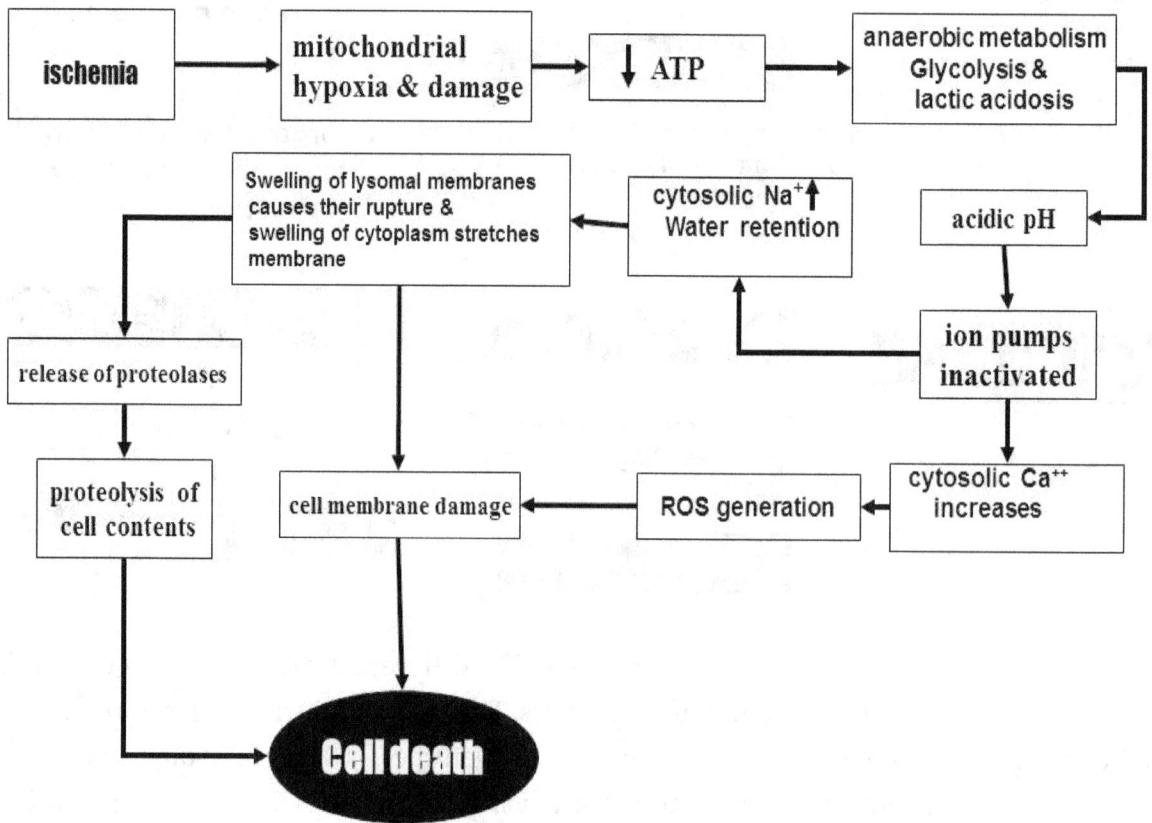

Fig. 3. Pathophysiology of Ischemic cell injury which may progress to cell death

Reperfusion injuries are linked to ROS. When ischemic areas are re-perfused, endothelial cells are activated, and produce more ROS but less NO. This imbalance leads to the production of inflammatory mediators like platelet activating factor, and tumor necrosis factor by the endothelial cells which inflict damage on the local microvasculature. Reperfusion injuries following ischemia are more common in pathology such as hypercholesterolemia, hypertension, and diabetes mellitus. Free radical induced damage has been associated with a large number of conditions such as aging, hyperbaric oxygen toxicity, atherosclerosis, ischemia of the heart and the brain, and radiation injuries.

Pathophysiology of Aging

Aging has been defined variously by various authors. Very simply put, it is a process during which there is a continuous wear and tear of the body with passage of time resulting over time in functional deterioration. The reader may like to look critically at this generated definition and try and find one's own definition of this biological phenomenon.

Successful (healthy) aging is a process by which deleterious effects are minimized and function is preserved until senescence makes continued life impossible. People who age successfully avoid experiencing many of the undesirable features of aging, and **remain functional both physically and mentally.** It is difficult to determine at what chronological age the process of aging actually begins. Within an individual, the rate of decline in function of different organs is variable. For example, kidney function may decline more quickly than heart or lung function in some, while it may be the other way around in others. With aging, many physiologic functions decline. Much of this a decline may be attributed to aging itself, and is therefore considered to be normal, not disease-related. For example, some aspects of respiratory function such as the age related decrease in vital capacity probably begins around the age of 27–30 years in most populations including Indians. A classical example of aging is presbyopia (decrease in accommodation of the lens of the eye). Presbyopia occurs in virtually everyone after the age of about 40 years, and no cause or explanation has been identified other than aging itself. In some situations though, statistical distribution may be used to define the thin dividing line between the normal and the abnormal. For example, some degree of glucose intolerance is considered part of normal aging, but diabetes is a pathological extension of the same glucose intolerance. Similarly, cognitive decline is commonly seen with aging and is considered normal; however, dementia, although common in late life, is a disease.

Theories of Aging

A number of theories to explain the phenomenon of aging have been proposed, but none seem to fit the bill fully. Two aspects that have been deliberated upon most are that it may be genetically pre programmed, and it may be a general response to cumulative injuries. It is possible that aging is a combination of both, a programmed process whose progress is modulated by environmental conditions with a potential for damaging or protecting cells. It is this interlocking of the two approaches that makes it difficult to arrive at a consensus of what exactly causes aging.

It is obvious that genetic inheritance, given that no external damaging force affects a person, determines his/her life span and aging process as per a set program. One such phenomenon is apoptosis, or programmed cell death, often called cell suicide. However the role which this process plays in aging is not clear. In the aging cell, DNA, RNA cellular proteins and the cell membrane are most prone to injury. The lack of DNA repair makes the cell susceptible to mutations. In some individuals, the process of aging is accelerated to an extent that even by the age of ten, the person appears to be extremely old. This is known as progeria, and may be a result of a deficiency of a protein called Lamin A in the nucleus. More recently the telomerase theory of aging has come to the forefront. Telomeres are sequences of nucleic acids found at the ends of chromosomes. These are thought to manipulate the clock that controls the life span of dividing cells. With each cell division, telomeres are shortened, and cellular damage is the result. Telomerase appears to repair the telomeres, and its availability may thus be responsible for controlling the rate of decline of cells. Free O_2 radicals may be responsible for collagen and elastin damage with advancing age. In the elderly, the naturally produced antioxidants may become deficient or less effective. Free radicals may also create mutant cells. They are also thought to hasten production of lipofuscin which interferes with the ability of the cells to repair and reproduce. Caloric restriction on the other hand has been shown to retard the aging process, possibly because this leads to a reduction in metabolic rate which in turn slows down the production of ROS. The proponents of the neuro-endocrine theory of aging feel that a decline in hormonal levels undermine the body's ability to repair daily wear and tear that the body undergoes. In spite of all the possibilities considered, the fountain of elixir of youth still remains elusive.

References

1. Best and Taylor's Physiological Basis of Medical Practice, 13th edn.; Ed. Best JB. William & Wilkins, London; 1990.

2. Braun CA, Anderson CM. Pathophysiology: Functional Alterations in Human Health. Lippincott Williams, London. 2007.

3. Cobbs JP, Hotchkiss RS, Karl IE, Buchman TG. Mechanisms of cell injury and death. British J Anaesthesia. 1996; 77: 3–10.

4. Dikshit MB, Raje S, Agrawal MJ. Lung functions with spirometry: An Indian perspective II: On the Vital Capacity of Indians. Indian J. Physiology and Pharmacol.2005; 49: 257–270.

5. Ganong WF. Review of Medical Physiology.21st edition. Lange Medical Books/McGraw-Hill, London. 2003.

6. Guyton AC and Hall JE. Textbook of Medical Physiology 11th edition. Elsevier, Philadelphia; 2006.

7. Harrison's Principles of Internal Medicine Edt. Longo D., Fauci A, Ksper D, Hauser S., Jameson J, Joseph Loscalzo J, 18th ed. MCGraw Hill, London. 2011.

8. Klatz R, Goldman R. Stopping the Clock. Keats Publishing, New Cannan, CT. 1997.

9. McCance KL, Huether SE. Pathophysiology: The Biologic Basis for Disease in Adults and Children. 4th edition. Mosby; St. Louis USA; 2002.

10. Text Book of Medicine. Ed. Souhami RL, Moxham J. 2nd edition. Churchill Livingstone, London; 1994.

11. Understanding Medical Physiology: A Text Book for Medical Students. Ed. Bijlani RL 2nd edition; Jaypee Brothers Medical Publishers, New Delhi. 1997.

CHAPTER - 2

Fluid, Electrolyte and pH Disturbances

Plan

1. Normal distribution of body water
2. ECF, ICF, osmolality
3. Pathophysiological consequences of reduction in body water
   i. maintenance of normal osmolality
   ii. water/fluid loss in infants and small children
   iii. water/ fluid deficit in the elderly
   iv. hypovolemia
   v. hypervolemia; excess of body water
   vi. electrolyte disturbances
   vii. pH disturbances

Normal Distribution of Body Water

Water is the most abundant constituent of the body. On an average, about 60% of the weight of an adult male is water. This is distributed between two main compartments: the intracellular which contains about 40% and the extra cellular which has about 20%. On the other hand, the female has slightly less amount of water, about 50% **(Fig 1)**. In the newborn, the percentage of water is high because of a relatively low cell mass which increases rapidly after the first year. Of the various tissues, fat has the least water content (about10%), the bone about 20%, while the kidneys have the highest of about 83%. The water content of other major tissues like the heart, muscle, skin, brain and liver ranges between 68–79 %.

Distribution of total body water (TBW) as the extra and the intracellular compartments **(ECF & ICF)** is given in Fig.1. (The reader at this point must revise the constituents of the ECF and the ICF).

TBW and the ECF may be measured by using the principle of dilution. (Volume = mass of substance injected/concentration of substance in the fluid compartment). **(*What should be the property of the substance which may be used to measure TBW?*)**.Dilution fluids cannot be used to measure the intra cellular and interstitial fluid volumes as no marker permeates exclusively through these two spaces. **(*If this is true, how can one estimate the volume of water in these two spaces?*)**

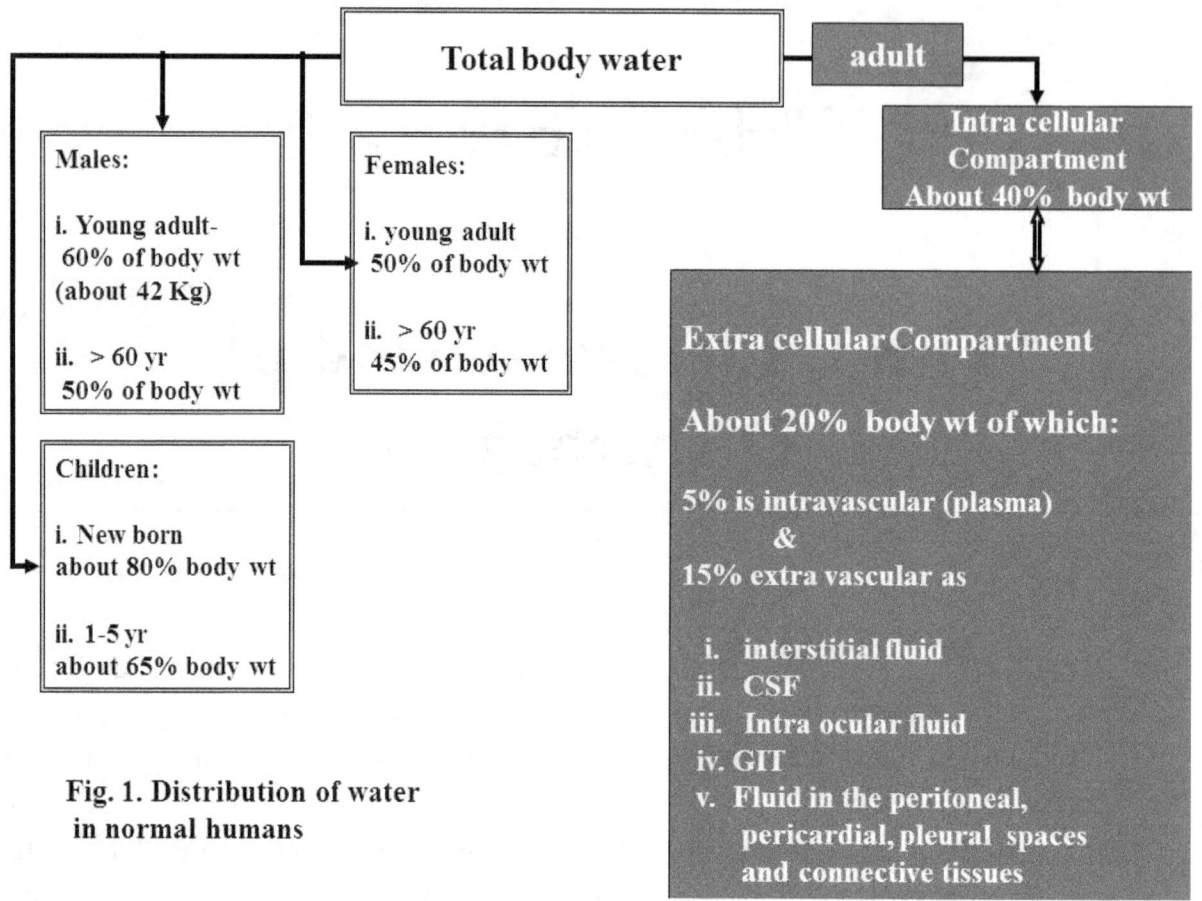

Fig. 1. Distribution of water
in normal humans

Table I: Normal 24 hour water intake and loss (in ml):
turn over approximately 2300 ml

Intake	Loss
by drinking about 1400	urine about 1400
via Food about 700	stools about 100
metabolism about 200	skin insensible about 300
sweating about 200	lungs 300

In 24 hours, the normal turnover of water is about 2300 ml (**Table I**). The ICF is protected by the ECF, and may exchange water only with the ECF. The latter on the other hand is directly exposed to the environment to which it may either lose water, or through which it may gain water (**Fig.2**). Acting via the ECF, the environment may indirectly influence the ICF. The ECF (internal environment)acts as a buffer between the two, and any disturbances in the ECF are bound to reflect on the ICF. Maintenance of the constancy of the ECF is therefore of paramount importance.

One of the best methods of estimating the constancy of the extra cellular fluid is to measure its osmolality (mosmol/kg of water).This takes into account not only the fully dissociable constituents like Na^+ and K^+ but also proteins, glucose and lipids, and hence is the more accurate estimate of the parameter. It may also be expressed as osmolarity (mosmol/l of water).Na^+ and Cl^- ions which are the most abundant, contribute maximally to this parameter. Osmolarity is affected by volume of the solution while osmolality is not as it is measured against weight

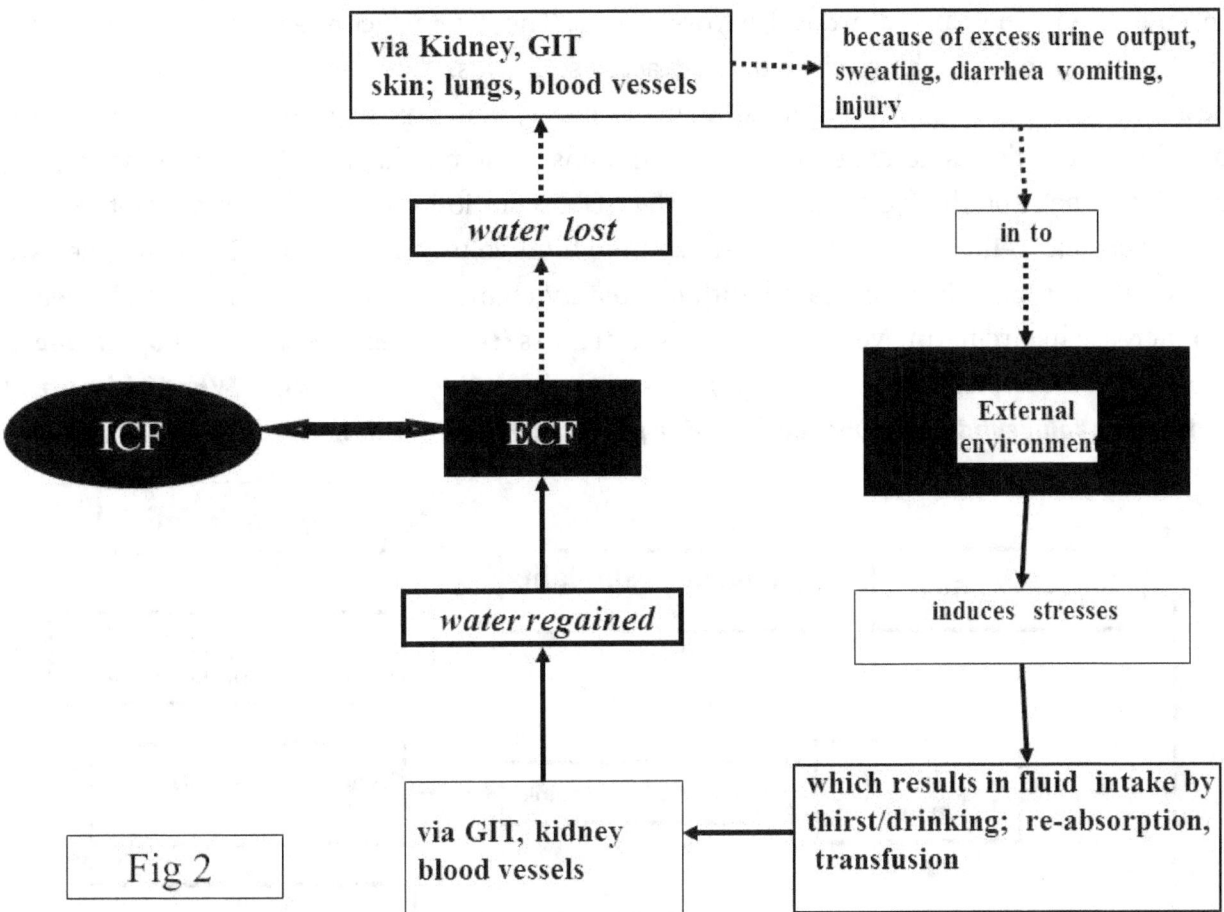

Fig 2

of the solution. Under physiological conditions the difference between the two measures is less than one percent, and one term may be used for the other. As osmolarity (mosmol/l) is a more practical estimate, conventionally, in physiology and medicine, it is this parameter that is estimated and expressed as osmolality/l. Normally osmolality of plasma (extracellular fluid) is approximately 290 mosm/l. At this point, the cells will not shrink nor expand because the water concentration in the ICF and ECF is the same, and hence solutes cannot move across the cell membrane. This is the state of **ISOTONICITY**. A reduction in osmolality to below the set point makes the solution **HYPOTONIC**, while its increase makes it **HYPERTONIC**. A change in osmolality by just 1–2%, invites physiological mechanisms to restore it to normal. Plasma osmolality may be measured with an osmometer which utilizes the principle of freezing point depression. In the absence of this special instrument, it may be calculated using commonly estimated plasma parameters as (**2Na$^+$+ 2K$^+$ + glucose+ urea**), all estimated in **mmol/l**.

*(**Work out what will happen to cells surrounded by i. hypertonic ECF and ii. hypotonic ECF. What sort of solution would you like to use intravenously to treat edema of the brain and why? Find out the name of this solution used in practice**)*

Normal osmolality is maintained by fine tuning both the water and Na$^+$ content of the ECF. A deficit of water makes the ECF hypertonic, and is met with a chain reaction which helps to restore the lost fluid (**Fig.3**). Anti Diuretic Hormone (ADH) secretion is also controlled by cardiovascular reflexes involving the low pressure cardiopulmonary receptors (Type B atrial receptors) and the carotid arterial baroreceptors. The afferent connections are made with the hypothalamus via the nucleus of the tractus solitarius. A decrease in circulating volume deactivates these receptors to reduce their discharge, which in turn increases the secretion of ADH. The reflex is activated by a non hypotensive hemorrhage (<15% loss in circulating blood volume), and this helps to augment the ADH secretion initiated by osmoreceptors. The cardiovascular reflexes are useful when isotonic fluid volume loss occurs. These reflexes are however less sensitive as compared with osmolality changes in bringing about ADH release. An increase in circulating volume increases water loss.*(How do low pressure cardiopulmonary (atrial type B receptors) help in reducing the excess circulating fluid volume. What is the atrial natriuretic polypeptide, and what do you think would be the role of this hormone in this situation?).*

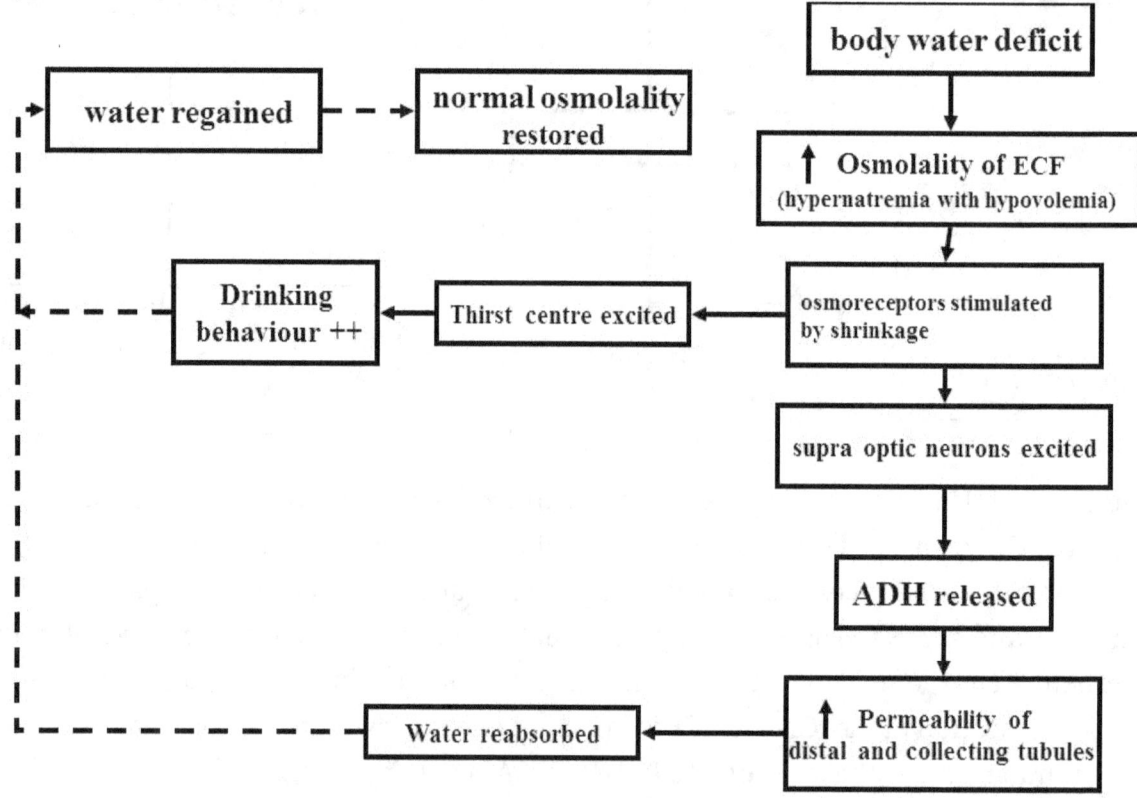

Fig. 3. Physiological adjustments to body water deficit

The most vital role in the maintenance of normal osmolarity, as mentioned earlier, is played by the Na⁺ balance in the body. Almost 80–90% of body sodium is present in the ECF. Changes in ECF Na⁺ generally reflect disturbance in body water homeostasis. An increase in ECF Na⁺ (>145 mmol/Liter) reflects a contraction of ECF while hyponatremia (Na⁺<135 mmol/L) is a manifestation of excessive body water retention.

Water/fluid Loss in Infants and Small Children

Early in gestation, almost 90% of a fetus's body weight is water. This ratio falls to about 70–80% in full term infants, 65% in young children, and approximately 60% in older children and adults **(Fig.1)**. The reduction in the water content occurs because of a shrinkage in the ECF compartment. Even then, 50% of body water remains in the ECF of neonates as compared with 20% in adults. Because of this, loss of ECF in the infant is proportionately greater. When this happens, its replacement must come from the intracellular compartment. The consequences of this are obvious.

The newborn's size is small, but the skin surface area is relatively very large. A full term neonate is estimated to have 2–3 times the surface area of an older child/ adult in relation to metabolically active tissue. The rate of metabolism in infants is significantly greater than in adults because of this large surface area. Any condition that increases metabolism causes a rise in heat production with it's concomitant insensible water loss. Hence, even if the absolute changes in body water and electrolytes may be small, they represent large proportionate changes in body fluid compartments.

Increased metabolism in the neonate demands more of the kidneys in terms of excretion of waste products. As such the concentrating capacity of an infant's kidneys is less than that of an adult. In response to water deprivation, the kidney of a full term baby can concentrate urine to only about 600–700 mosm/l as against 1200 mosm/l achievable by an adult's kidneys. Therefore even small losses of body fluids stretch the fluid and electrolyte handling capacity of infants. Volume overload on the other hand may be better handled by infant kidneys because they can excrete a very dilute urine (osmolality may be as low as 50 mosm/l) while the adult kidneys achieve a maximum dilution of about 70–100 mosm/l.

From the above, it is clear that even relatively small fluid losses in infants/small children will not be handled well by their body physiology. Cognizance of this must be taken by health professionals and care givers, and urgent fluid replacement must be given to infants suffering from conditions such as diarrhea, vomiting, and hyperthermia which occur commonly in this age group.

Water/Fluid Deficit in the Elderly

If infants have difficulty in handling water loss because they have about 80% of their weight as water, the elderly (usually > 65 yr) have problems because their water content amounts to only 50% of their body weight. Fluid deprivation, a hyper osmotic stimulus, or exercise in a warm environment (a combination of hypovolemia and hyperosmolality) in older adults is a challenge because this group tends to consume less fluids. Complete fluid restoration eventually occurs, albeit slowly. There is evidence to suggest that older men and women (i) have a higher baseline osmolality, and thus a higher osmotic operating point for thirst sensation (with little or no change in sensitivity), and (ii) exhibit diminished thirst and satiety. A reduction in their body water therefore may go unnoticed before it produces frank hypovolemia and its subsequent effects.

Hypovolemia is a reduction in circulating fluid volume as a result of excessive loss of fluid or inadequate intake. It is usually a combination of water and overall salt deficit. The more common causes, and the subsequent consequences are outlined in **Fig. 4.** (The reader may refer to a text book of Medicine/Surgery for a more detailed list of causes of hypovolemia). Physiologically, the reduction in cardiac output and blood pressure bring about sympathetic activation in an attempt to reverse the changes. **(*What is the reflex mechanism involved. How do the kidneys help?*)**If uncorrected using appropriate fluids, the situation may progress to shock. (Please also see chapter 3 on Pathophysiology of Shock and Burns). **[*What do you think is the "appropriate fluid" to be used in cases of a i. road accident; ii. burns; iii. excessive sweat loss?*]** As discussed earlier, infants and the elderly are prone to the adverse physiological changes produced by hypovolemia.

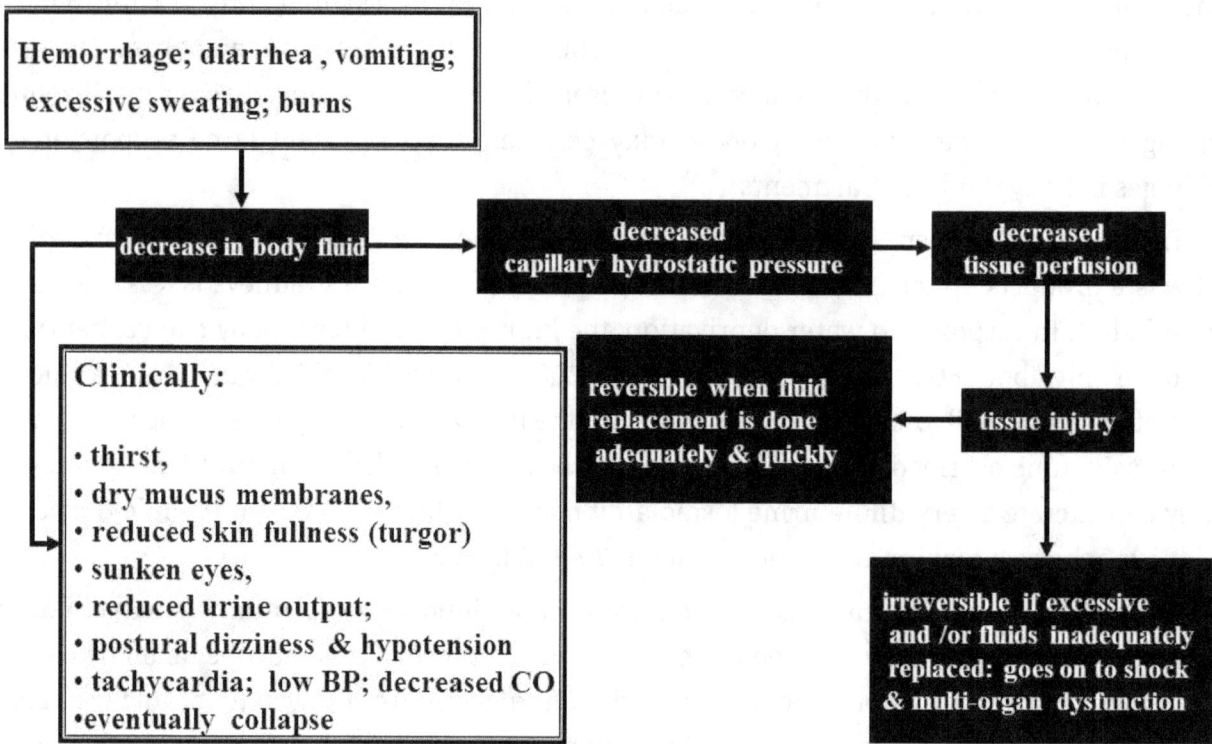

Fig 4: Hypovolemia and its effects

Hypervolemia on the other hand is a seemingly impossible physiological situation. It involves excessive water retention, and may happen because of compulsive water drinking in a patient with a psychogenic disorder. If after surgery, a patient is given hypotonic solution intra venously, hypervolemia may result because as such in such a patient the ADH levels are high. Any pathological condition which leads to a reduction of urine formation may produce a state of hypervolemia **(*Try and list out such situations*)**. In a condition known as Syndrome of Inappropriate ADH Secretion (SIADH), factors other than volume shrinkage or hyperosmolality are responsible for the increased circulating ADH so that the amount of the hormone secreted is inappropriate vis-à-vis the plasma Na^+ level. Bronchogenic carcinoma, and pulmonary tuberculosis are two of the diseases in which SIADH is known to occur. The usual water/electrolyte disturbance is a combination of hyponatremia with hypo-osmolality. The water tends to shift in to the intra cellular compartment because of this effect and leads to cerebral edema and raised intra cranial tension with its consequent effects on the level of consciousness, and convulsions. There is also the possibility of hypertension because of the excessive water load.

Electrolyte Disturbances

Only Na$^+$ and K$^+$ disturbances will be considered here.

Normal plasma Na$^+$ is maintained between 135 to 145 mmol/L. As 80–90% of Na$^+$ is extra cellular, its level is reflected by the ECF volume. With volume regulation, usually the Na$^+$ loss and gain get titrated, though individually, water and body Na$^+$ are independently controlled. Changes in water content therefore induce alteration in plasma sodium concentration, and imply abnormality in water balance while changes in the ECF volume are reflected as alterations in body Na content. *(Please revise the role of the renin-agiotensin-aldosterone system, the kidneys, GIT and the sweat glands in maintenance of sodium balance in the body).*

Hypernatremia is present when the plasma Na$^+$ level is > 145 mmol/L. This may occur as a result of increased Na$^+$ retention or increased water loss, the latter being more common. Urine output may vary depending upon the condition that causes hypernatremia (**Fig. 5**). When there is coexisting polyuria, the cause ishypothalamic diabetes insipidus or a nephrogenic diabetes insipidus. The distinction between the two may be made by use of desmopressin (**How?**).

The main physiological problem in hypernatremia is the increased osmolality which draws out water from the intra cellular space. The brain cells may get adversely affected because they shrink, and therefore major symptoms of this electrolyte imbalance are reflected as neurological disturbances such altered mental status, and more serious consequences such as convulsions and coma. On the plus side, shrinkage of osmoreceptors in the region of the hypothalamus induces thirst which promotes drinking behaviour aimed at correcting the water deficit, and also stimulates hypothalamic neurons to secrete more ADH for water retention by the kidneys.

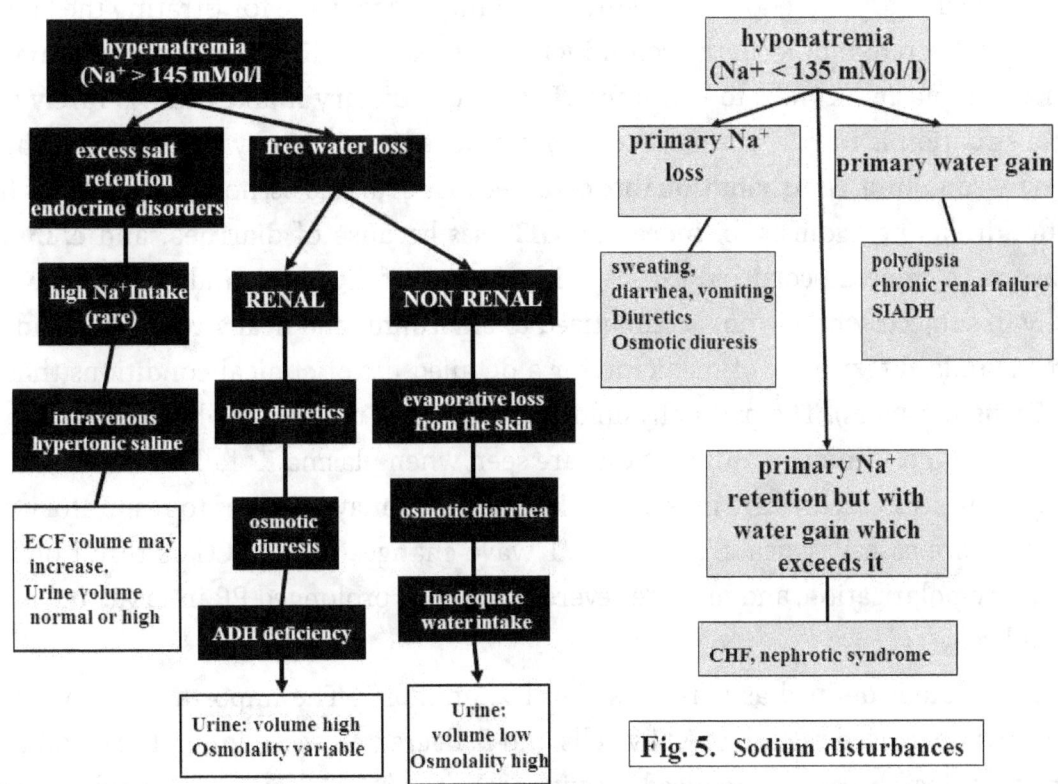

Fig. 5. Sodium disturbances

Hyponatremia means plasma Na$^+$ <135 mmol/L. It occurs as a result of excess water gain, or Na$^+$ loss, and results mostly in lowering of plasma osmolality. The dilution of the ECF results in a shifting of water into the intra cellular compartment. In the brain, this causes swelling and cerebral edema, resulting in neurological symptoms such as lethargy, headache, and confusion and stupor, which may progress to seizures and coma at plasma Na$^+$ level <120 mmol/L. It may be interesting to note that clinical situations such as heavy sweating, diarrhea and vomiting which result in hypovolemia, may be associated either with hypo orhyper-natremia, depending upon the relative losses of Na$^+$ and water.

Normal plasma potassium level ranges from 3.5 to 5 mmol/L. This is just about 2% of the total body K$^+$, the ICF containing about 38 times that found in the ECF. This happens because of the basolaterally located Na$^+$/K$^+$ ATPase pump which is mainly responsible for generating the trans-membrane electrical potential. The activity of this pump may be disturbed in heart and renal failure, and diabetes mellitus, resulting in the electrolyte imbalance problems often met with in these conditions. The large difference of K$^+$ between the two compartments may result in a low correlation between plasma levels of the cation, and the clinical symptoms and signs (especially ECG changes) produced by disturbances in K$^+$ metabolism. Hence it may be more prudent to be guided by clinical signs/ECG changes for initiating treatment for K$^+$ disturbances rather than await laboratory investigation reports.

Potassium is easily absorbed in the GIT. Excretion is mainly via the kidneys, and in sweat. The K$^+$ content of sweat of tropical populations like Indians is on the higher side, and is an indicator of their adaptation to environmental heat stress. In the kidneys, K$^+$ is reabsorbed in the proximal tubule, and is secreted via the principal cells in the distal convoluted tubules. Hence it is the urinary secretion of K$^+$ which is mainly responsible for titrating the body level of this cation. Secretion of K$^+$ in the colon increases greatly in diarrhea. **Hypokalemia** is said to occur when plasma K$^+$ falls to <3.5 mmol/L. Reduced dietary intake of K$^+$ is rarely a cause as such a situation is rapidly counteracted by a reduced secretion by the kidneys. The more common reasons are **a. redistribution into cells** because of alkalosis, hormones such as Insulin and beta adrenergic agonists; **b. increased GIT loss** because of diarrhea, and **c. increased renal loss** as in mineralocorticoid excess, with the use of diuretics, and excessive sweating, especially in subjects who are not acclimatized to environmental heat stress. (The reader may refer to a standard text book of medicine for a detailed list of clinical conditions that cause plasma K$^+$ disturbances). The pathophysiological feature is the relative hyper-polarization of the cell membrane. Clinical manifestations are seen when plasma K$^+$ falls to about 3 mmol/L. The commonly seen features are muscle weakness (which may progress to respiratory muscle paresis in severe cases), typical ST segment/T wave changes in the ECG as result of delayed ventricular re-polarization, and in more severe situations, prolonged PR interval and widened QRS complexes.

Hyperkalemia is defined as plasma K$^+$ level > 5 mmol/L. The important reasons for this disorder are i. increased release of K$^+$ by cells and ii. decreased secretion in the distal tubules. It occurs in acidosis because of reduced activity of the Na$^+$/K$^+$ ATPase pump. Insulin deficiency also affects this pump and induces hyperkalemia. In renal failure, secretion of K^+ into the distal tubules is reduced. Similarly, aldosterone deficiency results in retention of potassium. Sudden massive hemolysis releases large amounts of K^+ in to circulation causing hyperkalemia. Increase in dietary intake of K^+ is compensated for by an increase in secretion in urine, and hence in most situations, does not contribute to hyperkalemia. *(If you measure plasma K^+ immediately after a patient of epilepsy has thrown violent convulsions, what will you find?)*

Clinical presentation of hyperkalemia is some what similar to that of hypokalemia. The cardinal pathophysiology is a prolonged depolarization which impairs membrane excitability. Muscular weakness occurs, and this may progress to flaccid paralysis. Cardiac toxicity manifests as peaked T waves, followed by prolongation of the PR interval and the QRS duration. Ventricular fibrillation may develop. Though hyperkalemia is less common than hypokalemia, it is more dangerous because of its effect on the myocardium.

Acid-Base Disturbances

Table II: Common buffer systems

- Bicarbonate buffer $NaHCO_3/H_2CO_3$
- Hemoglobin Hb/HHb
- Phosphate HPO_4^{--}/H_2PO_4
- Other Proteins

Normal arterial blood pH ranges between 7.35 – 7.45

(Look up the pH values for urine, gastric juice and pancreatic juice, and reason out why they should be so).

The arterial pH is regulated by various buffer systems tabulated in Table II. These extra cellular and intracellular systems operate in concert with respiratory and renal regulatory mechanisms. The respiratory system makes rapid adjustments to the changes in arterial PCO_2 while the kidneys regulate plasma HCO_3^- by reabsorption of the filtered bicarbonate; formation of acid which is easily titratable; and excretion of NH_4 in urine. Buffer systems which operate in the blood and respiratory system act quickly between seconds to minutes. The kidneys may take between hours to days to regulate pH. These systems may correct pH to normal, or may only compensate partly by taking it as close as possible to the ideal. When the system **compensates**, the ratio of the numerator and denominator is retained at normal by bringing about adjustments in the values. For example the ratio of the $NaHCO_3/H_2CO_3$ buffer is normally 20/1 because the quantity of the numerator to the denominator is 24 mmol/l and 1.2 mmol/l. To maintain this ratio while adjusting for a pH disturbance involving this system, if the numerator increases, then the denominator will also change in the appropriate direction in order adjust the ratio to the desired normal level, there by adjusting the pH. As against this, when **correction** occurs, the adjustments are made so as to bring back the absolute levels of both $NaHCO_3^-$ and the H_2CO_3 to normal, as a result of which the ratio, and with it the pH, return to normal.

Shift of the pH to a level higher than 7.45 is alkalosis while a downward shift to <7.35 is acidosis. The disturbance which produces these changes is either alkalosis or acidosis respectively. The Acid-Base Nomogram is useful in interpreting the disturbance. This diagram may be found in any standard textbook of Physiology or Biochemistry. The diagram is made up of plots of bicarbonate concentration, $PaCO_2$, arterial blood pH values and arterial blood H^+ ion concentrations which intersect as per the Henderson-Hasselbalch equation.

Simple Acid-Base Disorders

Acid- Base imbalance may result in an Acidosis or an Alkalosis which in turn may be metabolic or respiratory (**Fig. 6**). Their definition evolves around change in the plasma bicarbonate, or the arterial carbon dioxide tension, with the appropriate change in the other parameter. An increase in $PaCO_2$ is normally accompanied by an appropriate increase in plasma bicarbonate (see below).Some times, a mixed disturbance may occur. For example, a respiratory alkalosis may co-exist with a metabolic acidosis.(These are highly complex disturbances, and beyond the purview of this book). Assessment of acid base disorders is supplemented by the estimation of the **Anion Gap.** This is the measurement of plasma anions such as anionic proteins (albumin), phosphate, sulfate and other organic anions which are not measured as a routine using the standard equipment. In certain situations, it is these anions that may determine the type of acid /base status in a patient. It is calculated as the sum total of plasma Na^+ - (Cl^- + HCO3$^-$). The normal value is usually about 12 $\pm$ 4 mmol/L. Addition of any acid radical other than HCl will increase the anion gap. Therefore estimation of this parameter becomes useful in detailed evaluation of patients with metabolic acidosis in conditions such as lactic acidosis, ketoacidosis, uremia, salicylate poisoning. Clinico-physiological manifestations of only the simple acid-base disturbances are outlined in **Fig. 7.**

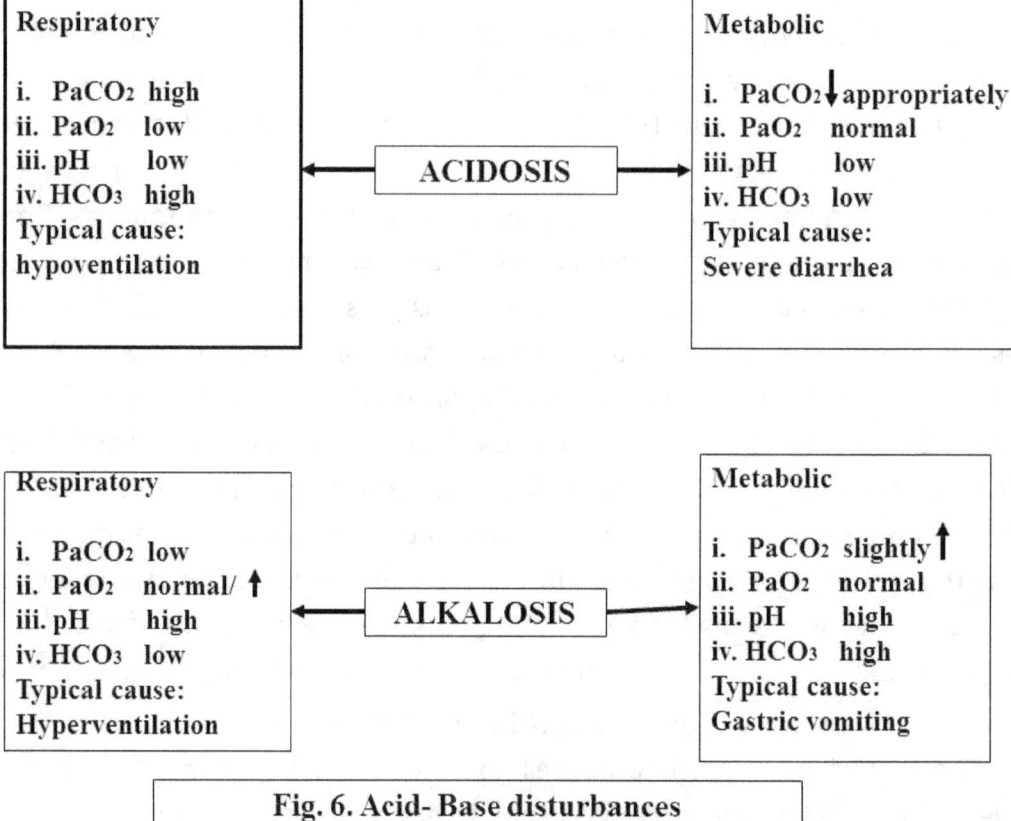

Fig. 6. Acid- Base disturbances

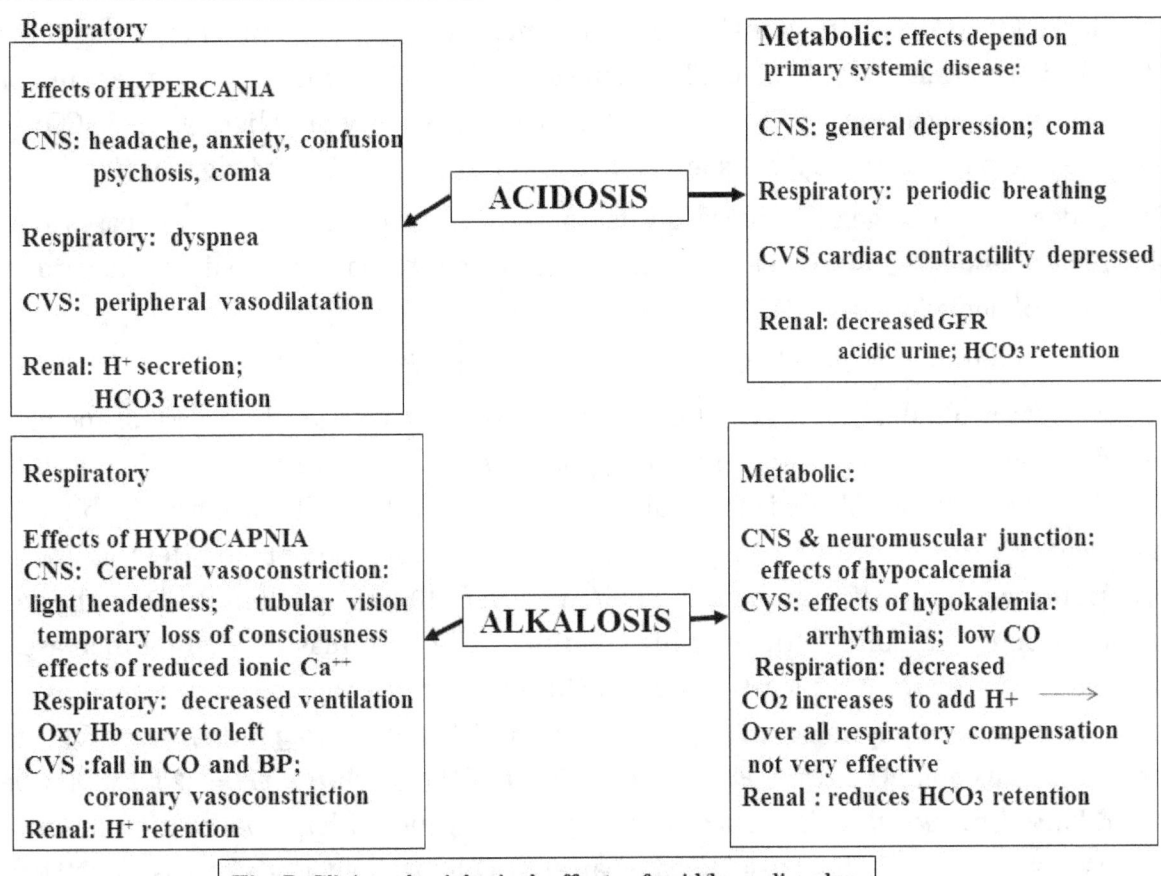

Fig. 7. Clinico-physiological effects of acid/base disorders

Acidosis: When the arterial pH falls below 7.35, acidosis is said to be present.

i. Respiratory acidosis occurs as a result of inability of the respiratory system to regulate blood CO_2 Any pathological situation which leads to HYPOVENTILATION *(see chapter on Physiological Disturbances in Respiratory Disease, and Fig 5 there in)* will lead to the development of respiratory acidosis. Here, an increase in ventilation cannot be offered as a clearing agent for the rising PCO_2 in blood. The inability of the respiratory system to do so may lie in the respiratory system itself, or outside the respiratory system. The latter may arise because of problems in the brain-for example effect of depressant drugs, or because of neuromuscular disorders of the chest wall. The typical cause of respiratory acidosis because of pathology in the lungs is Chronic Obstructive Pulmonary Disease. The main problem is the increase in $PaCO_2$. This adds H^+ to the blood, with a compensatory increase in HCO_3^-. The most distressing respiratory symptom is dyspnoea because of CO_2 retention. The kidneys help by secreting H^+ and retaining HCO_3^-.The "appropriate" increase in plasma bicarbonate is about 1 and 4 mmol/L per every 10 mmHg rise in PCO_2 in the acute and chronic stages respectively. The anion gap is within normal limits.

ii. Metabolic acidosis occurs when i. endogenously, excess amount of H^+ are added to the blood as in keto-acidosis, lactic acidosis; ii. kidneys fail to secrete adequately the naturally produced H^+ as in renal failure, or iii. there is a bicarbonate loss from the GIT as in severe diarrhea. In this situation, the respiratory system tries to compensate with an increase in ventilation (CO_2 washout) brought about by the stimulation of peripheral chemoreceptors by the high H^+ ion concentration in blood. The usual fall in $PaCO_2$ is about 1.25 mmHg per 1 mmol/L decrease in blood bicarbonate. The typical blood picture is a low pH, slightly low $PaCO_2$, low HCO_3^-. The anion gap is raised if the source of H^+ ions is organic acids as in ketoacidosis.

In acidosis, the oxygen-Hb curve shifts to the right because of an increase in 2–3 Biphosphoglycerate (2–3 BPG). In the chronic stage, the curve may swing slightly towards the left because of depletion of 2–3 DPG.

Alkalosis is said to be present when arterial blood pH is > 7.45.

i. Respiratory alkalosis. Hyperventilation (increase in ventilation in excess of metabolic demand) is the commonest cause of respiratory alkalosis. (Each and every increase in ventilation is NOT hyperventilation). There is invariably a washout of CO_2. The typical arterial blood gas and pH values are an increase in the pH; $PaCO_2$ falls to <35 mmHg; the blood HCO_3^- is low. The appropriate lowering of HCO_3^- is about 2mmol/L per every 10 mmHg fall in $PaCO_2$ in the acute stage, and about 4 mmol/L in the chronic condition. The PaO_2 may be slightly increased if hyperventilation occurs at sea level. Acute and chronic exposure to hypoxia of high altitude is also a potent cause of physiological hyperventilation. (*A healthy young man has the following arterial blood gas and pH picture: PaO$_2$= 20 mmHg; PaCO$_2$= 9 mmHg; pH =7.78. Explain how he could have developed this situation?*). The commonest cause of hyperventilation in human beings is psychogenic. Pathophysiologically, arterial CO_2 washout leads to vasoconstriction in the cerebral and myocardial blood vessels. The peripheral retinal vessels constrict, giving rise to symptom of tubular vision. Other vascular beds may dilate, and result in a transient and rapid fall in blood pressure. This in combination with the cerebral vasoconstriction, has been thought to be the cause of temporary loss of consciousness in susceptible individuals. The latter has been known complicate hyperventilation which occurs as a physiological response to certain situations in aviation such as exposure to severe heat stress, high acceleration forces, and acute accidental hypoxia, and may thus compromise flight safety. Hyperventilation may unmask seizure disorders. This may be explained by the fact that alkalosis induces a decrease in brain Gamma Amino Butyric acid (GABA), which is thought to function as an inhibitory neurotransmitter. This has been the principle of using hyperventilation as a "provocative test" while recording Electro Encephalograms in patients suspected to be having seizure disorders. Coronary vasoconstriction may precipitate an anginal episode in elderly patients with a history of ischemic heart disease.

ii. Metabolic alkalosis. Gastric vomiting which produces loss of HCl and extracellular fluid volume is by far the most common cause of metabolic alkalosis. Over enthusiastic naso-gastric aspiration may also be a cause, albeit accidental. Other less common reasons may be excess alkali administration, use of diuretics, excessive aldosterone secretion and hypokalemia. The arterial blood gas and pH picture is as given in **Fig. 6.** The $PaCO_2$ increases by about 0.75 mmHg /every mmol/l increase in HCO_3^-. The respiratory compensation is not very effective. The clinical representation of pathophysiology in the CNS is mainly directed by the reduction in ionized Ca^{++}. The most serious cardiovascular challenge is the development of arrhythmias as a result of the accompanying hypokalemia. At the kidney level, HCO_3^- reabsorption is slowed down.

References

1. Best and Taylor's Physiological Basis of Medical Practice, 13[th]edn.; Ed. Best JB. William & Wilkins, London; 1990.

2. Bijlani RL. Understanding Medical Physiology: A Text Book for Medical Students.2nd edn. 1997.

3. Cogan MG. Fluids and electrolytes: Physiology and Pathophysiology. Appleton and Lange; 1991

4. Ganong WF. Review of Medical Physiology. 21st edition. Lange Medical Books/McGraw-Hill, London. 2003

5. Gauer OH, Henry JP. Circulatory basis of fluid volume control. Physiol. Rev. 1963; 48: 423–481.

6. Guyton AC and Hall JE. Textbook of Medical Physiology 11th edition. Elsevier, Philadelphia; 2006.

7. Harrison's Principles of Internal Medicine Edt. Longo D., Fauci A, Ksper D, Hauser S., Jameson J, Joseph Loscalzo J, 18th ed. MCGraw Hill, London. 2011.

8. Iyer EM, Dikshit MB, Banerji PK, Suryanarayana S.100% oxygen breathing during acute heat stress: effect on sweat composition. Aviat Space & Environ. Med. 1983; 54: 232–235.

9. Kenny W. L., Chiu P. Influence of age on thirst and fluid intake. Med. Sci. Sports Exerc. 2001; 33: 1524–1532.

10. Lorenz JM. Assessing fluid and electrolyte status in the newbornClin. Chemistry. 1997; 43: 205–210[a]

11. Malhotra MS, Sridharan SK, Venkataswamy Y. Potassium losses in sweat under heat stress. Aviat. Space Environ. Med. 1976; 47: 503–504

12. McCance KL, Huether SE. Pathophysiology: The Biologic Basis for Disease in Adults and Children. 4th edition. Mosby; St. Louis USA; 2002

13. Phillips PA, Rolls BJ, et al. Reduced thirst after water deprivation in healthy elderly men. New Eng. J Med. 1984; 311: 753–759

14. Text Book of Medicine. 2nd edn. Ed. Souhami RL and Moxton J. Churchill Livingstone, London; 1994

CHAPTER - 3

Some Aspects of Deranged Kidney Function

Plan

1. The nephron
2. Renal failure:

   ARF

   i. Pathophysiology of pre-renal azotemia

   ii. Consequences of ARF

   iii. Intrinsic and post renal ARF

   CRF

   i. Pathophysiology ofCRF

   ii. Diabetic nephropathy and CRF

   iii. Hypertensive kidney disease

3. Nephrotic syndrome

   i. Definition

   ii. Pathophysiology of proteinuria

   iii. Pathophysiology of edema

4. Pathophysiology of common urinarybladder dysfunctions

   i. Dysfunction secondary to obstruction to urine flow

   Benign prostatic hypertrophy and bladder dysfunction

   ii. Urinary incontinence

   Automatic bladder

   Atonic bladder

   Stress incontinence

   Incontinence in the aged

Introduction

The myriad physiological functions performed by the kidney **(Table I)** fine tune the regulation of the milieu interior. The nephron, which is the functional unit of the kidney, selectively filters the plasma to form an ultrafiltrate through the glomerular membrane **(Fig. 1)** at a rate of about 125 ml/min (the glomerular filtration rate (GFR). The filtrate contains all the blood constituents except the proteins and the cells. *(At this stage the student should review the forces which determine the formation of the ultra filtrate, and recap its contents).* Though the cortical nephrons are the more numerous (approximately 70–80%), it is the juxta medullary nephrons that play the major role in concentrating urine. They act as "counter-current multipliers" and are more effective in the concentrating function because

Table I: Functions of the kidney

- concentration of urine
- water and electrolyte balance
- plasma volume maintenance
- regulation of pH
- blood pressure control
- excretion of waste products
- metabolism of chemicals, drugs
- endocrine activity
- gluconeogenesis

of their long loops of Henle which dip deep down in to the medulla. *(Which type of animals do you think will have long loops of Henle. State a plausible hypothesis for your answer?).* The functional histology of the tubule **(Fig.2)** explains the role that its various parts are likely to play in this. Most of the renal blood flow goes to the cortical nephrons while that to the medullary nephrons in their peritubular capillaries (the vasa recta) is very low (<5% of the total renal blood flow) and hence sluggish. This combination ("low and slow") helps the vasa recta to function as "counter current exchangers" in order to help retain the concentration achieved in the tubule. The renal blood flow is auto-regulated over a pressure range of about 75 to 160 mmHg, and this in turn helps to maintain the GFR also relatively constant over this pressure range. *(List sequentially the factors that determine the GFR).* After the age of about 30 years, the GFR declines normally at a rate of about 0.6ml/m²/min/year. In the females the GFR is slightly lower than in males.

Any disruption in kidney physiology affects the milieu interior adversely. In this chapter, the focus will be on the more common disruptions in kidney function which are reflected as pathophysiology.

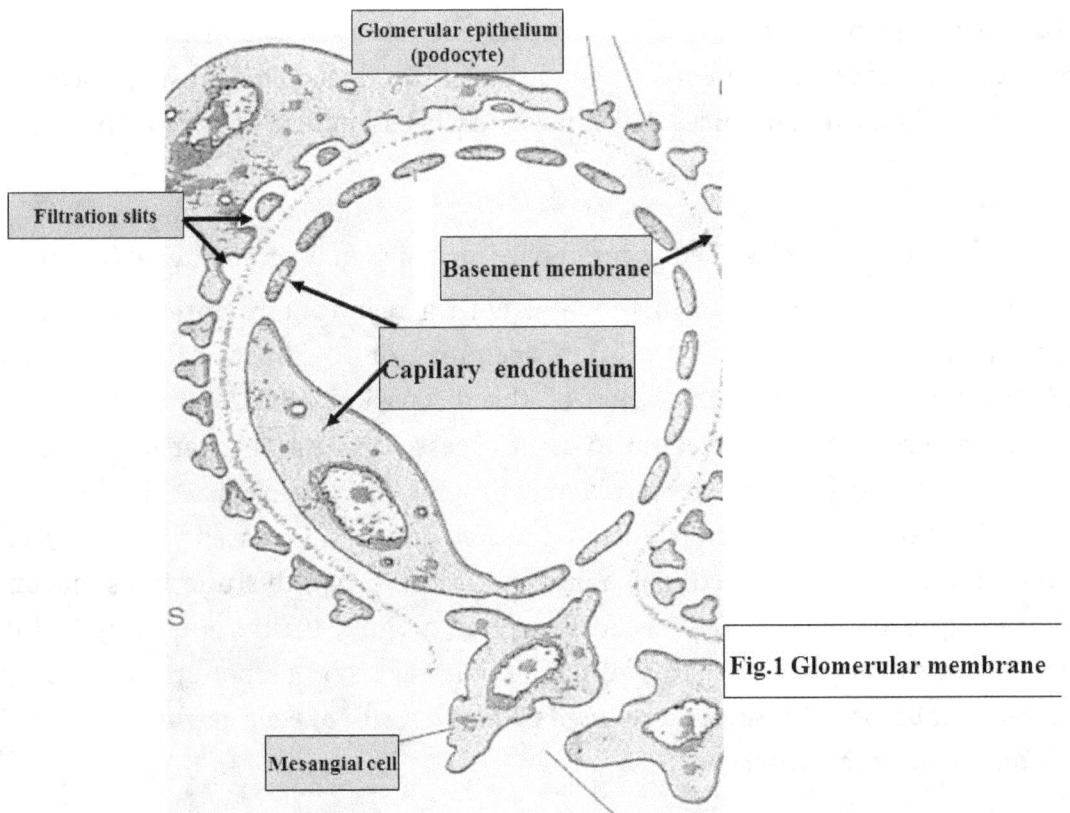

Fig.1 Glomerular membrane

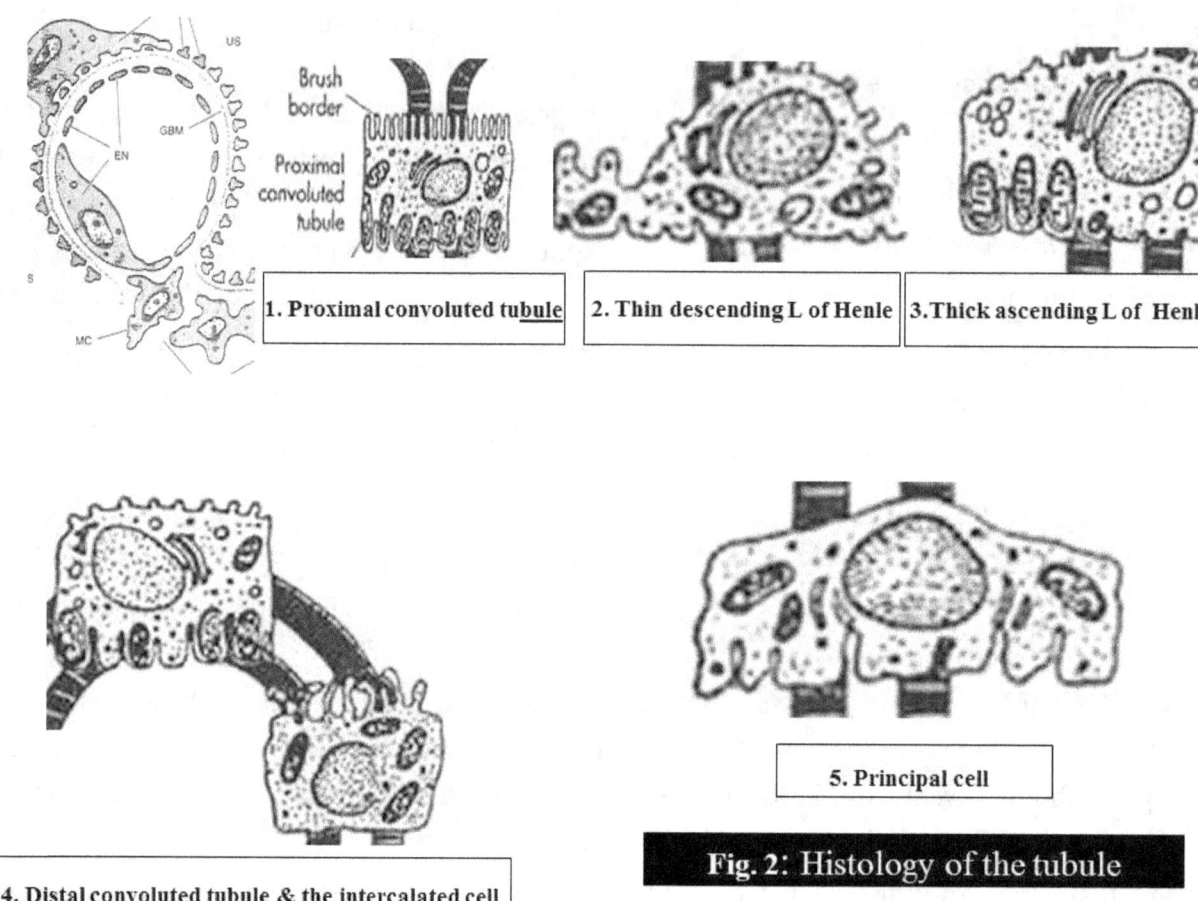

1. Proximal convoluted tubule | 2. Thin descending L of Henle | 3.Thick ascending L of Henle

5. Principal cell

Fig. 2: Histology of the tubule

4. Distal convoluted tubule & the intercalated cell

Renal failure

Renal Failure is the rapid decrease in the glomerular filtration rate which may occur over hours to days, with a resultant retention of nitrogenous waste products. This is usually accompanied by electrolyte and fluid disturbances. The failure may be either acute (ARF) or chronic (CRF)

Acute Renal Failure

Acute renal failure may be classified as pre-renal, intrinsic renal, or post renal **(Fig 3)**.

Pre-renal azotemia (reduction in the GFR with a resultant retention of nitrogenous waste products such as urea and creatinine) accounts for 40–80% of ARF. (It is this aspect of ARF that will form the main theme of the discussion). Biochemically, a reduction in GFR with a serum creatinine equal to or in excess of 44 μmol/lover base line value is considered diagnostic. Pre-renal azotemia may progress on to the classical ARF in which is chemictubular necrosis supervenes. ARF if detected early, and treated vigorously, is easily reversible. However, when it occurs as a part of a multi organ dysfunction syndrome, the mortality may be as high as 50–80%. ARF may be oliguric (urine output <400 l/day) or non oliguric (urine output = or > 400 ml/day), the latter having a better prognosis. (*Define anuria. Would it be correct to say that reduced GFR as a result of poor perfusion of the kidneys is in fact an adaptive response?*)

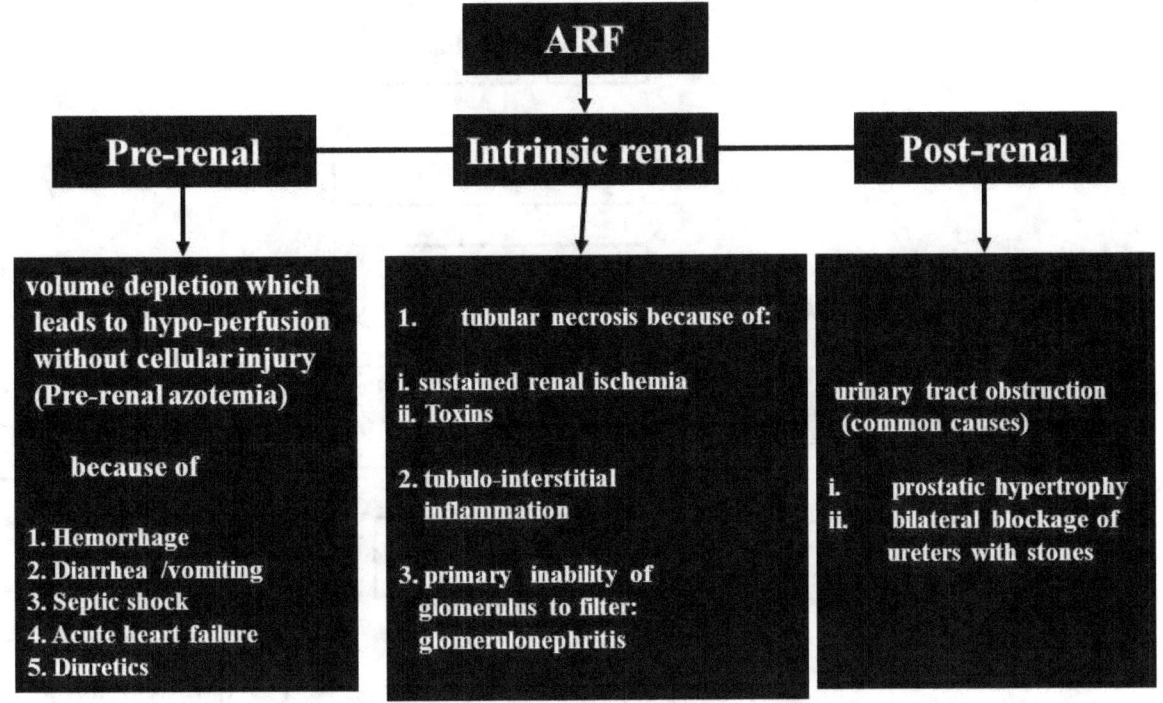

Fig. 3. Causes of Acute Renal Failure (ARF)

Pathophysiology of Pre-renal Azotemia

Pre-renal azotemia progressing on to the classical ischemic ARF is` a manifestation of renal hypo-perfusion. The early response to hypo-perfusion is a reduction in the GFR reflected as reduced urine output, and increasing blood creatinine and urea levels. In fact a monitoring of these biochemical parameters in a patient who is, or has been in circulatory shock, is an excellent method of diagnosing whether pre-renal azotemia is setting in. To begin with, there is an afferent arteriolar constriction and mesangial contraction because of sympathetic excitation, and the local release of angiotensin II which is initiated by Tubulo-Glomerular feedback mechanism. *(Please recapitulate this mechanism)*. Ischemia also induces formation of Endothelin I (ET I) by the capillary endothelium. When this occupies ET I receptors on the capillary endothelium, nitric oxide, prostaglandin E_2 and prostacyclin are produced. They help by dilating the afferent arteriole in order to increase perfusion, and with it the GFR. (**Fig. 4**). But when the compensatory mechanisms are overwhelmed, the GFR continues to fall, reaching a stage of pre-renal azotemia.

The **initial phase** of ARF lasts for hours to a few days, and is associated with a severe reduction in the GFR. Up to this point, the ischemic damage to the glomeruli is minimal but the volume of urine produced is low, though > 400ml/day (nonoliguric azotemia). The urine at this stage can be concentrated (osmolality > 500 mosmol/Kg), and the urine Na^+ concentration is <20 mmol/l indicating that the tubular function is intact, and that the nephrons are capable of producing normal urine, albeit the volume is well below normal. Only rarely does the situation lead to bilateral cortical necrosis and irreversible kidney damage.

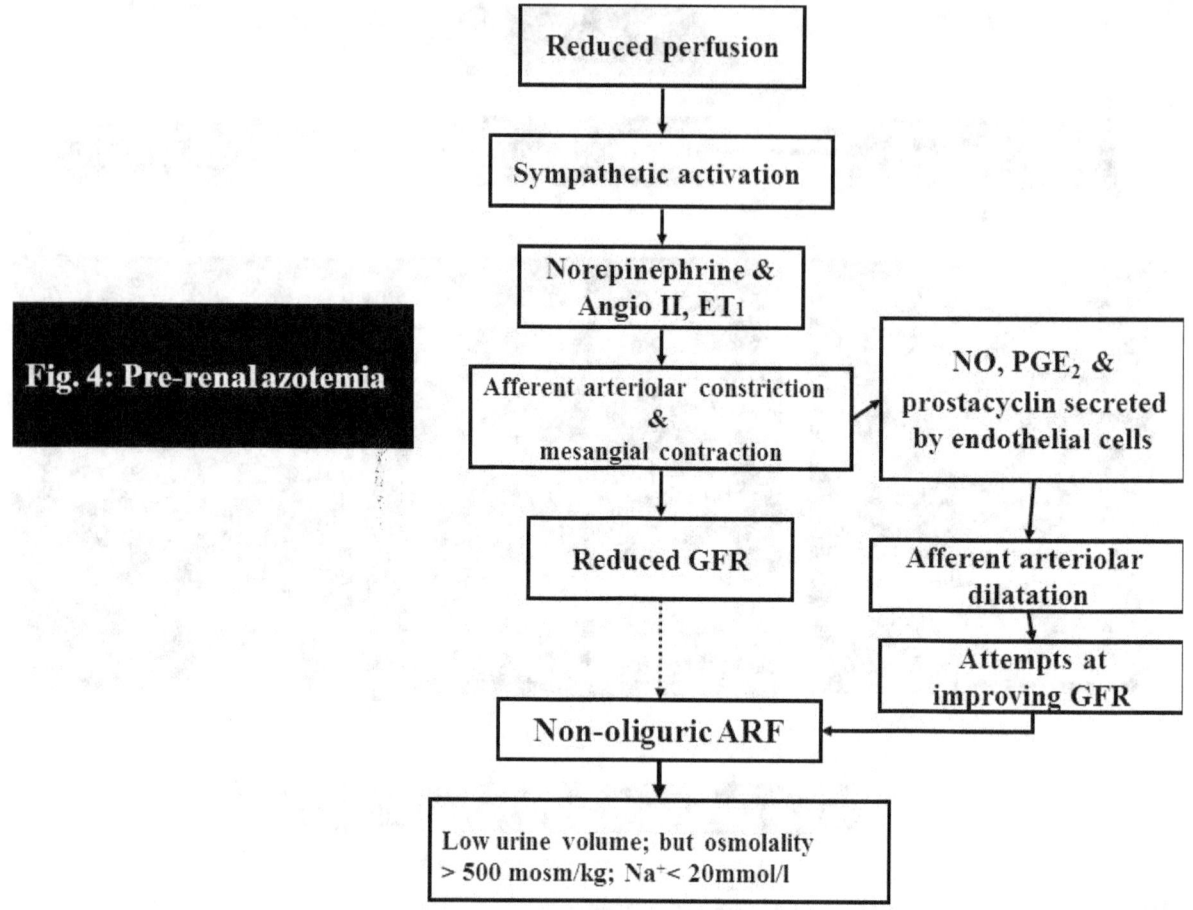

Persistence of severe ischemia may result in cellular ATP depletion which in turn leads to acute tubular cell injury. There is impairment of cell volume regulation, increase in intra cellular calcium and release of oxygen free radicals which further accentuate cell membrane disruption. The terminal portion of the proximal tubule and the medullary portion of the thick ascending loop are most affected (*Why are the terminal parts of the medullary PCT and the thick ascending limb maximally affected by ischemic injury*). Three possible patho-physiologic mechanisms affect the GFR adversely: i. shedding of the ischemia damaged apical membrane into the tubular lumen results in cast formation which blocks fluid flow in the tubules, and creates a back pressure ii. tight junctions between the lateral borders of the PCT cells open up, probably because of alterations in actin and microtubule cytoskeleton. This disruption allows the filtered fluid to leak back in to the extracellular space, and in to the blood to further reduce the GFR. iii. abnormal proximal Na⁺ reabsorption by the PCT results in presenting a large amount of this ion to the distal tubule, altering the tubulo-glomerular feedback mechanism into constricting the afferent arterioles. The damage is reflected as a decrease in the urine osmolality to <350 mosmol/Kg with a urinary Na⁺ concentration of > 40 mmol/l. Restoration of renal perfusion during this period will help in limiting the extent of cellular damage.

A **phase of maintenance** (1–2 weeks) follows. During this phase, cell damage is established, and GFR reaches its lowest value of about 10ml/min. It is important to note that this may happen if restoration of blood volume and with it the renal perfusion to normal has been inadvertently delayed. Possible mechanisms may involve persistent vasoconstriction of afferent arterioles because the ischemic damage has induced endothelial dysfunction resulting in compromised production of vasodilator metabolites such as NO and PGs, and an increase in the formation of Endothelin 1, and other leucocyte derived mediators. It is during this phase that effects of uremia become evident. During the **phase of recovery** which lasts for about 2 weeks, cellular and tubular integrity is re-established by tubular cell regeneration. Various growth factors such as Epidermal GF, Hepatocyte GF and Insulin like GF I may be involved. Thyroid hormone may help in the formation of epidermal GF. The outcome there after is a slow recovery of organ function. There is often an accompanying diuresis as the glomeruli have begun normal function, but the tubules have yet to regain fully their power of concentrating urine. The diuresis may also be caused by a prolonged excretion of the accumulated salt and water during the renal shut down phase. The important electrolyte disturbance during this stage is hypokalemia. The polyuria needs to be monitored closely to prevent dehydration.

(Can the kidney of a patient who has recovered from pre-renal azotemia be successfully transplanted in to another human being who requires the transplant? Explain your answer?)

Consequences of ARF

Disturbances in water and electrolyte and pH disturbances occur as a consequence of renal failure **(Fig. 5)**.

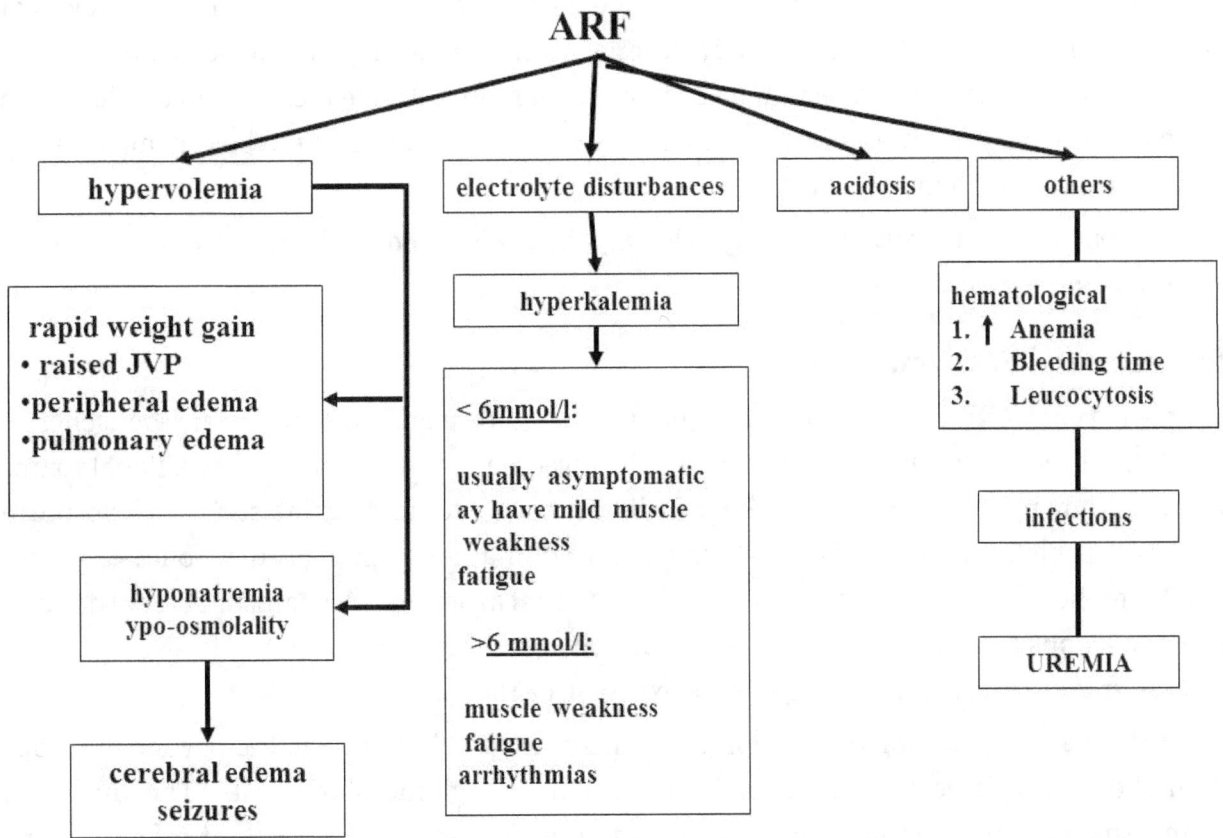

Fig. 5 Consequences of Acute Renal Failure

In the initial and maintenance phases of ARF, the inability of the kidneys to produce adequate amount of urine results in retention of water. In addition, parental and naso-gastric administration of fluids and nutrition is likely to add to the water load. This leads to hypervolemia and hypo-natremia with their consequent effects.

A common complication of ARF is hyperkalemia. The administered and ingested K^+ is not excreted adequately. In addition, cellular injury and the prevalent metabolic acidosis are additional sources, the latter because it affects Na^+/K^+ ATPase pump and forestalls the entry of K^+ in to the cells. Symptoms of hyperkalemia such as muscle weakness and fatigue appear when serum K^+ level > 6 mmol/l. Cardiac arrhythmias need to be detected in time by ECG monitoring.

The breakdown of ingested protein contributes to non-volatile acid which is not excreted as required, leading to metabolic acidosis with an increase in the anion gap. (*Recapitulate what is the anion gap. How may its measurement help in ARF*).

Anormocytic normochromic anemia may occur rapidly because the damaged kidneys are unable to secrete enough of erythropoietin. This is coupled with reduced RBC survival time, and hemodilution as a consequence of water retention. The bleeding time is prolonged because of reduced platelet count and function. Leucocytosis if it occurs indicates that secondary infection has set in. The latter may be a reflection of altered immune mechanisms, or repeated venepuncture or bladder catheterization. Other minor complications may be i. hypocalcemia because of relative tissue resistance to effect of parathyroid hormone as also reduced levels of calcitriol, and ii. hyperphosphatemia. In severe cases, uremia may be a major complication. (This will be discussed in the section on chronic renal failure).In the recovery phase, depletion of body water may occur as a result of extensive diuresis. This is likely to be accompanied by hypernatremia and hypokalemia.

(*Keeping the above consequences of ARF in view, what should be the approach to fluid management in such a patient*).

Intrinsic and Post renal ARF

Intrinsic renal ARF is most often a sequel to acute prerenal azotemia when the ischemic insult has been severe. The pathophysiology involved has been described above. Renal toxins are also known to cause such an injury. Radio-contrast dyes and agents such as cyclosporin are responsible for severe intra-renal vasoconstriction and the changes that follow are very similar to those brought about by the classical pre-renal azotemia. Mechanism of renal damage by endogenous nephro-toxins such as calcium, myoglobin and hemoglobin is also the same. The latter two are also known to generate oxygen free radicals.

Inflammatory conditions of the glomerulus may also precipitate ARF. (It is beyond the scope of this book to describe the various conditions which cause glomerulonephritis).The functional changesmay be triggered by a primary insult to the glomerulus as a result of inflammatory, toxic, metabolic or hemodynamic agents. The end result of this injury is determined by the secondary mediator activity that is induced. Leukotrienes, cytokines, oxygen free radicals and lysosomal enzymes may be associated in one way or another. Apart from a reduction in GFR, hematuria and protein urea not seen with pre-renal failure, are the hall mark of the acute nephritic syndrome. This happens because of damage to the capillary endothelium and its permeability, as also the basement membrane of the glomerulus.

The commonest cause of **post renal failure** in the male is benign prostatic hypertrophy (BPH) with bladder outlet obstruction. Any other cause of obstruction of the urinary tract at the level of the renal pelvis and below will also cause post renal failure. The obstruction to flow of urine filtered by the kidneys builds up a back pressure which finally increases the tubular hydrostatic pressure to a level equal to or higher than the filtration pressure in the glomerulus. The rapidity with which the tubular elements of the renal tract distend dictates the onset and severity of pain which may be felt as a renal or an ureteric colic. There may be azotemia or anuria in a complete obstruction. The ARF is easily reversed by release of the obstruction though the rise in GFR to normal is relatively slow. A post obstructive diuresisis likely and may lead to fluid depletion and electrolyte imbalance.

Chronic Renal Failure

Table II. Etiological factors in CRF

1. Hypertension
2. Diabetes mellitus
3. Autoimmune disease
4. Recurrent episodes of ARF
5. Polycystic disease of the kidneys
6. Nephrotic syndrome

Chronic Renal Failure (CRF) is the progressive attrition of the nephron population as a result of multiple etiologies, and which culminates in life threatening uremia. When the patient has to depend upon renal replacement therapy either by dialysis or transplantation, s/he is said to have reached **End Stage Renal Disease or ESRD.** The disease must last at least for about 3 months before this stage is reached. CRF may be the end result of a number of conditions, and the pathophysiology which initiates the process is determined by the initiating clinical condition. Thereafter, the progressive decline in nephron mass leads to a set of common progressive set of events which are independent of the original etiology. The penultimate representation of CRF is the occurrence of the uremic syndrome which is manifested when the GFR falls to <5% of normal. At this phase in time, dialysis or renal transplantation are required for survival. The compensatory hyperactivity of the undamaged renal tissue occurs as result of mechanisms brought into play in the presence of the initial renal damage inflicted. This is helpful in the maintenance of the GFR to the extent possible. *(What do you think happens if a person is left with only one kidney because the other has been removed surgically?)*This hyperactivity is modulated by cytokines, various growth factors, vasoactive substances, and the rennin-angiotensin system. The early adaptive role of these substances later changes to a maladaptive one. Of these, the rennin-angiotensin system has been associated with the development of sclerosis. Clinico-physiologically CRF is graded as various clinical stages, stage 1 starting with kidney damage but with a normally maintained GFR. When the GFR drops to <50% of normal, the stage of diminished renal reserve has been reached. Frank renal failure is said to have set in when the GFR is lowered to about 25% of normal. In the final stage, the GFR reduces to <5%, and ESRD is the end result. Survival at this stage of the disease is dependent upon renal dialysis or renal transplantation. This gradation is useful towards clinical management. CRF manifests as a multiple system involvement collectively called **uremia**.

Uremia

There is no exact definition of this entity. It is a **syndrome** of progressive deterioration of renal function with multi organ involvement. The syndrome involves the consequences of accumulation of protein break down products such as urea and creatinine as also disturbances in water and electrolyte regulation (hyperkalemia), and hormonal functions (more so in chronic renal failure). The clinical manifestations include anorexia, vomiting, headache, malaise and altered consciousness proceeding to coma, and have been attributed to high blood urea level. The various features which constitute the syndrome **(Table III)** are described below.

Water, Electrolyte and pH Disturbances

Table III. Pathophysiological manifestations of CRF

1. Impaired elimination of waste products
2. Water/electrolyte /pH disturbances
3. Hematologic and immune systems
4. Cardiovascular
5. Neurological
6. Metabolic and endocrine
7. Bone and musculoskeletal
8. Skin
9. GIT
10. Altered pharmacokinetics

The earliest sign of renal failure is azotemia (accumulation of urea, BUN and creatinine). The hallmark of CRF is the loss in flexibility in handling externally induced water and electrolyte changes. The ability of the tubules to reabsorb Na^+ gets impaired. This opens avenues for loss of the Na^+ ion, and along with it water. The problem is likely to be aggravated when there is vomiting/ diarrhea (often associated with uremia), and excessive sweating. This may lead to a further reduction of the GFR. The failing kidneys are unable to respond rapidly to these changes. Physiologically, external Na^+ supplementation should help in improving the GFR (water retention). However this principle of improving GFR may not be suitable if the patient is already hypertensive.

Most of the K^+ is excreted via the kidneys. When renal failure occurs, its fecal excretion increases markedly and helps in delaying the development of hyperkalemia as long as the intake of the ion is maintained normally. Dietary supplementation of K^+, use of beta receptor blockers, angiotensin receptor blockers, anti-inflammatory medications, and K^+ sparing diuretics, may cause hyperkalemia to occur rapidly, and must be guarded against. Classic retention of K^+, and the subsequent hyperkalemia occurs only when the GFR drops to about 10ml/min. At this stage of the disease, a careful monitoring of blood level of K^+ and its restricted intake is required. Hypokalemia is a rare occurrence in CRF.

Decreasing renal function limits the excretion of H^+ ions. Added to this, the limited ability of the tubular cells to generate NH_3 reduces the buffering capacity for organic acids. The overall effect is metabolic acidosis. The blood pH usually does not drop below 7.35, and is easily correctable. However the correcting fluids are likely to increase the Na^+ load. The severity of the acidosis may reach dangerous proportions if there is a sudden loss of alkali because of gastric vomiting or an increase in exogenous acid load.

Hematologic and Immune Involvements

A normocytic, normochromic anemia is by far the most important complication of CRF. Deficiency of erythropoietin production by the diseased kidney is the primary reason. Other contributory factors are i. Possible suppression of RBC synthesis because of circulating uremic toxins, ii. Shortened life span of RBC because of the same reason. Untreated anemia leads to lethargy, weakness, depression, and decreased cognitive functions. It is often associated with cardiovascular complications (discussed later in another section). It is thus obvious that treatment of this complication must be the priority in the management of CRF. ***What could be the best method of treating this anemia?***

Bleeding disorders which may manifest as epistaxis, menorrhagia, easy bruising of skin, and gastrointestinal bleeds have been attributed to altered platelet aggregation and adhesiveness because of uremia. Interestingly enough, anaemia may contribute to platelet function disturbance. Normally platelets are distributed close to the endothelial surface while the RBCs tend to move in the centre of the flow stream. In anemia, because of lowered viscosity which increases the velocity of flow, this normal "skimming" pattern of platelets is disturbed. They are no longer readily available for adherence to the endothelial surface whenever their activation is desired, the end result being a bleeding disturbance.

The high levels of urea and creatinine with other waste products of metabolism are known to impair inflammatory responses to infection. There is the possibility of reduced phagocytic activity, reduced cell mediated immune activity, and a decrease in the number of circulating granulocytes. All these factors increase the vulnerability of a patient of CRF to secondary infections.

Cardiovascular Complications in CRF

Patients with CRF are subjected to a manifold increase to the risk of developing cardiovascular abnormalities, and thus this greatly enhances their the morbidity and mortality. The more frequent manifestations of cardiovascular disease are i. hypertension, ii. Ischemic heart disease; and iii. congestive heart failure.

Hypertension is by far the earliest manifestation of cardiovascular disease in CRF. The main pathophysiology centres around increased circulating fluid volume **(Fig. 6)**.

Ischemic heart disease may follow hypertension. The coexisting anaemia reduces blood viscosity. This leads to an increase in the heart rate and cardiac output, thereby increasing myocardial work load. At the same time, myocardial hypoxia may occur because of the anaemia. These factors in combination with the dyslipidemia which develops, aggravate the IHD which has already been initiated by the left ventricular hypertrophy consequent to the hypertension. The end result of this pathophysiology is congestive heart failure. This may also take the form of pulmonary edema. Uremia is also known to produce pericarditis which may progress to a cardiac tamponade.

The neurological complications of CRF occur both, as peripheral neuropathy and central disorders. The neuropathy may be sensory and/or motor, and is a result of demyelination by uremic toxins. Uremic encephalopathy is the main clinical issue affecting the central nervous system. Exact mechanisms involved are as yet poorly understood. Delirium, coma and seizures are serious manifestations.

Metabolic disturbances usually involve impaired glucose metabolism. However the hyperglycemia is usually minimal. Blood insulin level may be slightly elevated as the kidneys which are responsible for insulin metabolism are non functional.Sex steroid function may be hindered. Menstrual disturbances in the female, oligospermia in the male, and altered sexual dysfunctions in either sex may be the end- result.

Bone and Musculoskeletal Involvement

Calcium, phosphate and calcitriol metabolism is affected early in the disease process. The problem starts with the disturbance in phosphate regulation. This results in the retention of phosphate. When this happens, there is a reciprocal fall in blood Ca^{++} level. In order to make up this deficiency, the parathyroid hormone secretion increases releasing excess Ca^{++} from the bones. The end-result is bone demineralization. As the formation of calcitriol is dependent upon the normal functioning of the kidneys, CRF reduces availability of the active form of the hormone. The absorption of Ca^{++}in the small intestines is reduced because of this, further accentuating non availability of the cation in blood- which in turn adds to the increased demand upon the parathyroid gland to make up the overall deficiency. The overall pathophysiology is collectively known as **renal osteodystrophy** of ESRD/CRF. There is a likelihood of abnormal Ca^{++} deposition in soft issues, and blood vessels leading to soft issue calcification (calciphylaxis). One of the manifestations of calcium metabolism abnormality is proximal muscle weakness, particularly in the lower extremities.

Skin involvement in CRF/ESRD is reflected as pallor because of anaemia, and dryness of skin because of reduced sweat gland activity. Hyperphosphatemia and phosphate crystal deposition in the skin leads to pruritis. Repeated skin pricks during dialysis subject such patients to skin infections. Deficiencies of platelet function may be seen as purpuric patches.

Gastrointestinal tract disorders include anorexia, nausea and vomiting. They probably result from decomposition of urea to NH_3 by GIT flora. GIT bleeding may occur.

Disturbed pharmacokinetics. As many drugs and their metabolites are eliminated via the kidneys, CRF results in derangement of their metabolism and excretion. The end result is a high incidence of drug toxicity.

Of the various causes of CRF **(Table II)**, due consideration is given here to hypertension and diabetes mellitus.

Hypertensive Renal Disease

(Fig.6). Hypertension (HBP) has a dual relationship with kidney function: It may be the end result of kidney disease, or kidney disease may be the end result of long standing HBP. The afferent arteriole is again the vessel involved. There is a deposition of an eosinophilic material which thickens the vessel and produces ischemic injury of the glomeruli and the tubules. The changes are often a part of involvement of other vascular beds such as retinal vessels and the myocardium. The GFR is usually maintained. However the consequences of malignant or accelerated hypertension are far more serious. Rapidly developed sustained HBP with a diastolic pressure of > 130 mmHg is accompanied by involvement of organ systems such as the retina, CNS, and the rapid deterioration of renal function.

Diabetic Nephropathy and CRF (Diabetic Glomerulosclerosis-Kimmelstiel-Wilson Disease)

Diabetic nephropathy is a leading cause of ESRD, and is known to occur in both type I and type II diabetes. By far the earliest manifestation of diabetic kidney damage is a micro-albuminuria. Pathologically, there is a thickening of the glomerular basement membrane and deposition of extracellular matrix in the mesangium. Factors associated with matrix deposition are hyperglycemia, glomerular hypertension, growth factors such as Growth Hormone, Angiotensin II, Insulin like Growth Factor (ILGF), and the Tumor Growth Factor (TGF)–β. The end result is the development of glomerular hypertension. Exact mechanism as to how this happens is not clear, but a possible sequence of events is given in **Fig. 7.** The principle of use of ACE inhibitors and Angiotensin II receptor blockers is also dictated by the set of events, and helps to delay the onset of diabetes induced nephropathy. More recently, it has been suggested that the podocyte foot process protein molecule nephrin has been found to be critical for the action of insulin on podocytes and may be involved in the pathophysiology of diabetes nephropathy. The reason as to why and how all diabetics do not develop these pathophysiological changes is not clear. It is possible that hemodynamic and metabolic factors act in combination with genetic predisposition as identical twins tend to develop such a complication more frequently.

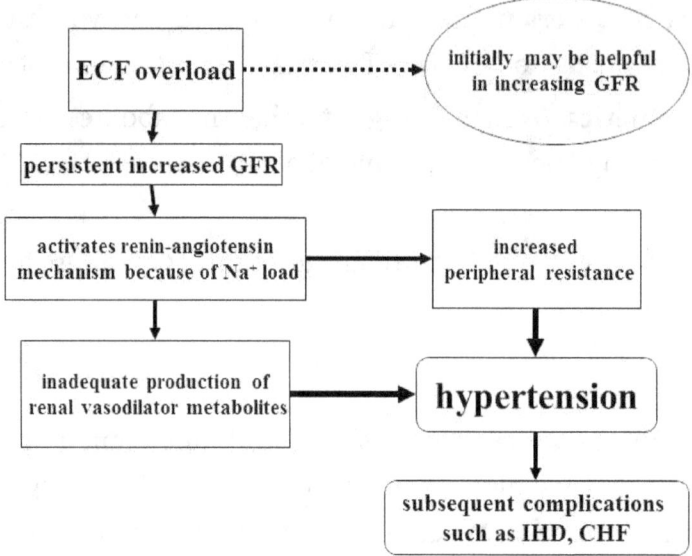

Fig. 6 : Possible algorithm for the development of Hypertension (HBP) in CRF

Table IV: Some Clinical conditions causing Nephrotic Syndrome

  i. Diabetes mellitus

 ii. Membranous glomerulonephritis

iii. Minimal change glomerulonephritis

 iv. Systemic Lupus Erythmatosis

  v. Amyloidosis

 vi. Rheumatoid arthritis

vii. Drug- induced nephropathy

Nephrotic syndrome is a clinical complex consisting of proteinuria (>3gms/24 hours), hypoalbuminemia because of *an increase in glomerular membrane permeability*, edema, hyperlipidemia and lipiduria, and hypercoagulability. This is to be distinguished from Nephritic syndrome in which there is an inflammatory response that *decreases the permeability of the glomerular membrane*, and which manifests clinically as hypertension, hematuria, azotemia and oliguria. The syndrome often develops as a complication of Diabetes Mellitus (common), Systemic Lupus Erythamatosis, or Amylodosis. **(Table IV)**. Sometimes it is a primary disorder.

Proteinuria is the pathognomonic feature of nephrotic syndrome (NS). This is mainly an albuminuria, and occurs because the permeability of the glomerular basement membrane to proteins is altered. Any disease that does this will result in the development of proteinuria. Normally, the endothelium of the membrane capillaries is negatively charged, and this easily repels the negatively charged albumin, preventing it from being filtered through. The mechanism because of which this situation is altered in NS is not yet worked out, but the defect is thought to lie somewhere in the basement membrane and its podocytes. Alteration in a number of protein molecules such as podocin, podoplanin, nephrin and alpha-actinin 4, found in the podocytes has been recently implicated in the pathogenesis of proteinuria. Even if the exact mechanism of action of these molecules is not yet known, these protein molecules are thought to be responsible for maintaining the selective filtration barrier in the normal nephron by controlling the cell signaling in the podocytes. There is a suggestion that nephringets relocated from the podocyte foot pores into the cell cytoplasm thereby disturbing the configuration of the slit diaphragm filaments of the podocyte. Sialic acid (an anion) content of the basement membrane may become deficient and this may make the membrane permeable to albumin passage. *(Why do you think these patients are more susceptible to infections? Also, why may the chances of iron deficiency anemia, hypercoaguability and some endocrine disorders increase?).*

Fig. 7: Possible pathophysiology of the development of glomerular hypertension secondary to Diabetes mellitus, & its consequences

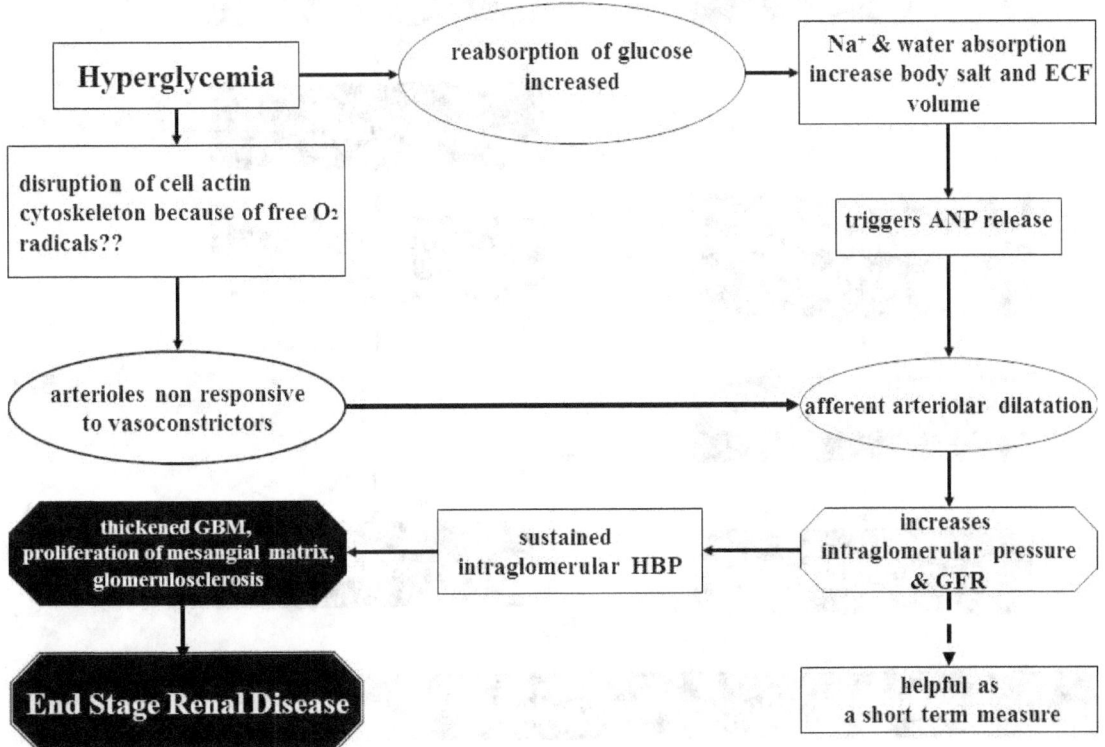

Edema formation in patients with NS was thought earlier to be a consequence of hypoproteinemia, and the resultant decrease in plasma oncotic pressure. This simplistic explanation is currently being debated. It has been proposed that increased renal catabolism and reduced hepatic synthesis of albumin may also be involved. Others are of the opinion that the decreased albumin synthesis by the liver is only relative because the liver is unable to keep pace with the amount of albumin lost. With leakage of fluid out of the blood vessels, the circulating fluid volume is compromised. To make up for this, the rennin- angiotensin system is brought into play resulting in salt and water retention. Increase in ADH secretion and suppression of the atrial natriuretic factor (ANP) *(How do you think these reactions are triggered?)* that occur also work towards retaining fluid. The end result then is an increase in hydrostatic pressure because of the accumulating fluid. This accentuates fluid leak in to the extracellular space to enhance the edema. Some experimental evidence however suggests that Na$^+$ and fluid retention may continue even after albumin infusion and blockage of the renin-angiotensin system. Here, blunted responsiveness to circulating ANP may play a role in maintaining a high circulating fluid volume. The pathophysiological consequences of nephrotic syndrome which result in proteinuria, edema, lipiduria are given in **Fig. 8.**

The hyperlipidemia of NS occurs because of increased synthesis of lipoproteins by the liver. This is in response to a relative deficiency of lipoproteins which also leak out of the kidneys. Defective lipid catabolism may also be involved. Increase in low density lipoproteins and cholesterol are the main contributors to the hyperlipidemia. It is possible that this metabolic complication may accelerate atherosclerosis and worsen the overall situation.

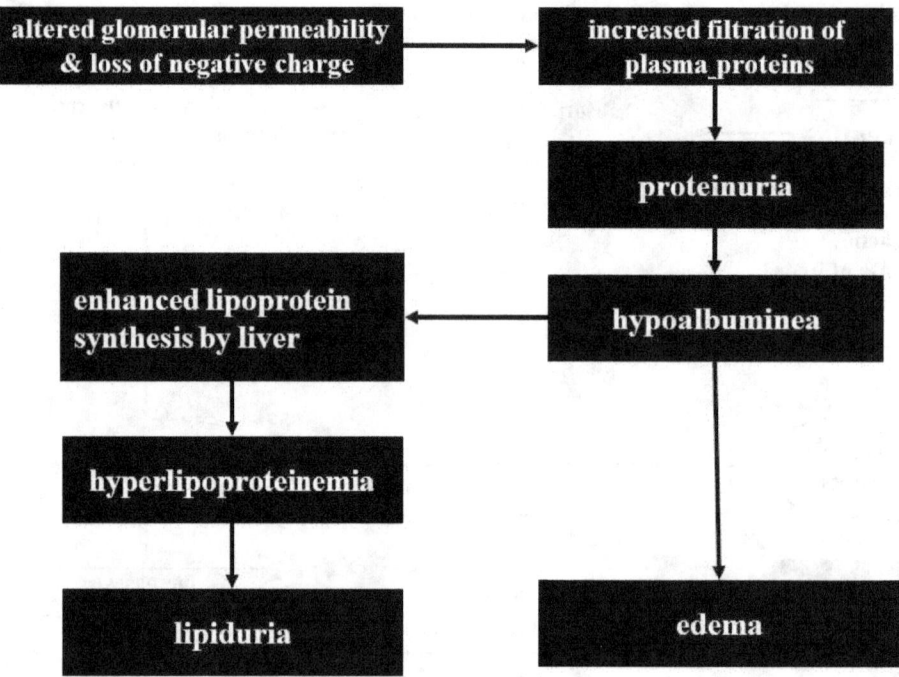

Fig.8 Pathophysiological consequences of nephrotic syndrome

Drug toxicity may increase because the proteins which normally bind to circulating drugs, regulating their effects, are in short supply. This increases availability of "free" drugs, enhancing their pharmaceutical effects.

(Now that you are aware of the pathophysiology of Nephrotic Syndrome, list its possible clinical complications).

Pathophysiology of common urinary bladder dysfunctions.

Prior to embarking on this study, it is suggested that the reader reviews the physiology of bladder function and the micturition reflex. The urinary bladder is a low pressure container with a high compliance. Collection of urine in the bladder progressively increases bladder volume from about 10 ml to 400 ml. The pressure change over this range of volume however does not exceed about 5cm/H_2O. The bladder sphincter, though, has a high tone with a pressure of about 45–60 cm/H_2O. Therefore if urine is to be voided, the contractions of the bladder must generate a pressure that exceeds the sphincter pressure. The micturition reflex takes care of this.

Dysfunction Secondary to Obstruction to Urine Flow

Extrinsic mechanical obstruction to urine flow is most commonly caused by benign prostatic hypertrophy. (Other causes of urethral blockage are not considered here. However the basic bladder function pathophysiology that emerges is similar).

The lobes of the prostate in the male surround the upper part of the urethra. When this gland hypertrophies with advancing age, the urethra is compressed, causing the extrinsic mechanical obstruction to the flow of urine.

The initial **compensatory** response of the bladder to obstruction of the neck is hypertrophy, as also an increase in its sensitivity to stretch secondary to accumulation of urine. Both help to overcome the obstruction. Functionally this results in an increase of frequency of micturition during the day as well as at night. As the obstruction progresses, the bladder needs to generate increasing force in order to void the urine. The bladder pressure exerted may be as high as 100 cm/H_2O (normal range 20 to 40 cm/H_2O). This degree of force generation during voiding is difficult to sustain throughout the process of micturition. The expulsion of urine therefore gets interrupted. In the meanwhile, urine continues to collect in the bladder, reactivating the micturition reflex. The consequence of this sequence of events is two fold: the frequency of micturition increases, and the bladder empties only partially every time reflex is activated. The end result is retention of urine. With continued obstruction, the smooth muscles of the bladder hypertrophy, producing folds in the mucosal tissue (Trabeculae). Some of these invaginate to form little pockets called cellules which extend outwards as diverticula. Urine easily stagnates in them making them a nidus for infection. The inter-ureteric ridges also hypertrophy, blocking the bladder-ward flow from the ureters. In the long run, the ureters dilate and the urine which is collected in them exerts a hydrostatic pressure which opposes the glomerular filtration pressure, gradually reducing the GFR. Finally the kidneys are subjected to a **post-renal failure**. At this point of time, compensatory mechanisms are no longer effective. The detrusor muscle contraction is too short lived and weak to expel the urine efficiently. Finally the overstretched bladder develops overflow incontinence. The classical symptoms at this stage are a distended, painful bladder, increased frequency of micturition, and urgency with overflow incontinence. The condition is very distressful for the patient and requires early surgical intervention.

Urinary Incontinence

The automatic bladder

If there is spinal injury of the spinal cord above the level of S2, an automatic bladder develops. In the immediate aftermath of the injury, there is the phase of spinal shock during which the bladder becomes atonic, and needs to catheterized. Following this phase, as the bladder fills, a threshold level at which the bladder stretch receptors are stimulated, is reached. This initiates the spinal micturition reflex, and the bladder empties itself. As there is no voluntary control of micturition, the process is repeated.

Atonic bladder

At the spinal level, the micturition reflex depends upon the afferent information going into the spinal cord, and efferent impulses from the sacral centres. The sensory input from the bladder may get interrupted because of neuropathy. If the stretch sensation from the bladder cannot be appreciated at the spinal level, or centrally because of this interruption, the bladder keeps on distending as the urine collects in it. At a particular point in time, the pressure which build up is enough to force open the bladder neck and allow the urine to dribble out. As soon as the pressure reduces even slightly after this decompression, the dribbling stops.

Stress incontinence

Normally, the urethro-vesical alignment is such that sudden changes in the urinary bladder pressure during a sudden bout of coughing, sneezing, weightlifting etc, is distributed all around the bladder neck. This maintains the intra urethral pressure at a level higher than that which is generated in the bladder, and prevents leakage of urine. In situations wherein there is a weakening of the pelvic floor muscles, or a reduction of tone in the urethrovesical sphincter, the normal alignment of the urethrovesical angle is disturbed, and urine gets voided involuntarily when situations mentioned above arise. This is more common in elderly women.

Incontinence in the aged

Incontinence is more commonly seen in the elderly because of a variety of reasons. This may happen because of a reduction in volume capacity of the bladder, weakening of the sphincter tone, degeneration of the detrusor muscle, and use of medications such as diuretics. Cerebral ischemia, or Alzeimer's disease may be responsible for diminishing attention to bladder filling. Mobility problems because of conditions like arthritis and sarcopenia (muscle wasting) may delay the individual's reaching the toilet in time. The various situations need sympathetic handling

References

1. Agraharkar M, Gupta R, Agraharkar A. Acute renal failure. E-medicine.com/Med/Topic 1595. htm. Sept 2007.

2. Baigent C, Burbury K, Wheeler D. Premature cardiovascular disease in chronic renal failure Lancet, 2000; 356:147–152

3. Best and Taylor's Physiological Basis of Medical Practice, 13[th]edn.; Ed. Best JB. William & Wilkins, London; 1990.

4. Bijlani RL. Understanding Medical Physiology: A Text Book for Medical Students.2nd edn. 1997.

5. Coward R, Welsh G, Koziell A et al. Nephrin is critical for the action of insulin on human glomerular podocytes. Diabetes. 2007; 56: 1127–1135

6. Davidson's Principles and Practice ofMedicine. Edt. Haslett C, Chilvers ER, Boon NA & Colledge NR. 19th Edn. Churchill Livingstone., London. 2002

7. Ganong WF. Review of Medical Physiology. 21st edition. Lange Medical Books/McGraw-Hill, London. 2003

8. Gauer OH, Henry JP. Circulatory basis of fluid volume control. Physiol. Rev. 1963; 48: 423–481.

9. Guyton AC and Hall JE. Textbook of Medical Physiology 11th edition. Elsevier, Philadelphia; 2006.

10. Gray's Anatomy: The Anatomical Basis of Medical Practice; 39thedn. Ed. Standring S; Elsevier, Churchill Livingstone, London; 2005. Kidney; Chapter 91; pp 1269–1284.

11. Harrison's Principles of Internal Medicine Edt. Longo D., Fauci A, Ksper D, Hauser S., Jameson J, Joseph Loscalzo J, 18th ed. MCGraw Hill, London. 2011.

12. Koop K, Eikman M, Baelde HJ et al. Expression of podocyte-associated molecules in acquired human kidney diseases. J. Amer. Soc. Nephrology. 2003; 14: 2063–2071.

13. Lorenz JM. Assessing fluid and electrolyte status in the newborn. Clin. Chemistry. 1997; 43: 205–210[a]

14. Mathieson PW. Editorial: Nephrin send us signals. Kidney International. 2003; 64: 756–757.

15. McCance KL, Huether SE. Pathophysiology: The Biologic Basis for Disease in Adults and Children. 5th edition. Mosby; St. Louis USA; 2006

16. Palmer BF. Nephrotic edema- pathogenesis and treatment. Amer. J. Med. Sci. 1993; 306: 53–67.

17. Porth C. Essentials of Pathophysiology: Concepts of Altered Health States. 2nd edn; Lippincott Williams and Wilkins; Philadelphia;2007.

18. Text Book of Medicine. 2nd edn. Ed. Souhami RL and Moxton J. Churchill Livingstone, London; 1994

19. Thadani R, Pascual M, Bonventre JV. Medical Progress: Acute Renal Failure. New Engl. J. Medicine. 1996; 334: 1448–1460

20. Travis L. Nephrotic Syndrome. E medicine.com/Ped/topic1564htm. 2005

CHAPTER - 4

Physiological Disturbances in Cardiovascular Disease

Plan

1. General layout of the CVS

2. Normal coronary circulation

 i. factors affecting coronary vascular diameter

 ii. collateral vesselsand myocardial protection

3. Pathophysiology of IHD

 i. myocardial oxygen supply demand ratio

 ii. altered situations which lead toIHD

 iii. effects of myocardial ischemia

 iv. markers of ischemia

 v. mechanical disturbances in the hypoxic myocardium

 vi. symptoms & signs of IHD

 vii. physiological basis for the management of IHD

4. Pathophysiology of hypertension

 i. Normal BP range and definition of HBP

 ii. Types of HBP

 iii. Cardiac output & peripheral vascular resistance

 iv. Factors affectingPathophysiology ofessential HBP

 v. Pathophysiology ofend organ damage with HBP

 vi. Clinical presentation of HBP vis-a-vis pathophysiology & physiological principles of treatment

5. Pathophysiology of Heart failure

 i. Definition

 ii. Normal pressure volume relationship in the ventricles.

 iii. Physiological adjustments and mal adjustments.

 iv. Systolic & diastolic dysfunction

 v. Types of heart failure

 vi. Clinical presentation of HF vis-a-vis pathophysiology & physiological principles of treatment

6. Pathophysiological consequences of valvular heart disease

General Layout of the Cardiovascular System

The cardiovascular system is designed exquisitely to transport oxygen and nutrients and various other materials to tissues, and is tailored to tissue demand. At the same time it removes metabolic waste products of the body for further disposal. The vehicle which is used is fixed at about 70 ml/Kg body weight or about 5 litres (as the blood volume) in a closed loop system enclosed in a network of conduits, the circulatory system. To propel this vehicle through the blood vessels which span the remotest outposts of the body, there is a pump (the heart) **(Fig. 1)**. At rest, this pumps out about 5 litres of blood every minute (the cardiac output) to various tissues. With every effort (beat) it pushes out about 70 ml of blood (stroke volume), and it does so about at a rate of about 70 times/minute (heart rate).Functionally, the pump is separated into 2 sections- a high pressure section (the left ventricle)which pushes the blood into the circulatory system, giving it enough momentum to make it go around fully, and return to the right side of the heart. The force it imparts (the arterial pressure generated) must be adequate enough to push the blood through various channels in spite of the variable resistance these channels offer. The right side of the pump however is required to operate against a relatively mild resistance of the pulmonary system, It does not therefore have to be a powerful pump.

With the volume available fixed at about 5 litres, it is impossible to "irrigate" every tissue with equivalent volume of blood per minute. So an internal distribution system in the form of resistances is used. This helps in directing the blood flow on an as required basis **(Fig. 2)**. Whenever a particular tissue demands more flow, diversion of blood flow is done by turning on/ partially closing off the taps (increasing/decreasing the resistances).

(At this stage the student must recall the functional anatomy of the heart chambers, of the muscles of the ventricles, and the resistance vessels).

The manner of distribution of the available blood volume is given in **Fig. 3 (A)**, while the pressure heads found at rest in the cardiovascular system are indicated in **Fig 3 (B)**. *(Look at the left atrial and left ventricular pressures. Why is the diastolic pressure in the LV so low?).*

After having considered the above, it might be realized at this stage that pathophysiology of the system will manifest itself if and when 1. The heart as a pump malfunctions. 2. The blood vessels are adversely affected (peripheral vascular disease) and 3. The blood volume that circulates within the cardiovascular system may either diminish or may increase beyond acceptable limits and change the pressure heads. In this chapter, the discussion will be limited to the effects of a combination of heart and peripheral vascular pathology which culminate into any one or a combination of the following conditions: Ischemic heart disease (IHF), heart failure (HF), hypertension (HBP) and valvular heart disease **(Fig.4)**. The loss of circulating blood volume and its manifestations as pathophysiology has been dealt with in the chapter on Shock.

Fig. 1 General layout of the CVS and its aims

- Perfuse tissues using 5000ml [80 ml/kg] of blood as per requirement on a continuous basis

- Rate at which this is done:

- At rest: 5 litres/min   (Cardiac Output)

- Increase in cardiac out put as and when needed

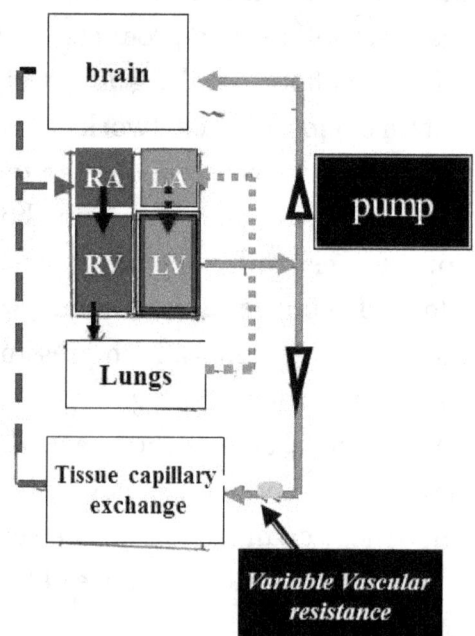

Fig. 2.   Distribution of cardiac output at rest

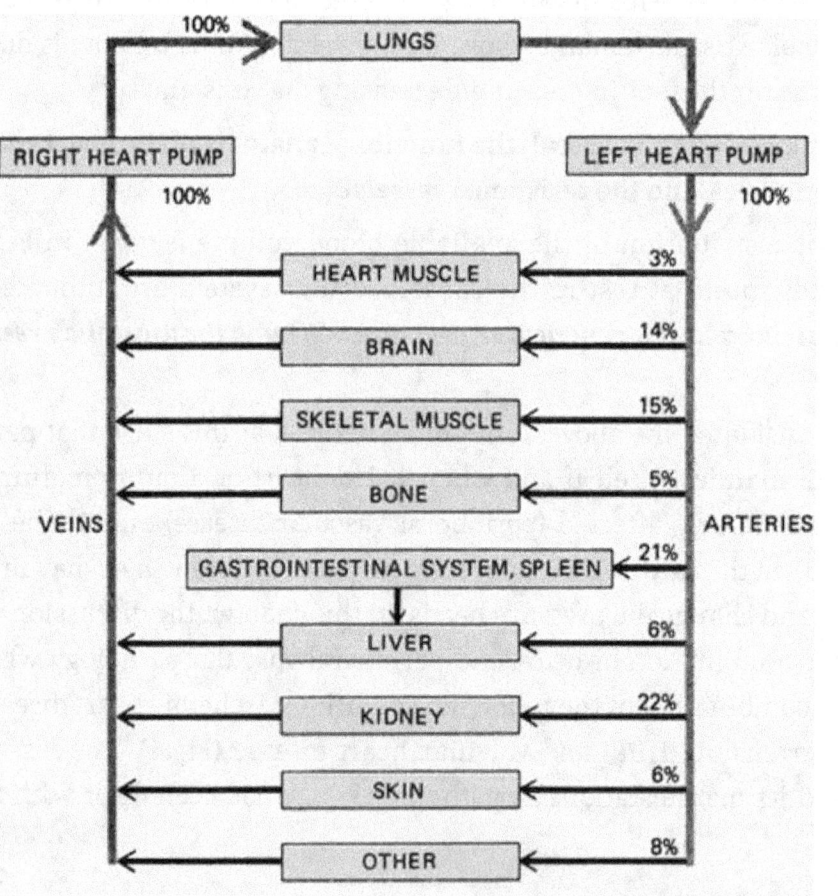

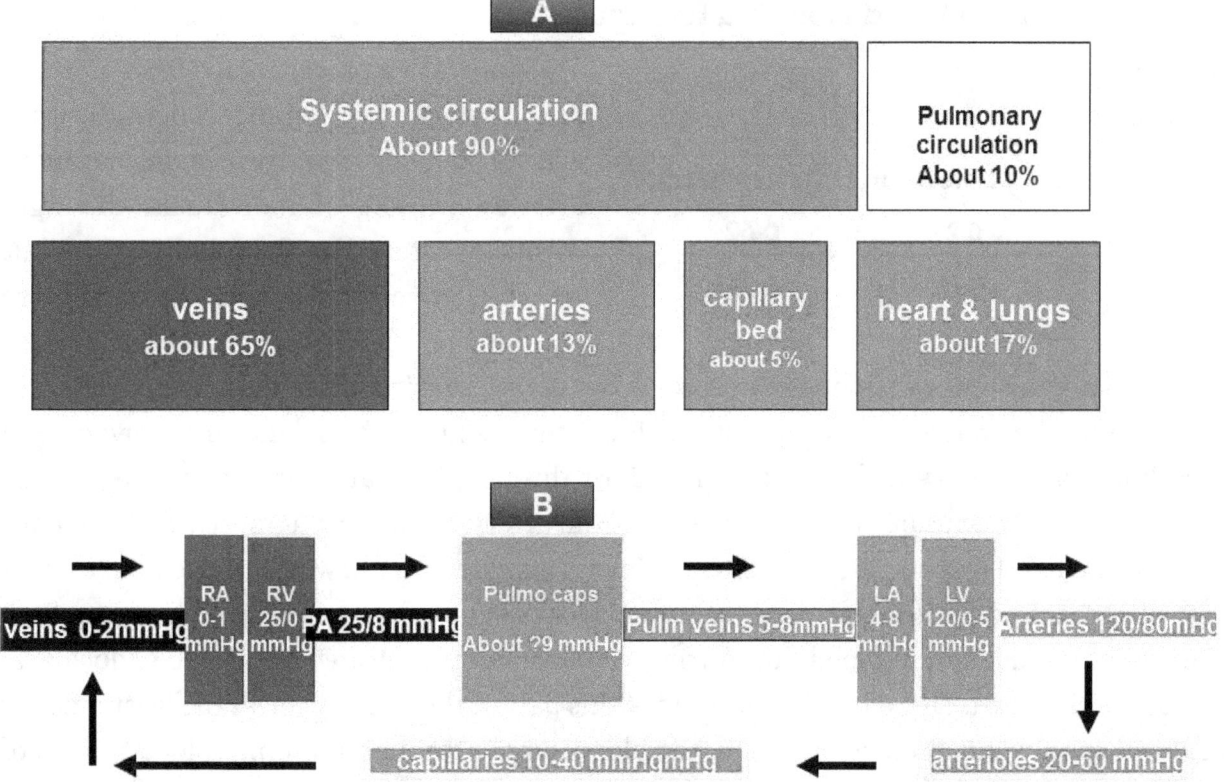

Fig. 3: A. Volume distribution & B. Pressure heads in circulation

Fig. 4: Consequences of pathophysiology of the CVS

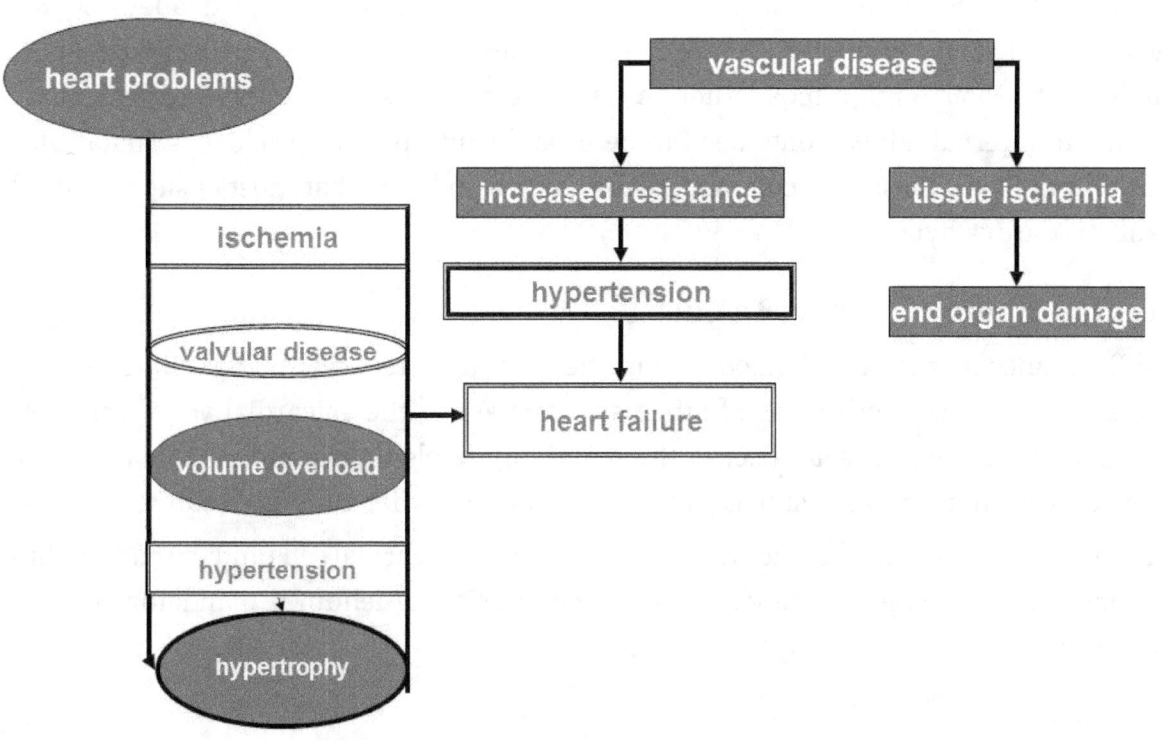

Normal Coronary Blood Flow and Its Regulation

Table I:		
HR /min	systole (sec)	diastole (sec)
65	0.27	0.62
200	0.16	0.14

The heart at rest receives about 5% (250 ml/min) of the cardiac output through a network arising out of the two main coronary arteries. In 50% of individuals, the right coronary artery has the most flow, in20%, it is the left vessel, while in 30%, the flow is equal in both the coronaries. This flow needs to increase many times during severe exercise. It is distributed in a three-tier system. The main blood vessels lie on the outside (epicardial). From these emerge the smaller intra myocardial vessels (pre-arteriolar) which run perpendicularly through the heart muscle to merge with the sub endocardial plexus of the heart emerging from arteriolar and intra- myocardial capillary vessels. The heart (particularly the left ventricle) is an extremely powerful muscle. So even at rest when it contracts, it compresses the intra myocardial muscles, reducing the coronary flow. This relative reduction in blood flow during systole is communicated to the sub-endocardial surface of the heart. Therefore during systole, this part of the heart receives the least amount of blood supply. When diastole occurs, the blood supply through the inner blood vessels of the myocardium and the endocardium obviously increases. During exercise, the heart muscle must contract even more powerfully (generating systolic pressure up to 200 mmHg or more). So the blood flow is even more compromised during systole. With increasing heart rate, the duration of both the systole and diastole, keeps on reducing. However the relative decrease in the systole period is less than that for the diastole (Table I). *(Why?).*This means that at high heart rates in a normal heart, there is a relative decrease in blood supply to the myocardium during diastole. Confounding this situation, the heart even at rest, extracts about 60–80% of the oxygen that is presented to it (the arterio-venous oxygen difference is about 15 ml as compared with about 5 ml in most other tissues). There is thus no reserve of oxygen supply to call upon. Internal adjustments need to be brought into play to produce a dilatation of the coronary blood vessels to accommodate the extra blood flow that must ensue in order to prevent myocardial hypoxia.

Factors Controlling Coronary Diameter

(Fig. 5) Resistance to myocardial blood flow is offered at all three levels of vascularization. Of these, as long as no serious degree of atherosclerosis exists, the epicardial vessel resistance is negligible. A number of factors act at the remaining levels to ensure that the myocardial oxygen demand supply relationship is suitably maintained under various circumstances.

A number of vasoactive substances are released by the myocardial cells and the endothelium in response to local oxygen demand. An active vascular endothelium is mandatory to make suitable regulatory effects to happen.

Fig.5. Factors affecting coronary vessel diameter

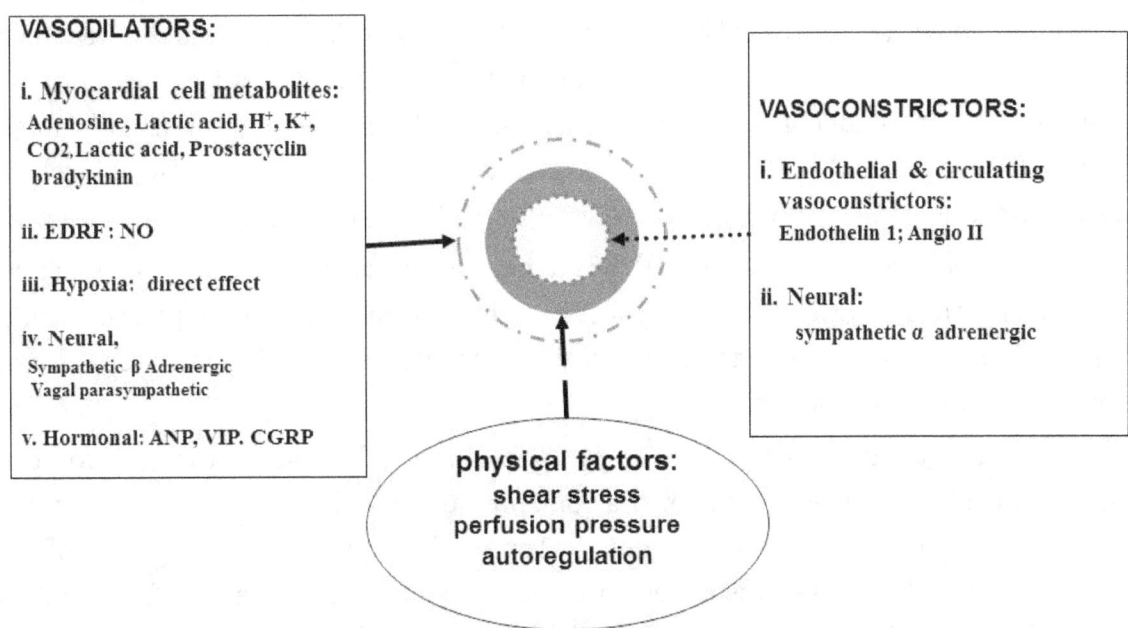

VASODILATORS:

i. Myocardial cell metabolites:
Adenosine, Lactic acid, H⁺, K⁺,
CO2,Lactic acid, Prostacyclin
bradykinin

ii. EDRF: NO

iii. Hypoxia: direct effect

iv. Neural,
Sympathetic β Adrenergic
Vagal parasympathetic

v. Hormonal: ANP, VIP. CGRP

VASOCONSTRICTORS:

i. Endothelial & circulating
vasoconstrictors:
Endothelin 1; Angio II

ii. Neural:
sympathetic α adrenergic

physical factors:
shear stress
perfusion pressure
autoregulation

Of the dilator metabolites released by the myocardial tissue in response to increased metabolic demand, adenosine has been considered a front runner. This substance may also be responsible for the reactive hyperaemia which follows vascular occlusion, and at the same time, probably helps to reduce the possible endothelial injury which may follow reperfusion. Hypoxia is known to have a direct vasodilator effect, and may also act via ATP sensitive K⁺ channels, as also by the release of adenosine. Nitric oxide (NO) is released by an intact endothelium of epicardial vessels, and is now identified as the classical Endothelium Derived Relaxation Factor (EDRF). Its release is stimulated by shear stress (induced by tachycardia), bradykinin, and acetylcholine which requires an intact endothelium to produce this effect. Damage to the endothelium may adversely affect vascular reactivity when needed as NO cannot be released. An Endothelium Derived Hyperpolarization Factor (EDHF) has also been identified in human vasculature. It acts as a vasodilator for mainly smaller arterioles, and is thought to act via Ca⁺⁺activated K⁺ channels. NO is known to inhibit its release. Hence this factor may have an important role to play when atheromatous changes in medium to small arteries prevent formation and release of NO. Angiotensin II is a vasoconstrictor which requires an intact vascular endothelium to act via AT1 receptors. It may also be responsible for increasing Ca⁺⁺ entry and the release of Endothelin 1 (ET1) which is a powerful constrictor. Thromboxane A₂ (TXA₂) from platelets is a vasoconstrictor, and also promotes platelet aggregation. The vasoconstrictor effect must be outclassed by the vasodilator effect if adequate blood supply is to be maintained whenever demand increases. Deficiencies in the production of these vasoactive substances, or an imbalance between them is classified as *Endothelial Dysfunction*. Demonstration of its presence often forms an important investigative tool in contemporary vascular research.

The coronary vasculature is freely supplied with sympathetic nerves, the arterioles having both α1 and β2 adrenergic receptors. The former have been located mainly in the epicardial vessels, while the deeper vasculature, fortuitously, is innervated by the latter. In any case, during sympathetic excitation, the overall effect is a vasodilatation because the constriction effect is overshadowed by the metabolically induced dilatation. Aiding this is the β adrenergic vasodilatation. More recently it has been thought that α adrenergic stimulation may play a preventive role in reducing the cardiac "steal" effect by restricting sudden diversion of myocardial flow to areas where the dilator effect becomes overtly prominent. There is relatively less parasympathetic innervation, and its role in controlling coronary diameter is minor.

Coronary blood flow is "auto regulated" over a mean arterial pressure range of about 60–140 mmHg. This is probably as a result of myogenic activity in response to the pressure changes in side the blood vessels combined with activity of the various dilator /constrictor substances that are being released constantly.

The myocardial oxygen supply/demand ratio determines the state of oxygenation of this tissue. The diastolic pressure time index (diastolic pressure mmHg x diastolic time in seconds) represents the supply status while the Tension Time Index (systolic pressure x systolic time in seconds). Normally this ratio is expected to be > 1, and if the value falls below 0.7, sub-endocardial ischemia is said to be the end result. A simpler guide to myocardial work load and oxygen consumption is the Rate Pressure Product (RPP: systolic BP x HR)/100). This index may be helpful in monitoring the clinical status of ischemic heart disease in a patient *(Try to explain how)*

Collateral Vessels and Myocardial Protection

Collaterals are blood vessels that interconnect epicardial coronary arteries. They are pre-existing, and remain closed because there is normally no pressure difference between the two arteries they interconnect. But if one of the two connected arteries is occluded, a pressure difference is set up, and the collaterals open. This is the process of arteriogenesis. After maturation, the collateral which when dormant has a diameter of 20–200 µm, may attain an intramural size of about 1 mm. These contain smooth muscle. Such collaterals are generally thought to emerge from pre-existing arterioles. *(What do you think is the significance of this?)*, The process of arteriogenesis may take anywhere between 3 weeks to 6 months. Shear stress, inflammation and secretion of growth factors, and the degree of obstruction (80% or more) promote collateral vessel formation. There are also factors that seem to inhibit angiogenesis. These are by products of plasminogen and collagen break down. Hypoxia of the myocardium stimulates formation of collaterals, probably because of secretion of Vascular Endothelial Growth Factor (VEGF). That is why patients with long standing IHD (chronic myocardial hypoxia) are likely to have better tributaries of collaterals which makes for a more viable myocardium when confronted with acute thrombosis. But this argument is partially nullified by the fact that in some patients, the collateral vessel may emerge away from the site of narrowing/ occlusion. Also, some subjects without coronary artery disease develop significant amount of collateral circulation. This makes the understanding of pathophysiological significance of development/presence of collateral circulation a bit of an enigma. Nevertheless, it has been found that in patients who suffer an episode of coronary blockage, viability of the myocardium that has already developed collateral circulation is better. Whether exercise training promotes collateral formation is debatable. If shear stress is a major stimulant for their development as is generally believed, regular exercise training should encourage this effect even in hearts that have normal coronary vessels. But this does not seem to happen. "Coronary steal" is an interesting phenomenon which has been associated with the presence of collaterals. In 10% of patients of IHD who have collateral circulation, there may be a diversion of blood from normal tissue towards the areas irrigated by collateral vessels. This may be a negative aspect of collateral circulation as it may induce a temporary relative ischemia in an otherwise normal myocardium.

Pathophysiology of Ischemic Heart Disease (IHD)

Table II: Possible causes of reduced O_2 supply to the myocardium

A. Reduced blood flow

    i. Atheroscleros is ii. Stenosis of coronary ostia iii. Aortic stenosis

B. Sudden complete blockage of blood supply

C. Excessive myocardial demand: hypertrophy

D. Combination of the above

E. Miscellaneous

    i. heart failure ii. hyperthyroidism iii. severe anemia

The corollary of the above is that if there is an imbalance between the demand of oxygen made by the myocardium as against the supply available, the myocardium will become hypoxic. This forms the basis of pathophysiology of Ischemic Heart Disease (IHD). Classically, IHD takes two forms: a chronic ischemia which manifests itself when there is a demand supply imbalance which is reversible, and a more serious, situation when there is an abrupt blockage of a coronary vessel where the tissue, if not re-perfused within approximately 20 minutes, undergoes irreversible damage (necrosis) followed by healing with scar formation.

Various situations which may compromise oxygen supply to the myocardium are tabulated in **Table II.** The commonest cause of IHD is atherosclerosis because of which the coronary vessel diameter is reduced progressively over a period of time. When the vessel diameter reduces to about 40–50% of normal, the stenosis is critical enough to restrict vasodilatation in response to increased metabolic demand. Even resting flow becomes inadequate when the narrowing is about 80–90%, at which vasodilatation to increased demand is almost unavailable. The ischemia may become obvious only when the demand increases greatly as during severe exercise. At other times it may be severe enough for symptoms to be present even at rest. Various other grades are obviously possible in between the two ends of the spectrum. Stimulation of the oesophagus by acid can cause coronary artery vasoconstriction and a reduction in coronary blood flow via a cardio-oesophageal reflex. There may be an abrupt blockage of the coronary vessel resulting in an immediate deprivation of oxygen to the heart tissue. The heart muscle may hypertrophy to the extent that the available blood supply is not enough to provide adequate nutrition even though there is no narrowing of blood vessels as in hypertensive heart disease. More often however, there is a combination of both, atherosclerotic narrowing of coronary vessels and myocardial hypertrophy. Of the miscellaneous causes, the raised left ventricular end diastolic pressure in combination with the lowered coronary perfusion pressure may compromise coronary circulation to precipitate IHD. **(What could be the link between severe anaemia and IHD?)**

Pathophysiological Effects of Myocardial Ischemia

There is a difference between myocardial ischemia and myocardial infarction. When ischemia occurs, the amount of oxygen supplied to the tissue is deficient. The end result is a myocardium that does not function optimally. If the surface area involved by this is large, an impairment in the relaxation and contraction of the myocardium may result. However, the effects are reversible if blood flow to the tissue is improved, whether by dilating the coronary vessels, or by reducing the work load (hence the demand of oxygen) of the myocardial tissue. When infarction occurs, the tissue has undergone irreversible death due to lack of sufficient oxygen-rich blood. The most frequent cause of infarction is a rupture of an atheromatous plaque with sudden blockage of blood supply at *any* stage of the spectrum of coronary heart disease.

Various mechanical, electrical, and chemical changes occur in the myocardial tissue which is exposed to ischemic hypoxia. The worst case scenario as noted above, is a sudden occlusion of the blood vessel. Initially the diffusible O_2, and O_2 stores from myoglobin are used up to make up for the deficiency caused by the ischemia, reducing further the oxygen tension inside the cells. This shifts cell metabolism from aerobic to anaerobic. ATP stores are rapidly depleted as its regeneration is compromised. There is an accumulation of adenosine, and adenosine monophosphate and diphosphate (AMP, ADP) in the cells as well as metabolites suchlactate, H^+ions, and K^+along with CO_2.In addition, mitochondrial calcium increases, making the myocyte more vulnerable to further damage, progressing to cell death if frank infarction has occurred. Vasodilatation of the arterioles following the accumulation of the various metabolites helps in the restoration of blood flow and nutrients to make amends to a certain extent. If severe obstruction persists for > 20 minutes or so, irreversible cellular damage with necrosis is the result. But if the blood flow is restored as early as possible, function may be fully restored. This forms the physiological basis of revascularization therapy using "clot busters" such as streptokinase, urokinase, or t-PA (tissue-type Plasminogen Activator).

Histopathology of the ischemic myocardium shows characteristic structural changes, one of the earliest being a decrease in the size and number of glycogen granules. In the first 30 minutes of ischemia, the available glycogen reduces significantly. Swelling of transverse tubules (T-tubules) and mitochondria also appears, the latter with reduced amount of matrix. Translated into functional terms, this means a reduction in the work capacity of the myocardial cells which accounts for the poor contractility and its subsequent effects. The point at which the ischemic cellular damage goes into an irreversible stage is as yet unclear. There are factors which tend to hasten this, as also those which try to delay/prevent the process. Clinically though it is well nigh impossible to judge the state of affairs as it unfolds.

The pathophysiology of acute myocardial infarction is complex. Loss of viable myocardium impairs cardiac function. The end result is a compromised cardiac output. If damage is severe, cardiogenic shock follows. Pulmonary congestion and oedema may occur when left ventricular function is severely impaired **(work out how such a patient will present clinically)**. The ischemia may also precipitate life threatening rhythm impairments such as ventricular tachycardia and ventricular fibrillation. Body physiology reacts to the reduced cardiac output by eliciting baroreceptor reflexes that lead to activation of sympathetic system and the renin-angiotensin-aldosterone system. Its aim is to restore cardiac output to as close as possible to normal. The pain and anxiety associated with myocardial infarction further activates the sympathetic nervous system, which causes systemic vasoconstriction and cardiac stimulation. Sympathetic activation is a double edged weapon: on one hand it helps to maintain arterial pressure, while on the other, it increases the myocardial oxygen demand. This can lead to greater myocardial hypoxia, and set up a vicious cycle which extends the infarcted area, precipitates arrhythmias, and further impairs cardiac function. It is also responsible for the sweating which is commonly experienced by the patient. Renal hypo perfusion and sympathetic activation stimulate renin release, which leads to increased plasma levels of angiotensin II and aldosterone that enhance renal retention of fluids to try and improve the filling state of the cardiovascular system, increase venous return, and thus enhance cardiac output. In the initial phases, this helps, but if the damage is extensive, it may actually worsen the situation by increasing the workload on the heart.

The infarction may be a "through and through" infarct involving the entire thickness of the left ventricular wall from the endocardium to the epicardium. This is known as "transmural infarction" and is associated with typical ECG findings with a typical evolutionary pattern over a period of a few days to a few weeks. Usually the anterior free wall and posterior free wall and septum are involved, with extension into the RV wall in 15–30%. Isolated infarcts of RV and right atrium are rare. Then there is the subendocardial infarct in which there are multifocal areas of necrosis confined to the inner 1/3–1/2 of the left ventricular wall. These do not show the same pattern of evolution of changes as seen in a transmural infarction. (The clinical student may like to review the classical ST /T wave and Q wave patterns in a patient with MI and distinguish between ST elevation MI (STEMI) and Non ST EMI.

Some patients may suffer from asymptomatic myocardial ischemia. The pathophysiology of this condition is at yet elusive. Some of these patients have been shown to have a higher threshold for pain, higher levels of circulating endorphins which may suppress pain triggered by a cardiac ischemic event, or may be having diabetic autonomic dysfunction. Such patients are obviously more likely to have recurrent cardiac events. The condition may get unmasked during routine exercise stress testing, or more tragically, as a first time unexplained cardiovascular catastrophe.

Markers of Myocardial Cell Damage

When severe/irreversible damage is incurred by the myocardium and its mitochondria, a number of chemical substances are released in to circulation. These substances are then detected in plasma to form the basis for markers for diagnosis of myocardial damage. These include enzymes such as lactate dehydrogenase (LDH), creatine kinase (CK), and isoenzymes of CK such as CKMB (specific to myocardial tissue, serum glutamic oxaloacetic transaminase (SGOT), and other intracellular proteins, troponins T and I, myoglobin and cardiac-specific myosin light chains. However, as yet there is no "single best" marker, particularly in the early hours after ischemic heart damage.

Of the many named above, an increased total CK is a simple and inexpensive test, but it is not specific for myocardial injury. However an isoenzyme of CK, namely CKMB is much more specific to cardiac muscle making it a good marker for acute myocardial injury. It rises in serum within 2 to 8 hours of the onset of acute myocardial infarction. Serial measurements made every 2 to 4 hours for about 12 hours after the patient is first seen provides a reasonably good indicator of myocardial injury. What is even more relevant is its utility in diagnosing subsequent infarctions because after the initial early rise, it dissipates in 1–3 days. Any further rise in its level then means a new event of infarction. The ratio of **total CK to CK-MB ("cardiac index")** was thought to provide a useful indicator for early MI. However its main drawback is that a false positive result may be given in patients with heart failure. Also, in elderly patients, because of their relatively low muscle mass, total CK estimate may below, and may contribute to a false increase in the CKMB/total CK ratio. It has thus been given up as a biomarker of myocardial ischemia.

Cardiac troponins are regulatory proteins found in both skeletal and cardiac muscle. The 3 known subunits are troponin I (TnI), troponin T (TnT), and troponin C (TnC). The genes that code for the skeletal and cardiac isoforms of TnC are identical. So there are nostructural differences between them, hence ruling out Tn C as a useful marker for cardiac tissue injury. However, the skeletal and cardiac isoforms for TnI and TnT are clearly differentiated by immunoassays. That is why cardiac troponins I and T show a unique specificity. They begin to rise within 3–10 hours, though the initial rise may not be as rapid as that of CK MB, but remain elevated for longer period of time (troponin I for 5–9 days and troponin T for up to 2 weeks. Because they remain for a long period after a single episode of infarction, they are not useful in the diagnosis of a re-infarction. Level of troponin T may also rise in skeletal muscle myopathies and renal failure, and is therefore not as sensitive as troponin I. Occasionally, reperfusion after occlusion results in rapid washing out of these markers making them peak earlier than expected. This may confound their interpretation. There is a direct relationship between troponin level and mortality.

Myoglobin may leak out of damaged myocardial cells, and can be a sensitive indicator of myocardial injury. Its level increases in blood even before CKMB, and has been related to the size of injury, but its interpretation becomes confounding in the presence of co-existing striated muscle injury.

Lactate dehydrogenase begins to rise in 12 to 24 hours following MI, and peaks in 2 to 3 days, gradually dissipating in 5 to 14 days. There are 5 isoenzymes (1 through to 5). Ordinarily, isoenzyme 2 is greater than 1, but with myocardial injury, this pattern is reverses. LDH-5 from liver may be increased with centrilobular necrosis from passive congestion with congestive heart failure following ischemic myocardial injury. With the availability of other specific markers, LDH has been relegated to a lower order as a marker for myocardial injury.

Now that inflammation, vascular remodeling after occlusive coronary disease are being understood, other potential markers of myocardial injury are being looked at. To name a few: C-reactive protein, adiponectin, and a type of natriuretic peptide. Their appearance has also been related to patient risk stratification.

Mechanical Disturbances in the Hypoxic Myocardium

Myocardial musculature in the region of sudden severe ischemia may remain functionally depressed for a few hours after the episode. This is known as myocardial *"stunning,"* and is thought to develop initially as a consequence of the ischemia, following which is its association of reperfusion. The physiological consequences of this are obvious, and are likely to contribute to the gamut of presenting symptoms and signs in a patient *(Physiologically what will such a patient present with?)*. The cellular pathophysiology most likely to be associated is a combination of release of free O_2 radicals, excess calcium accumulation and the inability of the myofibrils to respond to calcium, all of which culminate into a disturbance of excitation contraction coupling. Improvement in contractile function may take many days to weeks. It is felt that it is best to pre-empt it by administering antioxidants and calcium channel blockers, but once it sets in, inotropic support is required to help the myocardium to recover function.

In some patients of IHD, the left ventricular function at rest may be below par though there are no overt indications of frank IHD. When revascularization is carried out in such patients, the function improves considerably suggesting that the heart had been *"hibernating"* to cater to the reduced nutrition. At the cellular level, it is possible that the mitochondria sense the partial hypoxia because of reduction in activity of cytochrome oxidase and as a precautionary measure "reduce the output of its factories."

Symptoms and Signs as Reflections of Pathophysiology of IHD

The more common symptoms and signs which manifest when CVS pathophysiology occurs are given in **Fig. 6.** Some commonly done investigations which are likely to help in deducing the pathophysiology which has resulted in producing these signs and symptoms are also mentioned.

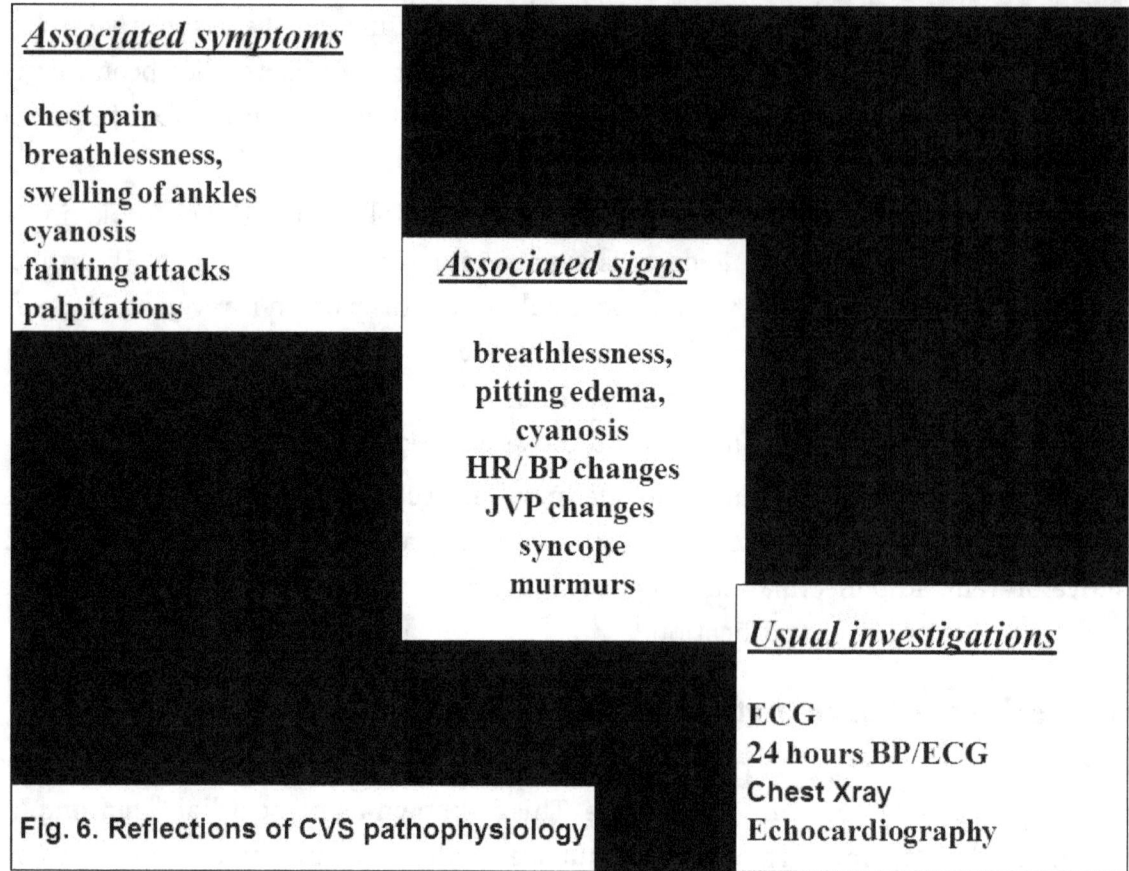

Associated symptoms

chest pain
breathlessness,
swelling of ankles
cyanosis
fainting attacks
palpitations

Associated signs

breathlessness,
pitting edema,
cyanosis
HR/ BP changes
JVP changes
syncope
murmurs

Usual investigations

ECG
24 hours BP/ECG
Chest Xray
Echocardiography

Fig. 6. Reflections of CVS pathophysiology

Chest pain is the cardinal feature of IHD. (The student is advised to review the clinical presentation of chest pain caused by ischemic heart disease). Briefly, definite acute coronary ischemia is indicated by sub-sternal discomfort precipitated by exertion which radiates to the shoulder, jaw or inner aspect of the arm, and is easily relieved by rest or nitro-glycerine in less than 10 minutes. Radiation of the pain to various areas is explained by the fact that common dermatomes are shared by the heart with these areas. More often than not, cardiac pain is substernal because developmentally, heart is a midline viscus. Postprandial angina results from a diversion of blood flow away from areas supplied by severely stenosed coronary arteries to those areas supplied by less diseased or normal arteries This may be due to sympathetic activation from food ingestion, and norepinephrine-induced vasoconstriction in diseased vessels. The pain generally lasts for about 5–10 minutes, but does not exceed 20 minutes. If it at all does, then it indicates a far more sinister consequence such as infarction. Anginal pain often has a circadian disposition, and is known to occur more frequently in the morning because of enhanced sympathetic activity and increased platelet adhesiveness.

The chest pain is likely to be coronary insufficiency if most of the features of angina are present, but may not be typical in some aspects. On the other hand, if the history is questionable as to the relationship of the pain to exertion, and when the pain is not relieved by nitroglycerine, it is clearly non-cardiac in origin.

Mechanisms responsible for myocardial ischemic pain are complex and still under elucidation. Chronic ischemia due to atherosclerotic lesions is responsible for the classical angina of effort -acute coronary insufficiency (ACI). Whenever the oxygen supply is deficient in response to an increase in demand as during exercise, the myocardial tissue "cries out" in protest.

Ischemia (which in effect causes tissue hypoxia)reduces the formation of adenosine triphosphate (ATP), resulting in the development of acidosis, and the release of substances such as lactic acid, serotonin, bradykinin, histamine, reactive oxygen species, and adenosine.. The released chemicals are thought to stimulate chemo- and mechano-sensitive receptors, probably free nerve endings, found within cardiac muscle fibers, and around the coronary vessels. More recently, adenosine is considered to be a strong contender. It is also possible that veno-dilatation as a response to ischemia can activate these receptors. The sensory impulses travel up the sympathetic afferent pathways from the heart and enter the sympathetic ganglia in the lower cervical and upper thoracic spinal cord (C7-T4). Impulses are then transmitted via the ascending spinal pathways to the thalamus and ultimately to the cerebral cortex. There is obviously a strong psychogenic element that is associated with this variety of pain as patients often describe it as "oppressive," "applies pressure on the chest wall," "spells imminent death." The latter is an important part of the clinical presentation in patients with acute myocardial infarction which is an extension of IHD.

Myocardial ischemic pain in women is more often atypical as compared with that which is experienced by men, and occurs 5–10 years later in age. Coronary micro vascular endothelial dysfunction is prevalent in women with chest pain in the absence of frank obstructive CAD. Although their long-term prognosis is uncertain, these women are at increased risk of having inducible subendocardial hypo-perfusion. Inducible metabolic ischemia is also prevalent in women with chest pain not attributable to CAD. The pathophysiologic mechanism of this endothelial dysfunction is not clear. It does not appear to be related to atherosclerosis risk factors or inflammation.

Breathlessness is often associated with anginal pain. Hypoxia of the myocardial muscle may result in diastolic dysfunction which then increases the left ventricular end diastolic pressure. The resultant back pressure most probably causes an increase in pulmonary interstitial fluid pressure to excite J receptors described by Paintal. Stimulation of J receptors has been related to the sensation of dyspnea. Mild bronchospasm accompanies the clinical presentation as a part of the vagally mediated reflex. There are occasions when patients with passing left ventricular failure complain of breathlessness which is accompanied by a mild but audible rhonchi, and may be diagnosed erroneously by an inexperienced physician as an attack of mild bronchial asthma. However the history, the relatively higher age group of such patients and the presence of fine basal crepitations spell out the true pathophysiology.

Other symptoms may include belching, nausea, indigestion, diaphoresis, dizziness, light-headedness, clamminess of skin, and fatigue. These are a reflection of the involvement of the autonomic nervous system in this syndrome. Dizziness suggests a sudden drop in brain perfusion because of a rapid decrease in myocardial contractile function. Sometimes this may induce a frank syncope which some patients with myocardial infarction may experience. The blood pressure and heart rate usually increase inpatient having an anginal attack and is a result of enhanced sympathetic activation. In a patient with the more serious form of the condition (infarction), a fall in the blood pressure indicates a serious situation *(explain why?)*

The physiological basis for the management of IHD is to limit the discrepancy between the O_2 available to the myocardium, and the oxygen demand made by the myocardium. Various tools which may help in reaching this goal are outlined in **Fig.7**, the major approaches being i. to reduce the oxygen consumption by the myocardium, and ii. to improve/restore blood flow. Pain relief with the use of powerful analgesics (morphine still remains the drug of choice) is of paramount importance, particularly in patients with infarction. The severe oppressive pain precipitates anxiety and with it, sympathetic excitation which accentuates O_2 requirement in an already compromised system. Use of supplemental Oxygen does not seem to be of much relevance if the oxygen saturation is normal, though theoretically the hypoxic myocardium should benefit by its use.

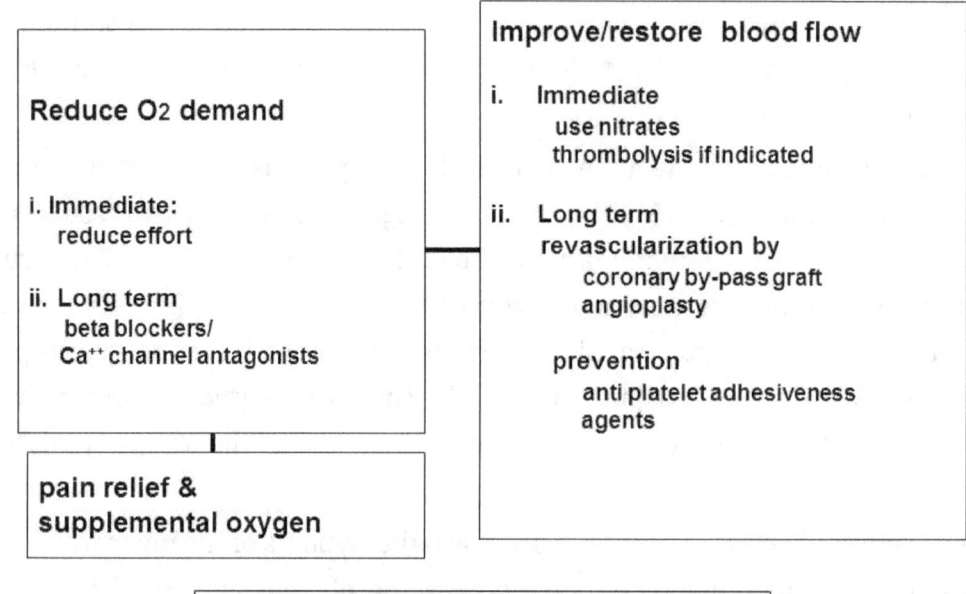

Fig. 7 Physiological basis of management of IHD

Pathophysiology of Hypertension (HBP)

Table III The norms for blood pressure(mmHg) in adults

	Systolic BP	Diastolic BP
Optimal BP	<120	<80
Normal	<130	<85
High normal	130-139	85-90
Hypertension	>140	>90

The current concept of what constitutes normal blood pressure for adults irrespective of age is given in Table III, and is based on the WHO classification. The cut off level for frank HBP still remains at 140/90 mmHg, and 180/110 and more is classified as severe HBP. However there is a grey area from120–139 systolic, and 80–89 diastolic which is currently the focus of much debate. It has been proposed that those so called normal individuals who have a BP in the high normal category are more likely to develop frank HBP with the passage of time. Hypertension poses a serious health problem as it may remain undetected for long periods of time, and gets revealed only when the end organ damage it has inflicted surfaces as a serious clinical condition.

Types of HBP

Almost 90–95% of those who have hypertension (BP 140/90mmHg and more) do not have definable cause for their condition, and are said to have developed Essential or Primary HBP. The remaining 5–10% who have HBP (secondary HBP) have an identifiable cause such as endocrine or renal diseases, pregnancy induced, long term use of medications such as oral contraceptives, or coarctation of aorta. A small percentage may have only high systolic BP (isolated systolic HBP). Here the discussion centers around the primary variety. However the end organ damage inflicted –the end result of prolonged hypertension, remains basically the same. The only difference is that identification of the 2ndary HBP may be relatively more easy as the basic pathology which results in the condition may have surfaced as a distinctive clinical entity of which HBP is an accompanying feature.

Normal BP maintenance is dependent upon two main factors: the cardiac output and the resistance offered by the tone of the blood vessels to this parameter. Some of the more important factors which affect this are given in **Fig. 8.** The student at this stage must review the physiology of maintenance of normal cardiac output and the peripheral resistance.

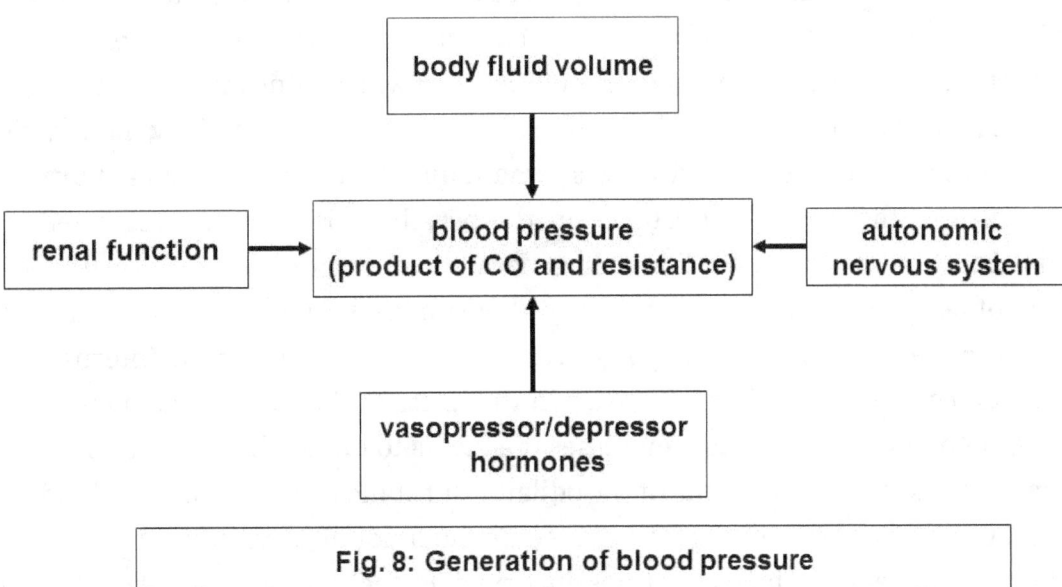

Fig. 8: Generation of blood pressure

Logically therefore, factors that increase cardiac output and peripheral resistance must lead to HBP. Review what happens to BP in a patient with hypovolemic shock. The BP falls. Replacement of the lost volume brings up the blood pressure. In the same context, if the body mechanisms retain more than the required amount of fluid, the blood pressure should increase. But in normal individuals this is not the case because body regulatory mechanisms come into play. (***what are*** *these mechanisms?*). These mechanisms are obviously wanting in patients who develop essential HBP. A number of possible reasons as to why this may happen are discussed below.

Cardiac Output and HBP

An increase in cardiac output may be the initial finding in patients who develop HBP. This increase in the CO could be attributable to i. An increase in circulating fluid volume (Reducing ECF volume by using diuretics still remains an important armamentarium in the treatment of essential HBP), and ii. An increase in the contractility of the ventricular muscle because of neural stimulation. At a later stage in the course of the disease, this parameter is known to revert back towards normal, and in fact in well established cases, may actually be lowered. The corollary to this is that there is an increase in the peripheral vascular resistance. It has been postulated that the initial increase in CO increases the blood flow. This needs to be auto-regulated. As a consequence, the peripheral vascular resistance increases, adjusting the blood flow. In the process the peripheral vessels undergo a rapid thickening on a permanent basis. With this the heart must generate more pressure in order to sustain blood flow. Thus HBP develops. At the same time, because of the more than warranted resistance offered to the flow, the CO starts to fall, initially to normal, and later in some patients, to below normal.

Vascular Resistance and HBP

The site of maximum resistance is at the pre-capillary level in vessels of internal diameter of 500 μ and less and is related inversely by a power of 4 ($1/r^4$). The resistance is offered because of the presence of a certain amount of circular smooth muscle in the tunica media of the vessels. This muscle is under neural control, and therefore a change in vascular diameter brought about by the contraction/relaxation of this smooth muscle layer will significantly alter the resistance offered. The two other factors which determine the resistance offered to flow, namely viscosity of blood and the length of the blood vessels remain unaltered for all practical purposes. The essentials of how the vascular resistance in the periphery is normally regulated is given above in the section on IHD (**Refer Fig. 5**). In general the role of vagal/parasympathetic neural regulation of peripheral vascular resistance is almost non-existent. Two main characteristics determine the presence of abnormally high resistance. i. A decrease in the internal diameter of the resistance vessels, and ii. An increase in thickness of the tunica media constituted by circular smooth muscle layer which increases the media to lumen diameter ratio. The former might happen because the amount of vasodilator substances eg. NO, prostaglandins (PGs) produced by the vascular endothelium reduce or the amount of vasoconstrictor substance viz endothelin increases. This constitutes in principle the pathophysiology of endothelial dysfunction. On the other hand growth factors, and hormones such as AII and NE may induce a thickening of the smooth muscle layer increasing the media/ lumen diameter ratio, or rearrange the existing structures without a change in this ratio (vascular remodeling). In both situations, there will be a sustained increase in peripheral resistance, and this in turn will be reflected as essential HBP.

Factors which affect Pathophysiology of Essential HBP

Genetic influences have been associated with the development of essential HBP. Hypertension is twice as common in children whose parents (one or both) have hypertension. The association may be as much as 60% in twins. On the other hand, children adopted by hypertensive parents have less predilection to develop HBP. Multiple genes have been implicated. These influences may act in combination with factors such as obesity, alcoholism and smoking which are known risk factors for the development of HBP.

Babies born with **low birth weight** have been known to develop HBP in adult life. They may also be more prone to coronary artery disease, insulin resistance, diabetes mellitus, and abdominal obesity. These observations suggest that fetal influences such as intra uterine under nutrition may be contributing to the development of hypertension, though the mechanism by which this occurs is not clear.

One of the possibilities that has been considered in the pathophysiology of HBP is the role played by Na^+. There is experimental evidence that an increase in dietary sodium results in HBP and a reduction does the opposite. High Na^+ intake may activate increase in intracellular Ca^{++}, increase the level of catecholamines, and may worsen insulin resistance. Increase in salt sensitivity may be brought about by a defect in its excretion by the kidney, impaired NO synthesis, and increased activity of Na^+/H^+ exchanger pumps. Increase in the incidence of essential HBP in the elderly has been related to increased Na^+ sensitivity in this age group.

There is a dual relationship between the **kidney and HBP** (*There is this perpetual debate as to what comes first-the chicken or the egg?*). Altered renal physiology and pathology may be present before changes become manifest in other organs.

It has been postulated that the risk of development of HBP is inversely proportional to the number of nephrons at birth. The reduction in the surface area available for Na^+ excretion may be responsible for the HBP. This hypothesis finds support in the fact that women, the elderly, as well as blacks in whom incidence of essential HBP is more, have smaller kidneys and less number of functional nephrons.

The most sensitive, long-term mechanism for control of BP is dependent on the fine tuning of fluid volume control by the kidney. At an average perfusion pressure of 100mmHg, about 150 mmol of Na^+ are excreted /day. A shift to the right of this pressure diuresis relationship may be the starting point for essential HBP.

Renin AngiotensinII Aldosterone (RAA) System in the Pathophysiology of Essential HBP

The Renin –angiotensin II-aldosterone system (R-A-A system)is the major determinant of body Na^+ and water regulation. (*The student must review the R-A-A system*). In principle, when the Na^+ presented to the distal convoluted tubule reduces, the RAA gears up towards conservation of this ion. The two main target tissues involved are the adrenal gland and the renal vasculature which need to respond suitably to circulating levels of Angiotensin II (AngioII). Normally, if dietary Na⁺ is reduced, the aldosterone secretion by the adrenals in response to AngioII is enhanced, while the renal blood flow is reduced in order to minimize sodium loss in urine. The opposite happens when the sodium availability increases. If on the other hand, **non-modulation** of this mechanism occurs, the abnormally regulated AII does not allow the appropriate response of aldosterone secretion when Na⁺ intake is low, nor that of renal vasculature when sodium loading occurs. This may then be responsible for the development of essential HBP. Use of angiotensin converting enzyme (ACE) inhibitors/ Angiotensin II receptor blocking agents help in the management of this variety of HBP.

Some patients of essential HBP (approximately 20%) seem to have low level of plasma renin activity (PRA). They have high salt sensitivity, low circulating aldosterone levels, and blunted response of aldosterone secretion to AngioII infusion. Theoretically, such individuals should have high levels of circulating aldosterone which might explain the blunted response to stimulation by AngioII, and increased extracellular fluid volume, but seem not to. It is possible that the AngioII receptors in such people are down regulated. These patients tend to respond well to diuretics.

In some patients, the PRA is high. When this happens, the circulating AngioII level will also be high and the consequences of this are relatively more easy to understand. Here, the non- modulation of PRA seems to be the bottom line in the pathogenesis of essential HBP. Theoretically then, use of AngioII blockers/receptor antagonists should be by far the most effective method of control of this HBP. But surprisingly this does not happen in up to 50% of such patients. Hence the apparently simple logic in explaining pathophysiology of this variety of HBP is after all not that simple. All the above points suggest that pathophysiology of essential HBP seems to pivot to a large extent around Na⁺.

Neural Mechanisms in the Pathophysiology of Essential HBP

Psychogenic stress has been associated with the development of essential HBP. An interaction between the nature of the stress, the way it is perceived by a given individual, and the susceptibility of an individual to stress is likely to be involved in determining the final outcome. However, the exact role of this type of stress in the pathogenesis of HBP is not clear. **Stress induced sympathetic nervous system** over activity may be one of the important factors involved **(Fig. 9)**, particularly in the early phases. Sympathetic activity prompts the kidneys to initiate more AngioII secretion. Apart from AngioII, norepinephrine and other growth factors derived from the endothelium and the platelets may also promote vascular thickening (remodeling). Interestingly, offspring of hypertensive parents show sympathetic hyperactivity in response to stress tests long before they may become hypertensive.

Baroreceptors resetting. High pressure (arterial) baroreceptors are slowly adapting stretch receptors which are a physiological means of adjusting BP when it changes during day to day activity. The most studied baroreceptors are the carotid ones. A sudden rise in BP activates these receptors to set up an inhibitory reflex which produces bradycardia with vasodilatation, while their deactivation does the opposite. (*Recall the carotid baroreceptor reflex pathway*). Though the correction of the BP in either direction may be relatively incomplete, the response is rapid. The baroreceptor reflex gets temporarily inactivated during exercise **(Why so?)**. In other words, there is always a chance that its protective function may get compromised. It has been postulated that resetting of this reflex may be associated with the development of hypertension. A change in the baroreflex sensitivity (BRS) defined as a reflex modulation of heart rate with a change in the BP, may be involved. Earlier, this parameter was measured invasively as a prolongation of the RR intervalin response to changes in BP induced by bolus injection of a vasopressor (phenylephrine) or a shortening of the same in response to a vasodilator such as nitroglycerine. More recently, a non-invasive continuous method of measuring beat to beat BP allows estimation of BRS even at rest. The exact mechanism as to how BRS is reset is not clear, but there is a possibility that the cardiovascular centres show a blunted response to inhibitory input from the baroreceptors. This may then be responsible for letting the vasoconstrictor centre have a greater sway, resulting in greater sympathetic tone, and hence an increase in peripheral resistance.

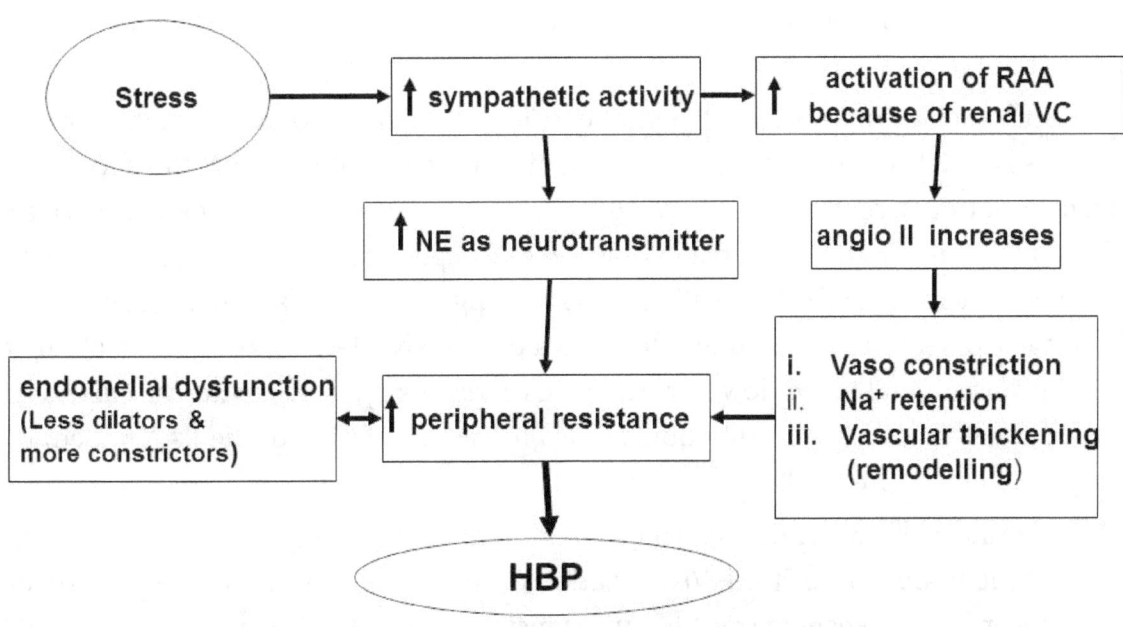

Fig. 9: Sympathetic activity in the pathogenesis of essential HBP

Obesity has been associated with hypertension. Cardiac output is higher in obese individuals, and may be responsible for initiating the hypertension as described earlier. Dyslipidemia in obese people may promote thickening of blood vessels and contribute to an increase in peripheral resistance. In addition, obese people develop insulin resistance which in turn has been blamed for the pathogenesis of HBP. Insulin resistance is thought to i. enhance BP and aldosterone responses to AngioII; ii. stimulate secretion of growth factors which in turn cause a thickening of vascular smooth muscle; iii. increase endothelin release by the vascular endothelium; and iv. reduce the synthesis of vasodilator substances, particularly prostaglandins. Insulin resistance has also been noted in non-obese hypertensives.

Pathophysiology of End Organ Damage by HBP

Essential HBP is often referred to as the "Silent killer" because it remains asymptomatic while organ system damage inflicted by it progresses until one day it may precipitate as a medical catastrophe. Early detection of hypertension is more often accidental during a routine visit to the doctor. In this context, the importance of annual medical examination as carried out in certain organizations such as the Armed Forces, cannot be overlooked.

The heart is the most frequently affected organ in hypertensive disease. Initially it responds enthusiastically by enhancing its pumping activity to generate more than normal pressure to overcome the increase in peripheral resistance. This results in left ventricular hypertrophy *(Why only LVH?)* which, unlike the hypertrophy of skeletal muscle, is irreversible. When this is associated with an atherosclerotic narrowing of the coronaries, the stage is set for the onset of ischemic heart disease as a complication of hypertension. *(Explain why?).* With passage of time, if the heart continues to work against a high resistance, the hypertrophy progresses to a dilatation followed by failure. The process involves a phenomenon called *remodelling* which will be discussed in the section on heart failure. Because of the two pronged attack on the physiology of the myocardium, heart disease is by far the commonest cause of death in hypertensives.

Recollect from the chapter on renal pathophysiology that the major manifestation of chronic renal failure is HBP. Reciprocally, hypertension is often the cause of development of renal failure. Persistent elevation in renal vascular pressure induces the development of glomerulosclerosis. The classical lesion is in the afferent arterioles, with a deposition of eosinophilic material with a subsequent thickening of vessel walls and narrowing of lumen. As a consequence, hypoxic injury to the glomeruli and the tubules leads to a progressive deterioration of renal function. The compromised renal blood flow in turn begins a vicious cycle of further vascular damage because of renal endothelial dysfunction, as also further activation of the RAA system which in turn perpetuates more HBP.

Damage inflicted by systemic hypertension on to the nervous system may manifest in the retina or in the brain matter itself. Visualization of the retinal vasculature by fundoscopy is an invaluable means of keeping track of hypertensive changes that may occur. The pathology starts with a sclerotic hardening of the retinal arterioles followed by splinter haemorrhages, and small necrotic patches in the nerve fibres seen as cotton wool spots. Damage to the endothelium leads to formation of exudates because of release of lipid and fluid in to the macula, and may finally result in frank papilloedema.

The CNS manifestations of hypertension include transient ischemic attacks, frank hemorrhagic stroke and hypertensive encephalopathy. The latter is associated with dilatation of cerebral arteries and a fibrinoid necrosis. The cerebral arterial dilatation results from a failure of auto-regulation due to high sustained HBP. There is a leakage of fluid out of the blood vessels with cerebral oedema and raised intra cranial tension which is responsible for the clinical signs and symptoms of severe headache, transient blindness, vomiting, transient paralysis and convulsions. These form a part of the syndrome of malignant hypertension in which there is a sustained increase in diastolic pressure to > 130 mmHg. Retinal haemorrhages and exudates, papilloedema and oliguria are a part of this dangerous condition which may occur in <1% of hypertensive patients, especially in those who are not on regular treatment.

Clinical Signs and Symptoms vis-à-vis Pathophysiology, and Physiological Principles of Treatment

Generally, essential hypertension may remain silent until the target organ damage induced symptoms announces its presence. The popular belief that generalized headaches are a common clinical presentation has not been substantiated. However, morning occipital headache is often felt by patients with severe essential hypertension, and is most probably caused by arteriolar dilatation as a consequence of failure of auto-regulation. Factors such as age (earlier the onset, lower is the life expectancy if not adequately treated), sex, obesity, smoking, alcohol consumption, and atherosclerosis, are all known to affect pathophysiology of essential HBP.

Physiological principles of management revolve around i. Lowering of the circulating fluid volume ii. Reduction of workload of the myocardium iii. Prevention of the development of cardiovascular remodeling.

Pathophysiology of Heart Failure

Table IV: Causes of Acute heart failure

Acute failure

i. Myocardial infarction

ii. Sudden arrhythmias: Ventricular fibrillation

iii. Sudden loss of valve function as a result of rupture of papillary muscles/chorda tendinae

iv. Cardiac tamponade

Chronic failure

i. Hypertensive heart disease

ii. Ischemic heart disease

iii. Dilated cardiomyopathy

iv. Valvular heart disease

v. Severe anemia

vi. Hyperthyroidism

Heart failure may be defined as the inability of the myocardium to function effectively as a pump. The consequence of the pump failure is the deprivation of tissues of the much needed cardiac output and perfusion for sustaining normal physiological function. Of the two parameters which constitute cardiac output, it is mostly the stroke output that suffers when failure occurs. *(The student at this point must review the factors that affect myocardial contractility with particular reference to the Frank Starling law).* The failure may follow some sort of a preceding event which may be an abrupt one such as acute myocardial infarction **(Acute failure)**, or the result of on going set of insults delivered to the myocardium over a period of time (volume overload, increased peripheral resistance (**chronic heart failure-CHF**). It is the latter aspect which will constitute the main theme of this write up. At times, the acute failure may supervene the existing chronic variety. ***(Can you think of any examples at this stage?)***. The more important clinical causes of acute and chronic heart failure are given in **Table IV.**

Even if each one of the ventricles is a "pump," in its own right, it is the left ventricle that gets due recognition as the more important functional entity. ***(Why)?***. At this stage, it will be helpful to review the sequence of events that occurs when the normal ventricle at rest pumps out the stroke volume **(Fig. 10)**. At point A when the end systolic volume has been reached, the LV pressure is close to 0 mmHg, and that at the end of the diastolic filling stage (B), it remains almost the same. During isovolumic contraction (BC), sufficient force must be generated by the contracting ventricle to open the aortic valve and push out the bolus of about 70 ml of blood (the stroke volume -CD). This may be best estimated by measurement of the ejection fraction (stroke volume/end diastolic volume) x 100. Normally this is about 55% or more. The ejection phase during which the SV is pushed out lasts for about 250 msec. In an athletes heart, the point A of the rectangle ABCD will move towards the left, while the point B will shift to the right. The pressure generated (BC) will be higher than that of a normal person, and the stroke volume CD will now also be greater than the normal. A somewhat similar situation, though at a lower level, will also be present in a normal person who changes posture from upright to supine. Consequently the failure of this sequence of events would be of major pathophysiological concern. **(At this point, just look at the rectangle abcd of Fig 10, and note how it differs from ABCD. Also, draw ABCD for an athletic LV)**. The inability of the LV to perform the assigned role is brought about by an internal damage to the myocyte which may be combined with an overload attributed to excessive volume and/or higher than normal peripheral vascular resistance. By far the most recognized cause of heart failure is myocardial ischemia. As this damages the functional unit of the myocardium, some authors label this variety as "Myocardial" failure, while "Heart failure" may be used to encompass a larger spectrum of pathophysiology where myocardial muscle damage is not the primary event.***(Which of the causes listed in Table IV do you think by pass myocardial muscle damage as the primary cause of failure?)***.

Physiological adjustments to failure. Initially, the left ventricular (LV) dysfunction may remain masked because of various compensatory mechanisms **(Fig 11)**. Depending upon various factors such as genetic disposition, gender, age and environment, these measures may help to support LV function over a variable period of time by maintaining myocardial contractility. Sympathetic activation, and an increase the circulating fluid volume are the two main props of the compensatory process. The latter invokes the beneficial effects of the Frank Starling mechanism, and this added to the increase in adrenergic activation helps to restore myocardial contractility to some extent. Venous constriction aids venous return. Thus during the compensatory stage, the left ventricular dysfunction gets masked either fully or partially. The peripheral vasoconstriction induced by sympathetic stimulation is partly offset by vascular dilators such as nitric oxide and various prostaglandins which are released. The atrial stretch which occurs as a result of volume overload is responsible for the release of

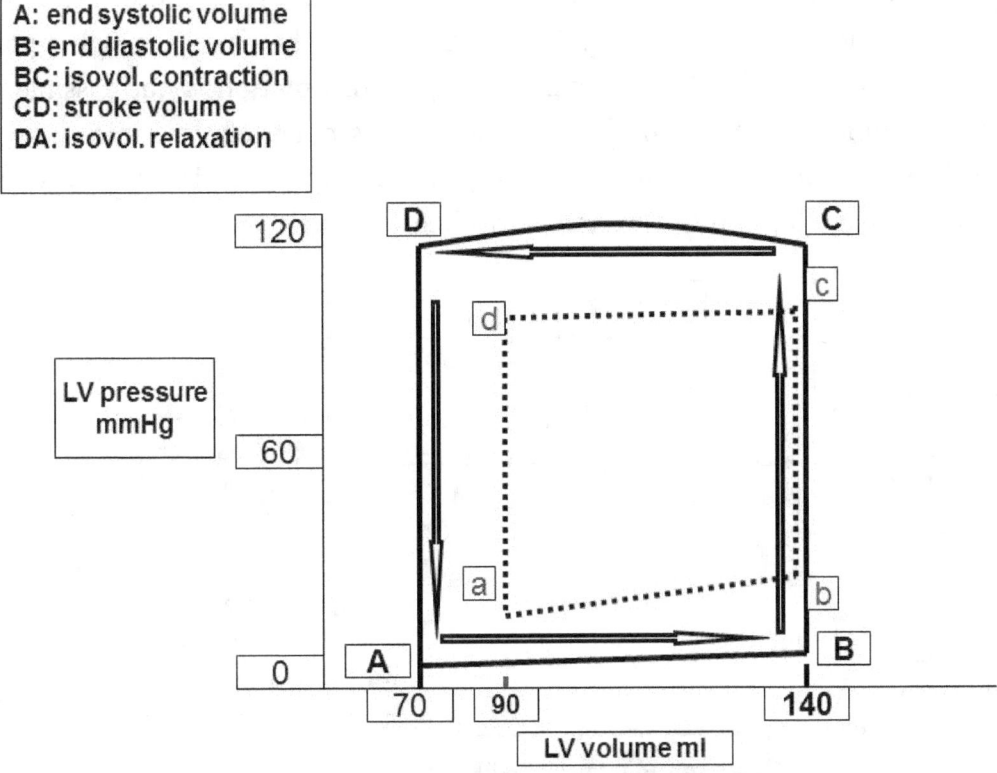

A: end systolic volume
B: end diastolic volume
BC: isovol. contraction
CD: stroke volume
DA: isovol. relaxation

LV pressure mmHg

LV volume ml

Fig.10: The pressure-volume relationship in the left ventricle as it functions as a normal pump (ABCD), and when failure sets in (abcd)

atrial natriuretic peptide (ANP) which may reduce the extent of volume overload initially, but is not adequate enough to completely reverse the process. At some stage during this process, as yet ill defined, decompensation sets in **(Fig. 12)**. The high levels of circulating AngioII and aldosterone which had initially helped in hypertrophy and compensation, stimulate fibrosis of the myocardial tissue. The various neuro-humors (norepinephrine vasopressin and ANP) are assisted by activation of the dormant embryonic myocardial growth factors which also cause hypertrophy of the undamaged myocardial cells. The release of adrenergic mediators increases afterload, inotropy and chronotropy, and hastens myocardial cell damage because of the increase in workload in an already damaged myocardium. The occurrence of sudden arrhythmias with sudden death as a consequence have also been associated with high circulating levels of catecholamines. High levels of endothelin detected in circulation are thought to be responsible for the increase in afterload in heart failure. Tumor Necrosis Factor (TNF)α, a cytokine, is released by the damaged myocardial cells. This has been associated with hastening of cell death and apoptosis through inducible isoform of nitric oxide synthase. The exact mechanism of its occurrence however is still elusive. The sequential events involved in the progression of myocardial cell damage are i. The hypertrophy of unaffected myocytes with alterations in their contractile properties; ii. Abnormal myocardial metabolism and energy utilization; iii Ongoing loss of myocytes because of necrosis, accelerated apoptosis and autophagia. iv. Reduced responsiveness to β adrenergic stimulation. v. Reorganization of extracellular matrix and its subsequent replacement with collagen matrix and fibrosis. This process of progressive degeneration of myocytes is known as **remodeling**. These events may be reversible if the treatment is initiated early. But left unheeded, the changes reach a point of no return –"the terminal threshold" after which salvaging of function is not possible. As yet the exact point at which terminal threshold is reached has not been determined. Apart from a possible role in remodeling of cardiac tissue and hastening apoptosis, TNFα has been associated with the cachexia that is often seen in patnic heart failure.

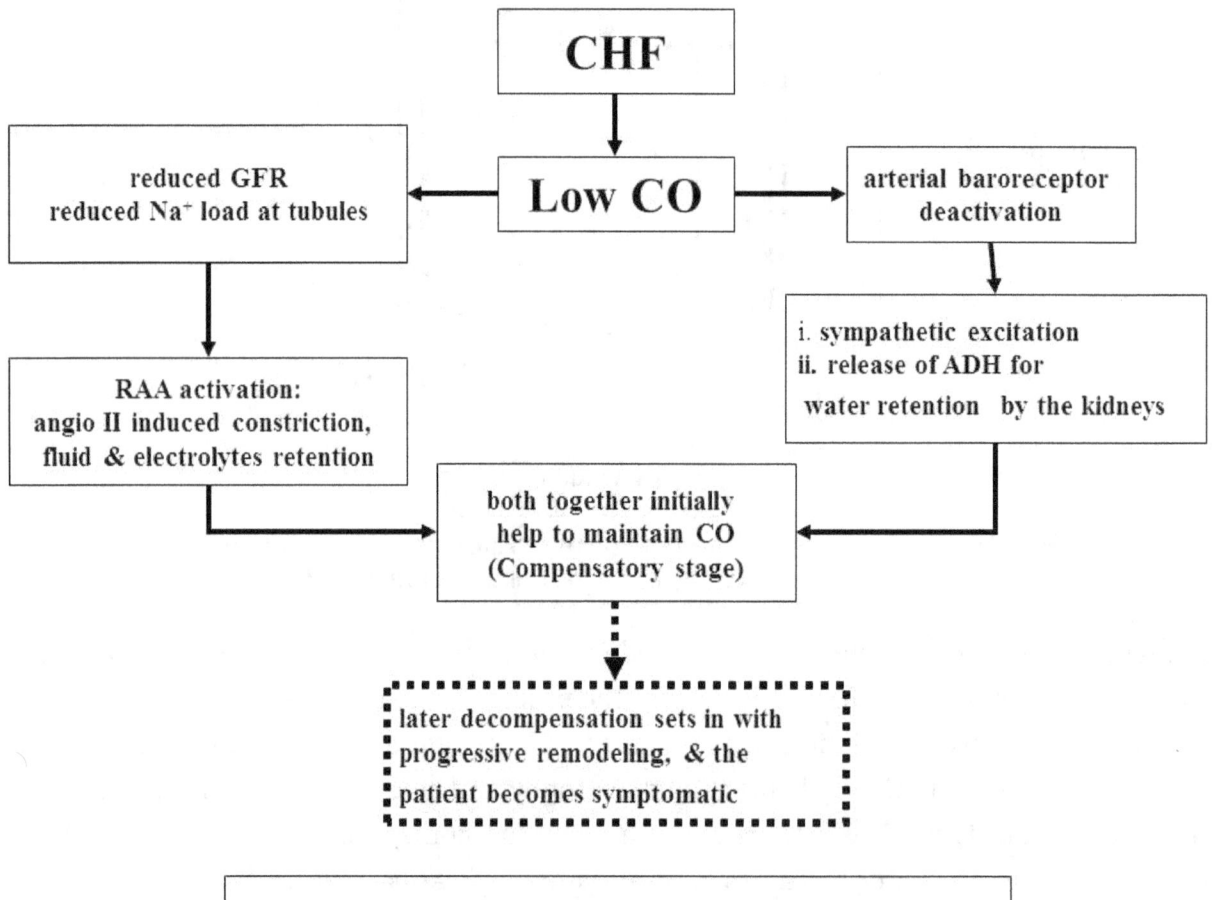

Fig. 11: Compensatory mechanisms in heart failure

The commonest end result of progressive heart failure is the **systolic dysfunction**. Simply stated, it is the inability of the ventricles to push out an adequate stroke volume. Referring to **Fig 10**, under the circumstances, the pressure generated during the isovolumic contraction phase will be lower than the normal. As a result the ability of this force to propel out a normal stroke volume will reduce. Diagrammatically, in a simplistic form, this is represented in fig. 10 as the rectangle **abcd**. Note that at the point of "a" of abcd (the end systolic volume) "a" is greater than the normal end systolic volume at A. **(Explain why?)**. As the myocardial dysfunction progresses, the situation just described will worsen, and the end systolic volume of the LV will continue to increase because of stagnation of the uncleared stroke volume. As the remodeling consolidates, the ventricle will become stiffer with a gradually increasing left ventricular end diastolic pressure (LVEDP), and the inability of the heart to relax and fill normally (**Diastolic dysfunction**). The end result is the development of cardinal symptoms and sign of heart failure viz. dyspnea and cyanosis **(Fig. 13)**.

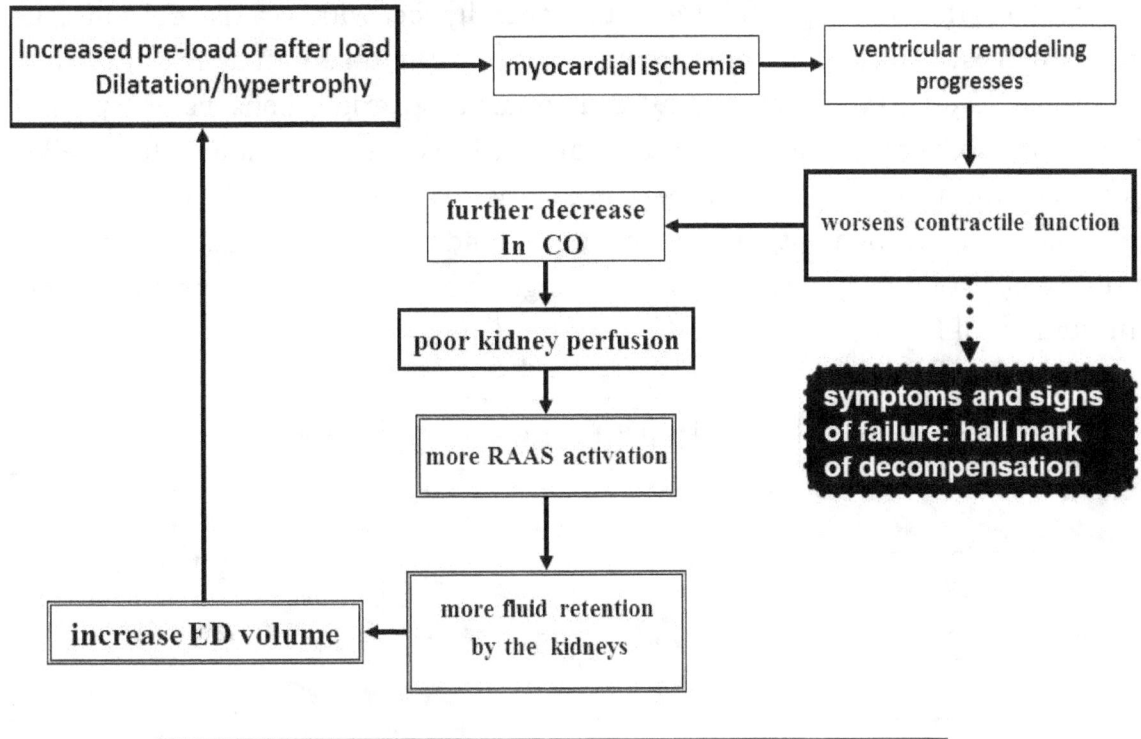

Fig. 12. Propagation of failure: decompensation

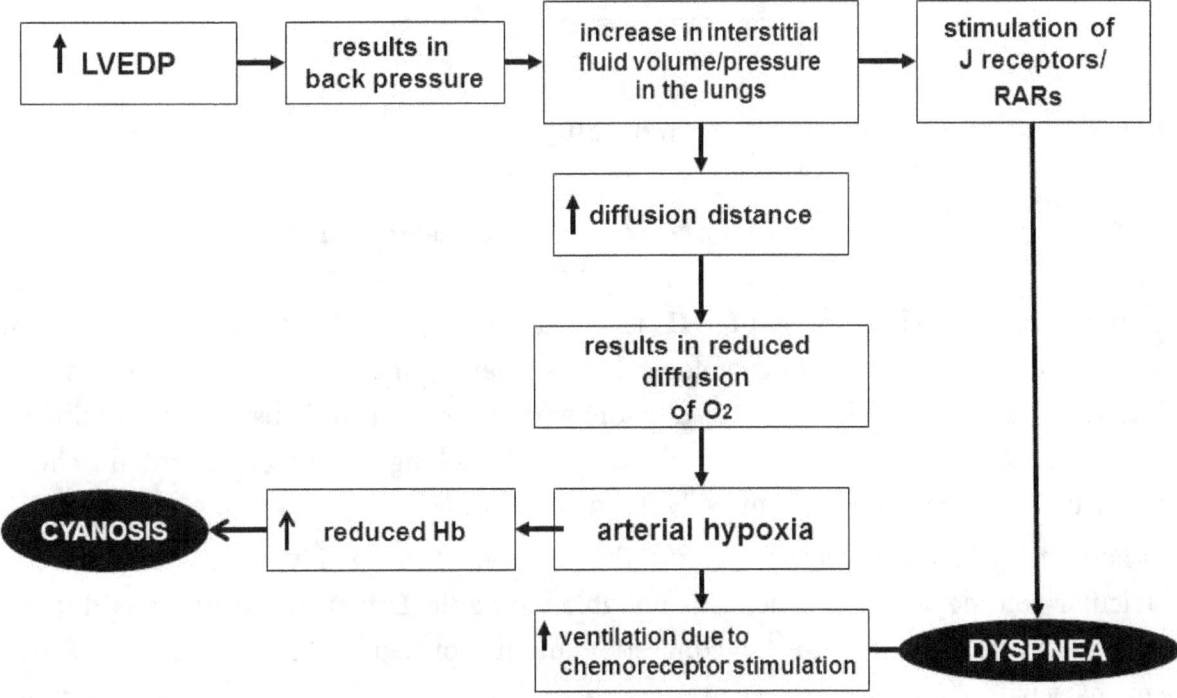

Fig. 13 Projected mechanisms of dyspnea & cyanosis in heart failure. J receptors and RARs (Rapidly Adapting Receptors) are located in the pulmonary interstitial space and airways.

Types of heart failure. Apart from the above, other descriptors of HF may be helpful in classifying the clinical manifestations of heart failure **(Fig. 14)**. However these variations merge as the pathophysiology, and at best they may be useful in the early clinical settings of the disease. It is the **low output** failure that is generally met with as a natural consequence of myocardial ischemia. In certain circumstances, the cardiac output may rise to more than normal levels (severe anemia, hyperthyroidism, arterio-venous fistula, pregnancy, Beriberi being some of the more important causes). The common denominator in all these situations is the markedly lowered peripheral vascular resistance. By the time the HF sets in, the originally high output usually reduces, and may be recorded as normal or just slightly elevated. Clinically, it is difficult to distinguish between the high out and low output manifestations of HF.

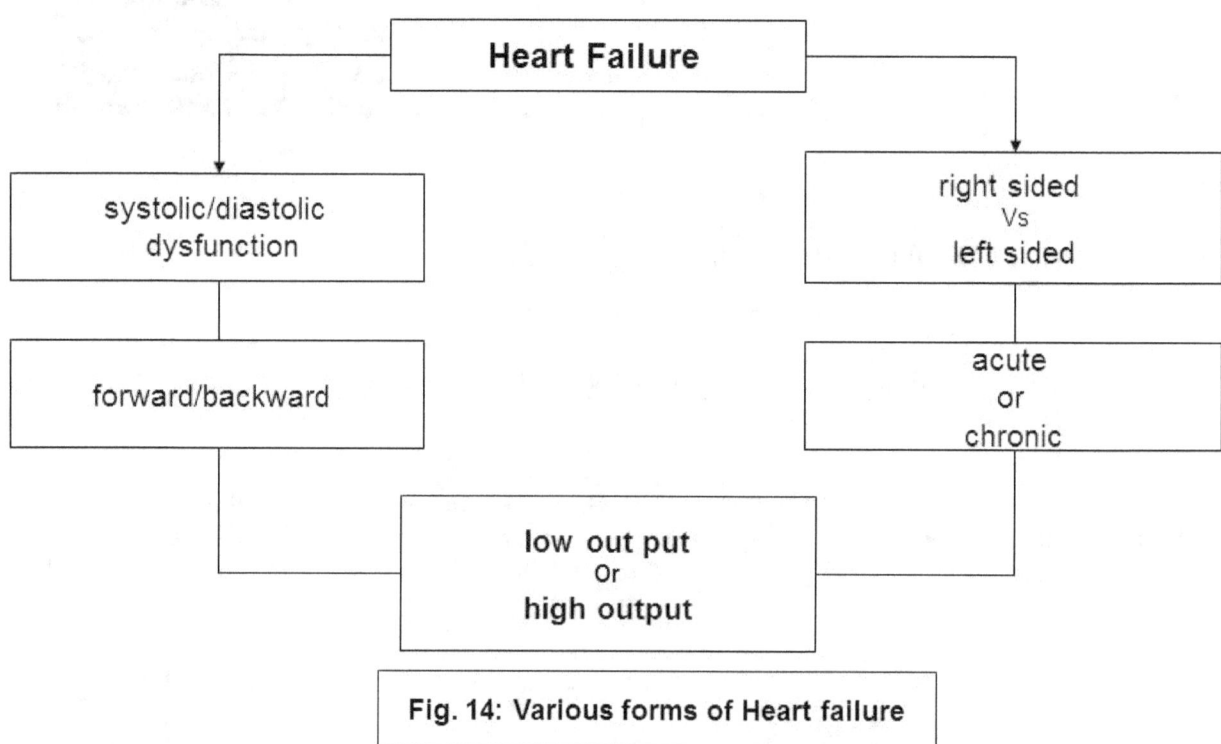

Fig. 14: Various forms of Heart failure

The more common cause of **Acute HF** is a massive myocardial infarction. The sudden hypotension is the main pathophysiological consequence, and acute complications such as pulmonary edema may follow. In the **Chronic** variety, arterial BP is usually maintained till late in the progression of the disease, and usually follows long standing myocardial ischemia, dilated cardiomyopathy and chronic valvular heart disease.

When the primary pathology affects the left ventricle eg. Myocardial ischemia, left ventricular volume overload which it is not able to handle, **Left sided failure** is said to have occurred, and results in pulmonary congestion and its consequences. In **Right sided failure**, the primary pathology affects the right ventricle eg. Pulmonary hypertension of long standing, chronic obstructive pulmonary disease (COPD). In this situation, the pathophysiological manifestations seen are as a result of back pressure building up behind the right ventricle-increased JVP, liver enlargement which may progress on to ascites, oedema and cyanosis. Over a period of time, both ventricles finally succumb.

The concept of the so called **forward** versus **backward** failure is not fully resolved. It is the contention of some that in the former, fluid retention occurs as a result of decreased renal perfusion which in turn activates the Renin-Angiotensin-Aldosterone System, while in the backward failure, the back pressure which develops on the atria and the venous system because of ventricular hypo function causes a transudation of fluid into the extracellular spaces and promotes fluid retention. The controversy at this point of time, is more a matter of semantics rather than an understanding which may dictate a therapeutic strategy.

Clinical Signs and Symptoms vis-à-vis Pathophysiology of Heart Failure, and Physiological Principles of its Treatment

The clinical manifestations of altered physiology as a result of HF are illustrated in **Fig. 15.** **Dyspnea** is by the most common derangement. The basic pathophysiology involved is described earlier, and the principle hold good for any variety of dyspnea listed in Fig. 15. In the initial stages of failure, breathlessness occurs on exercise which is not severe enough to produce the same symptom/sign in a normal individual. The weakened myocardium is unable to clear the increased venous return which occurs during exercise, while at the same time, the existing myocardial ischemia worsens because of the increased O_2 demand which the circulation is unable to meet. This further reduces the pumping ability of the heart, resulting in a backpressure which congests the lungs, setting up the events described in Fig. 13. Orthopnea is posture related. The myocardium is weakened to the extent that in the supine posture it is unable to cope with the venous return, but in the upright posture, gravity helps by redistributing the fluid towards the lower limbs, thereby reducing the lung congestion. Paroxysmal nocturnal dyspnea is an extension of the same where in there is a sudden pulmonary congestion which excites the reflex activity via the J receptors producing dyspnea. The patient reports that s/he feels better by "Standing near a window and taking in deep breaths" which actually means that the vertical posture has redistributed the accumulating fluid in the lungs to the periphery, thereby reducing the central blood volume, and ameliorating the stimulation of the J receptors in the lung interstitial spaces.

Periodic breathing (Cheyne Stoke's breathing) is a form of breathing pattern which occurs in the presence of reduced sensitivity of the respiratory centres to $PaCO_2$. The respiratory effort waxes and wanes, producing a cyclical pattern of breathing. Prolongation of circulating time in patients with HF has been blamed for the occurrence. This form of breathing is typically seen in individuals exposed to high altitude hypoxia.

Cyanosis of HF is a central cyanosis and indicates the presence of arterial hypoxia severe enough to increase de-saturated hemoglobin to > 5 gms%.The hypoxia is the end result of an increase in the diffusion distance for O_2 because of the accumulating fluid in the interstitial space. The sluggish circulation in HF is seen as peripheral cyanosis, and delayed filling of nail beds.

Pitting edema as a sign in patients with HF occurs because of transudation of fluid out of the veins because the increase in hydrostatic pressure. Classically, in early failure, the patient develops an edema around the ankles as s/he goes about the daily chores. The inadequate pumping of the heart results in a backpressure on the venous side, with the fluid accumulating in the dependent parts –around the ankles. The increase in the hydrostatic pressure produces a transudate which appears as the pitting edema. During the night, when the patient rests in the supine posture, this fluid which has accumulated in the peripheries, gets redistributed, and the edema is relieved. In more severe forms of HF, the edema occurs in the sacral region which becomes the dependent area. **Pulmonary edema** is a manifestation of a similar pathophysiology in the lungs when the pulmonary capillary hydrostatic pressure can no longer be contained, leading to an exudation of fluid out of the capillaries in to the interstitial spaces, and later in to the alveoli.

The increase in **JVP followed by hepatomegaly** are obviously the effects of venous back pressure –the end result of the inability of the ventricles to clear the volume load they encounter. An extension of the hepatomegaly is the development of the ascites when the back pressure exceeds the capacity of the hepatic veins to handle it.

The decrease in cardiac output to the brain which causes cerebral hypoxia, is probably the reason for the CNS effects of HF. In patients who are handicapped with cerebral atherosclerosis, these effects are likely to be more obvious. The **oliguria** is the manifestation of low cardiac output to the kidneys.

Patients with severe CHF suffer from weight loss, and muscle wasting with a loss of appetite and generally thrive poorly. This gamut of symptoms and signs is collectively termed as cardiac **cachexia.** Various reasons attributed to its occurrence are i. Presence of circulating cytokines such Tumor Necrosis Factor α. ii. Impaired intestinal absorption because of congested veins. iii. Relatively high metabolic rate because of the over worked respiratory muscles and iv. The anorexia and abdominal discomfort with loss of appetite which follows hepatic enlargement. The cachexia, loss of appetite and the often reduced cardiac output to the working muscles accounts for the fatigability of these patients.

The management of HF is complex. The physiological basis includes the reduction of circulating fluid volume by the use of diuretics, and restriction of salt intake to control the volume overload which the failing heart is unable to cope with, use of cardiotonics to maintain cardiac contractility, and prevention (and if instituted early enough in the course of the disease, reversal) of ventricular remodeling.

Fig. 15. Clinical manifestations of pathophysiology in HF

Respiratory

i. Dyspnea
ii. Orthopnea
iii. Paroxysmal Nocturnal Dyspnea
iv. Periodic respiration
v. Cyanosis

Cardiovascular

i. Hypotension
ii. Pitting edema/Ascites
iii. Raised JVP
iv. Hepatomegaly
v. Cold clammy skin

Renal

i. Oliguria
ii. Pre-renal azotemia
iii. Albuminuria

CNS

i. Slow cerebration
ii. Mental confusion
iii. Depression

General

i. Fatigue
ii. Anorexia
iii. Cachexia

Pathophysiological Consequences of Valvular Heart Disease

Heart valve disease may be congenital or acquired. Here only some functional aspects of acquired valvular disease will be dealt with. The commonest clinical condition which affects heart valves is Rheumatic heart disease *(It licks the joints but bites the heart).*

Table V: Normal functions of heart valves

  i. separation of atria from the ventricles

  ii. separation of the great arteries from the ventricles

  iii. ensure one way flow between atria and ventricles

  iv. prevent backflow into atria during ventricular contraction

  v. generation of heart sounds

  vi. help generate required pressures in the heart chambers

Normal functions of the heart valves are listed in **Table V**. From this table it becomes obvious that valvular damage becomes an impediment to the normal flow of blood between the various compartments that the valves separate from one another. The two main categories of mechanical defects in the valves which are likely to disrupt normal functioning are i. Stenosis (narrowing) and ii. Regurgitation (back flow) through a partially open valve. It may be deduced that any point of time, the valves may suffer such a fate resulting in Aortic, Mitral, Tricuspid, Pulmonary stenosis or regurgitation. Sometimes a multiple valvular involvement may be seen, classically so in Rheumatic heart disease. The typical appearance of systolic and diastolic murmurs in context of the first heart sound (S1) and the second heart sound (S2) are seen in **Fig.16**.

The possible hemodynamic effects of an aortic stenosis murmur are also illustrated. **(The reader may deduce the effects of other stenotic and regurgitation murmurs)**.The genesis of other varieties of murmurs such as those produced as a result of hyper-dynamic circulation, have not been considered.

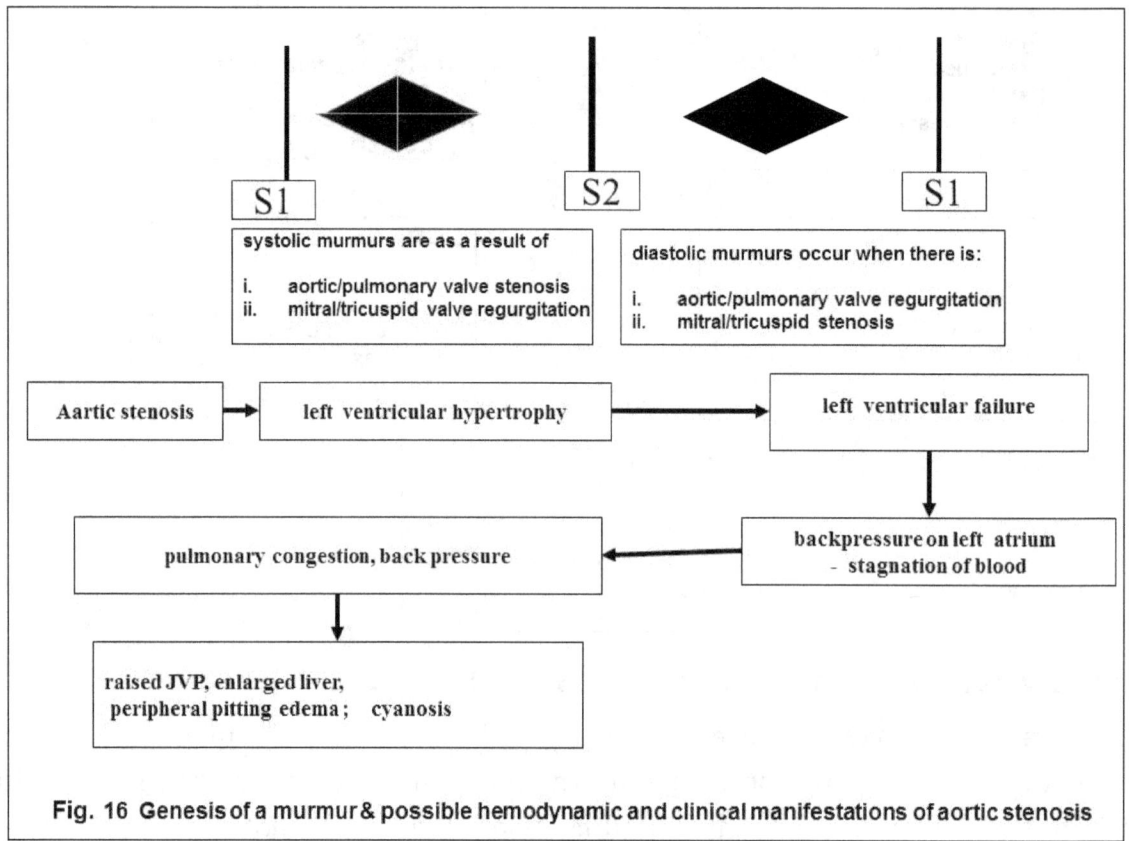

systolic murmurs are as a result of

i. aortic/pulmonary valve stenosis
ii. mitral/tricuspid valve regurgitation

diastolic murmurs occur when there is:

i. aortic/pulmonary valve regurgitation
ii. mitral/tricuspid stenosis

Aartic stenosis → left ventricular hypertrophy → left ventricular failure

pulmonary congestion, back pressure ← backpressure on left atrium - stagnation of blood

raised JVP, enlarged liver, peripheral pitting edema ; cyanosis

Fig. 16 Genesis of a murmur & possible hemodynamic and clinical manifestations of aortic stenosis

Mitral valve prolapse syndrome is an interesting aspect of valvular pathophysiology. It produces a wide variety of symptoms, which include palpitations, tachycardia, dysarrhythmia and syncope which are very often alarming for the patient. But this is generally an innocuous problem. It occurs because of a ballooning of the anterior and posterior cusps of the mitral valve into the atrial cavity during systole because of a collagen disorder. It has been associated with autonomic dysfunction and excessive secretion of catecholamines, both of which may account for the various symptoms. Added to this is the possibility of development of a mitral regurgitation because of the ballooning of the valve.

References

1. Beevers G. The pathophysiology of hypertension. British Med. J.2001; 14 April.BNET.com: 74435271

2. Best and Taylor's Physiological Basis of Medical Practice, 13th edn.; Ed. Best JB. William & Wilkins, London; 1990.

3. Davidson's Principles and Practice of Medicine. Edt. Haslett C, Chilvers ER, Boon NA & Colledge NR. 19th Edn. Churchill Livingstone., London. 2002

4. Delagado IIIRM, Willerson JT. Pathophysiology of heart failure. Texas Heart Institute Journal. 1999; 26: 28–33

5. Ganong WF. Review of Medical Physiology. 22nd edition. Lange Medical Books/McGraw-Hill, London. 2005

6. Ganz P, Ganz W. Coronary blood flow and myocardial ischemia. Chapt. 34 in Heart Disease: A Text Book of Cardiovascular Medicine Edt. Braunwald E, Zipes D, Libby P. 6th edn. WB Saunders, London; 2001; pp. 1087–1113.

7. Gray's Anatomy: The Anatomical Basis of Medical Practice; 39th Edn. Ed. Standring S; Elsevier, Churchill Livingstone, London; 2005.

8. Guyton AC and Hall JE. Textbook of Medical Physiology 11th edition. Elsevier, Philadelphia; 2006.

9. Harrison's Principles of Internal Medicine Edt. Longo D., Fauci A, Ksper D, Hauser S., Jameson J, Joseph Loscalzo J, 18th ed. MCGraw Hill, London. 2011.

10. Kappagoda CT, Ravi K. The rapidly adapting receptors in mammalian airways. And the irresponses to changes in in extracellular fluid volume. Exp. Physiol. 2006; 91: 647–654.

11. Klabunde R. The Pathophysiology of Myocardial Infarction - Causes and Effects in "Cardiovascular Pharmacology Concepts. CV Physiology.com. March 2007

12. Kirk ES, Factor S, Sonnenblick EH. Newer concepts in the pathophysiology of ischemic heart disease. G. Ital. Cardiol. 1984; 14: 881–891

13. McCance KL, Huether SE. Pathophysiology: The Biologic Basis for Disease in Adults and Children. 5th edition. Mosby; St. Louis USA; 2006

14. Packer M. New concepts in the pathophysiology of heart failure: beneficial and deleterious interaction of endogenous hemodynamic and neurohormonal mechanisms. J. Internal Med. 1996; 239: 327–333.

15. Paintal AS. Vagal sensory receptors and their reflex effects. Physiol. Rev. 1973; 53: 59–227

16. Pant S, Deshmukh A, Neupane P, Kavin Kumar MP, and Vijayashankar VS. Cardiac Biomarkers. Chap. 2, pp 19–42 in Novel Strategies in Ischemic Heart Disease. Edited. Lakshmanadoss U. InTech Shanghai.2012

17. Porth C. Essentials of Pathophysiology: Concepts of Altered Health States. 2nd edn; Lippincott Williams and Wilkins; Philadelphia;2007.

18. Ramanathan T, Skinner H. Coronary blood flow. Continuing Education in Anaethesia, Critical Care & Pain. 2005; 5: 61–64: ceaccp.oxfordjournals.orhg/cgi/content/full/5/2/6

19. Ruwende C, Visovatti S, Pinsky D. Myocardial and cellular mechanisms of myocardial ischemia: reperfusion injury. Chapt. 58 in "Hurst's the Heart. Edit. Fuster V, O'Rourke R et al. 12th edn.

20. Struthers AD. Pathophysiology of heart failure following myocardial infarction. Heart. 2005; 91: suppl. ii.14-ii.16.

21. Vikrant S, Tiwari SC. Essential hypertension-Pathogenesis and pathophysiology. Indian Academy of Clinical Medicine 2001; 2: 141–161.

22. Werner GS. Editorial: Collaterals: How important are they? Heart 2007; 93: 778–779

23. World Health Organisation 2003 (WHO)/ International Society of Hypertension (ISH) statement on management of hypertension. World Health Organisation, International Society of Hypertension Writing Group. Journal of Hypertension. 2003;21:1983–1992.

CHAPTER - 5

Pathophysiology of Shock and Burns: Multi Organ Dysfunction

Plan

Shock & MODS

1. Definition and classification
2. Loss of circulating blood(fluid) volume) in shock: non-hypotensive & hypotensive haemorrhage
3. Cellular changes during shock: role of mediators
4. MODS

Burns

1. Classification
2. Extent of burns
3. Pathophysiology
    i. burn zones
    ii. systemic pathophysiological effects

a. cardiovascular b. respiratory; c.gastrointestinal; d. renal; e.hyper metabolic state f. hemolysis; g. immune suppression

General Introduction

Shock is defined as a state of inadequate tissue perfusion because of a relative or absolute deficiency of cardiac output, which if unresolved, may lead to a state of cellular dysfunction, and may finally progress to a *Multiple Organ dysfunction.*

Amongst the causes listed in **Table I,** this chapter will mainly consider hypovolemic shock. In principle, the basic pathophysiology of other variants of shock is similar to that seen in hypovolemic shock: hypoperfusion induced imbalance of oxygen, and substrate delivery at tissue level.

Of the many causes of hypovolemia, loss of blood to the external environment is an important one. This may be caused by trauma or bleeding from the gastrointestinal tract. *(What other causes of hypovolemia can you think of?).*

To comprehend the sequence of events which follows blood (fluid) volume loss, consider that there is a ratio of the blood volume available to the volume capacity of the cardiovascular system. Ideally this ratio should be 1. If the ratio falls below one, there could be i. a decrease in circulating blood volume (haemorrhage) or an increase in volume capacity because of vasodilatation **(Fig.1).**

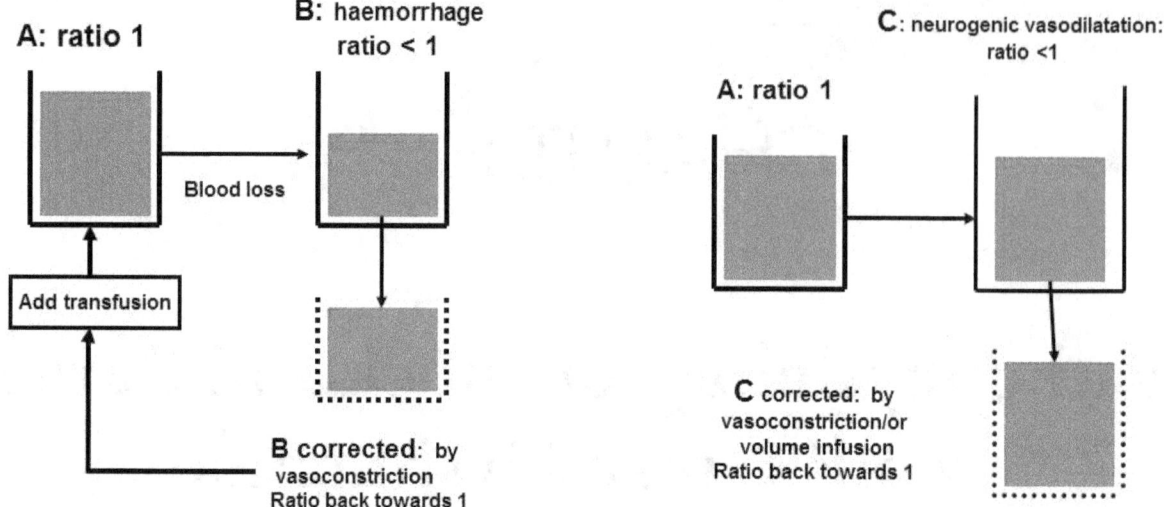

Fig. 1. Concept of volume availability/volume capacity ratio and shock

Shaded area is the volume available, and incomplete rectangle is the volume capacity of the cardiovascular system.

A represents normal ideal ratio of 1.

B denotes loss of circulating blood volume: classical hypovolemia in shock.
C denotes an increase in volume capacity because of sudden vasodilatation, while the available volume remains unchanged.: neurogenic variety of shock.

In both situations B & C, the ratio of volume availability /volume capacity is reduced. Coorections are attempted by vasoconstriction (.....). Additional support by transfusion will help to revert situation to normal (A)

Non-hypotensive Haemorrhage

> **Table I: Classification of shock**
>
> 1. Loss of circulating fluid volume-
>
> 2. Hypovolemic shock
>
> 3. Increase in vascular space availability-
>
> 4. Neurogenic shock: vasodilatation
>
> 5. Combination of 1 & 2 above:
>
>   i. Cardiogenic shock- heart failure
>
>   ii. Septic shock

Loss of blood which is <15% of the total blood volume is often known as non-hypotensive haemorrhage as there is no fall in the blood pressure. The main cardiovascular compensatory mechanism is a slight tachycardia, and an increase in peripheral vascular resistance in the forearm blood vessels. (the latter may be clinically evidenced as a delay in capillary filling of > 3 seconds). Both the changes are a consequence of deactivation of the low pressure cardiovascular receptors causing an increase in sympathetic excitation. In reference to **Fig 1**, this is a successful attempt to revert situation B (the lowered volume available/volume capacity ratio) towards the optimum (A). Deactivation of low pressure cardiopulmonary receptors is also associated with the early increase in circulating vasopressin (ADH) in response to non-hypotensive haemorrhage, and may thus assist in trying to recover the lost volume. With greater blood loss (> 15%), more stringent measures are brought into play by body homeostatic mechanisms. **(Fig. 2).** The decrease in pulse pressure, a result of an increase in the diastolic blood pressure (DBP), indicates a more widespread vasoconstriction. Even though systolic BP (SBP) is maintained in the supine posture, disturbance of cardiovascular homeostasis is reflected by postural hypotension if the patient attempts standing. It is obvious that the attempt now is to restore the volume/(volume capacity ratio), not only by adjusting the volume capacity by vasoconstriction, but also by attempting to increase the circulating fluid volume by water and salt retention. *(Students should work out as to how this is organised by body mechanisms)*.

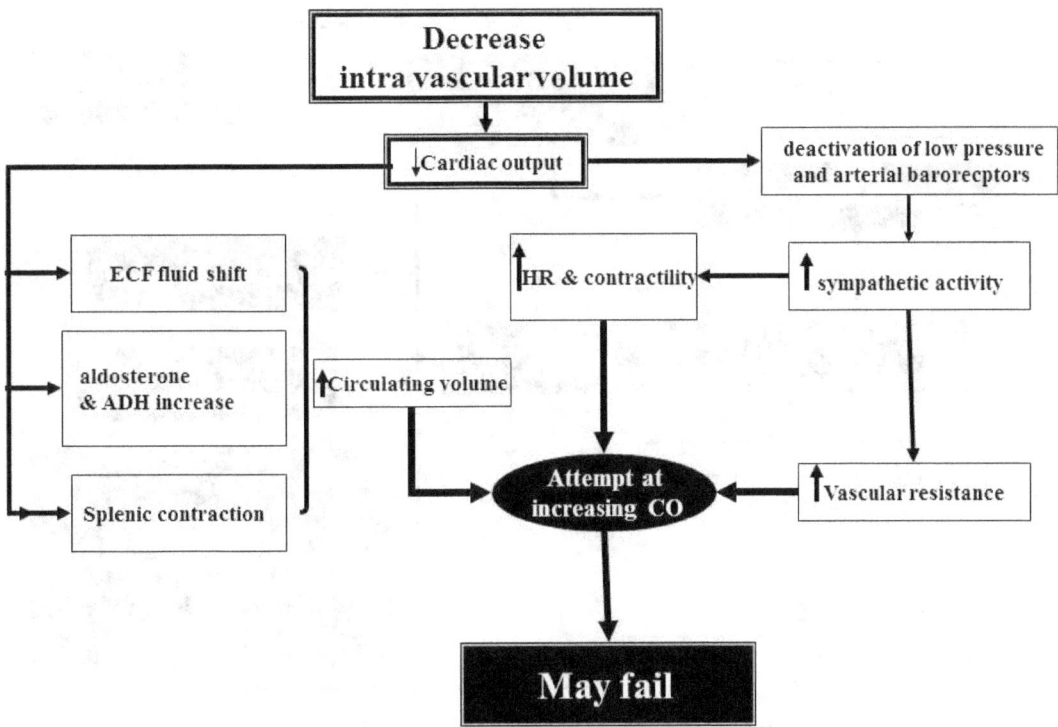

Fig. 2. Compensatory mechanisms in haemorrhagic shock

Hypotensive Haemorrhage

A blood volume loss of > 30 % takes the patient into the realm of hypotensive haemorrhage. The mean arterial pressure is likely to fall to about 60 mmHg. This is usually accompanied by oliguria (which may progress on to acute renal failure), and mental confusion /agitation as a result of cerebral hypoxia. If the condition is not reversed by enthusiastic volume supplementation and supportive measures, the situation is likely to deteriorate and develop into multiple organ dysfunction (irreversible shock).The time frame between reversibility of shock with vigorous measures, and progression to the irreversible state is very narrow. The sequence of likely pathophysiological events is outlined in **Fig. 3,** and the cellular changes that may occur are shown in **Fig. 4**

Apart from the adrenergic neurotransmitters and hormones, other vasoconstrictor substances which are released during haemorrhagic shock include angiotensin II, endothelin-I, and thromboxane A_2 Vasodilators too are secreted. The list includes nitric oxide, prostacycline, and local metabolic products such as adenosine and lactic acid. Local tissue perfusion depends upon the interaction between the vasoconstrictors and vasodilators. When tissue perfusion becomes inadequate, there is a shortage of O_2. Anaerobic metabolism in the cells results in a build up of lactic acid leading to metabolic acidosis. This causes the sodium potassium pump to fail resulting in rushing in of Na^+, and with it water, making the cells swell. Mitochondrial swelling occurs, followed by a cessation/reduction in ATP production. Intracellular disruption releases lysosomes which break up the cell membrane. There is a release of various inflammatory mediators **(Table II).**

Fig. 3. Pathophysiology of development of MODS

| If compensation fails/ blood loss continues | → | Further fall in blood pressure & CO | → | impaired O_2 use |

tissue perfusion reduced further

mitochondrial damage

further tissue hypoxia & Tissue damage → Impaired cellular metabolism

impaired glucose use lack of ATP

cell membrane disruption & cell death → **MODS**

- Generation of Free O_2 radicals by inflammatory mediators, vasodilators,
- Microthrombi formed & lysed,
- Adhesion molecules on endothelium
- Neutrophil activation

further microvascular injury & hypotension

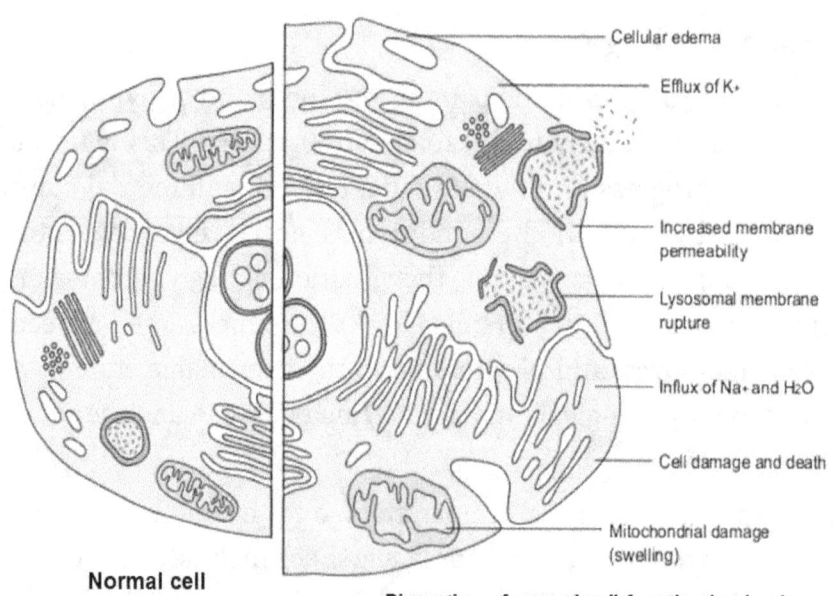

Cellular edema

Efflux of K+

Increased membrane permeability

Lysosomal membrane rupture

Influx of Na+ and H2O

Cell damage and death

Mitochondrial damage (swelling)

Normal cell

Disruption of normal cell function in shock

Fig. 4: Cellular changes in shock

Table II: Inflammatory mediators in shock

A. Eicosanoides

    iii. Cycloxygenase derived: prostaglandins PGI_2, PGE_2

    iv. Lipoxygenase derived leucotrienes:

LTB_4 (releases free oxygen radicals), LTC_4, D_4

B. Platelet activating factor: primes macrophages

C. Macrophage derived mediators:

    i. Tumour necrosis factor α

    ii. Interleukins (IL) 1, 8

D. Endothelial factors: toxic nitric oxide (NO) which induces free O_2 radicals

---

Activated coagulation proteins give rise to cycles of micro-thrombi production and their lysis leading to repeated occurrence of ischemia and re-infusion episodes. The latter may accentuate formation of oxygen free radicals which further enhance cell membrane damage. Inflammation causes expression of adhesion molecules on endothelial cell surface and activation of neutrophils. This also promotes micro-vascula rinjury and escape of fluid into interstitial space further aggravating hypotension. If the systemic release of various inflammatory substances cannot be controlled by body mechanisms, an auto-destructive generalized inflammatory reaction is the end result. This has been designated as a "Cytokine Storm." However the pathophysiology of injury-induced organ dysfunction secondary to shock is as yet poorly characterized.

Shock as a Multi Organ Dysfunction Syndrome (MODS)

Uncontrolled inflammatory response during shock may lead to a progressive failure of twoor more organ systems: MODS. The term arose from the American College of Chest Physicians/ Society of Critical Care Medicine Consensus Conference Document: Definitions for Sepsis and Organ Failure and Guidelines for the Use of Innovative Therapies in Sepsis. The mortality in the condition is about 45%, and reaches 100% if there is a failure of 3 or more organ systems for 3 days. Direct injury to an organ system produces primary MODS. Later, if an exaggerated host response to primary injury occurs, secondary MODS is said to have set in. Apart from severe traumatic shock, major surgery, extensive burns and systemic sepsis may also result in MODS. Some of the risk factors associated with the syndrome are age > 65 years; pre-existing baseline dysfunction and chronic diseases such as cancer and diabetes; patients on steroid treatment; coma on admission; multiple blood transfusions; persistent focus of infection. The development and consequences of MODS are outlined in **Figs.3 to 5.** Release of pro-inflammatory mediators in various tissues, and their influence in generation of free O_2 radicals is the centre point of the pathophysiological process. The latter in turn disrupt cell membranes, and cause cell death in various organ systems. (*Acute renal failure (ARF) is a single organ failure syndrome which may occur as a complication of hemorrhagic shock. How may it change into a MODS?*)

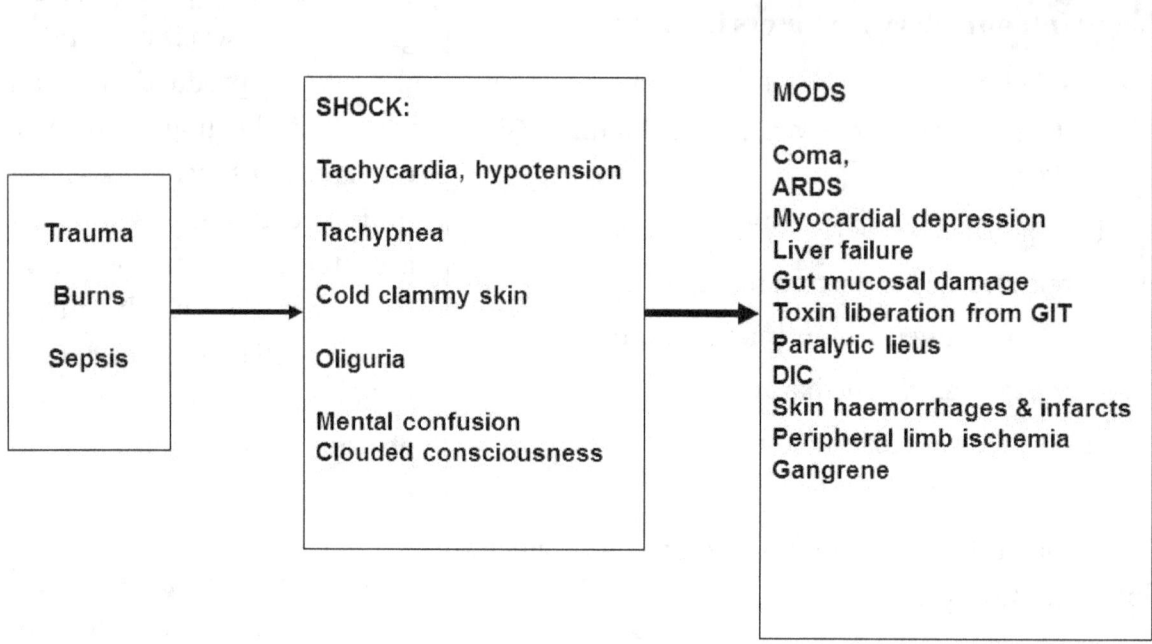

Fig. 5: The clinical consequences of MODS
(DIC is disseminated intravascular coagulopathy)

The more recent understanding of the pathophysiology of hypovolemic shock has introduced some interesting insights in the physiological principles of management. It is imperative that circulatory support using fluid transfusion and vasoactive agents must be introduced early and energetically in order to prevent onward march of the condition to MODS.

1. **Intubation and mechanical ventilation.** Respiratory muscle fatigue tends to set in rapidly because of the tachypnea caused by the hypoxia and lactic acidosis.. The inflammatory reactions may involve the lungs and enhance the hypoxia. Use of mechanical ventilation therefore helps to improve oxygenation, reduce oxygen consumption by respiratory muscles, and divert the available oxygen for perfusion of other needy tissues.

2. **Circulatory support.** Adequate cardiac output is required in order to make oxygen and nutrients available to the tissues, while there has to be a certain minimum pressure head [mean blood pressure (MBP)] which must be applied to the cardiovascular system if this cardiac output is to be delivered suitably.

Critical Closing Pressure and Its Relevance to Tissue Perfusion in the Hypovolemic State

At all times, the blood column must generate a certain amount of pressure inside the blood vessels in order to keep them open. If this pressure falls below the required minimum, the vessel closes down. This is the "Critical Closing Pressure." It is determined by a). The vascular smooth muscle tone of the arterioles and meta-arterioles. This is controlled predominantly by sympathetic innervation, the other factors being myogenic, metabolic, and flow dependent dilatation. b) Rouleaux formation by the RBC; c) Externally applied pressure (such as tissue edema) which compresses the peripheral blood vessels. *(In a normal situation, will the tissue pressure help to keep the blood vessel patent, or close it?)*. Of these, it is the vascular tone that is most amenable to modulation by both physiological and pharmacological means. Normally the homeostatic mechanisms "Auto regulate" blood flow in most tissues so that under a fairly wide range of pressure, blood flow remains constant. In a situation such as hypovolemic shock, the increase in the sympathetic drive induces a powerful vasoconstriction, while at the same time, the intra vascular volume that is now deficient, is unable to generate enough pressure to prevent a closure of the vessel because of the constriction. When this vessel is to be opened to restore blood flow, the intra vascular pressure needed to overcome this closing pressure will have to be higher than normal. On the other hand a withdrawal of sympathetic drive should lower the critical closing pressure and ease out the situation. The initial physiological principle of management then must aim to restore the MBP to autoregulatory level in order to restore tissue perfusion. This is best done by a controlled and rapid volume expansion by using various plasma expanders which may range from blood and blood products to crystalloid solutions so that the volume available/volume capacity ratio is brought as close to 1 as possible **(Fig.1)**. *(What physiological indices do you think indicate that the cardiac output delivery to the tissues is improving?)*. This ensures that in spite of a powerful sympathetic vasoconstriction, critical closing pressure is not reached earlier. If the source of bleeding is not arrested, the resuscitative measures will prove futile. The vasoactive medication will to help restore autoregulation as assessed by the clinical evidence. Vasoconstrictors and inotropic agents help to achieve this.

During severe hypovolemic shock, use of extraneous vasoconstrictors may not enhance the already available vasoconstriction because the maximum possible effect may have since been achieved by sympathetic activation, and the α adrenergic receptors may have been downgraded. It has been recently suggested that use of Vasopressin (ADH) acting throughV1 receptors may be of some assistance in enhancing vasoconstriction. *(Theoretically, do you think under such circumstances, vasodilators could be used to restore tissue perfusion?)*. On the other hand, vasoconstrictors may be most suitable in neurogenic shock as in this situation, the reduction of the volume available/(volume capacity) ratio to <1 is a result of an increase in the denominator (vasodilatation) **(Fig. 1-C)**. Blanket use of vasoconstrictors therefore in all circumstances where shock is present, is not recommended as this may further upset pathophysiology and accelerate the gravity of the condition. A careful clinical assessment of the perfusion status, the circulating volume availability, and its replenishment, must therefore precede use of vasoconstrictor agents.

Need for inotropic support. This is required when the cardiac output is low with high cardiac filling pressures and is best done using beta-adrenergic agonists. Such goal directed approach should help to arrest/ reduce the progress of pathophysiology of shock to MODS. More recently, use of anti-inflammatory drugs is being explored in order to prevent onset of severe inflammatory response which heralds multi organ dysfunction.

Pathophysiology of Burns

A burn can be defined as the condition resulting from direct or indirect action of heat on the human organism. Burns may be classified as thermal and non thermal, and may be 1st degree, 2nd degree or 3rd degree. **(Fig. 6).**

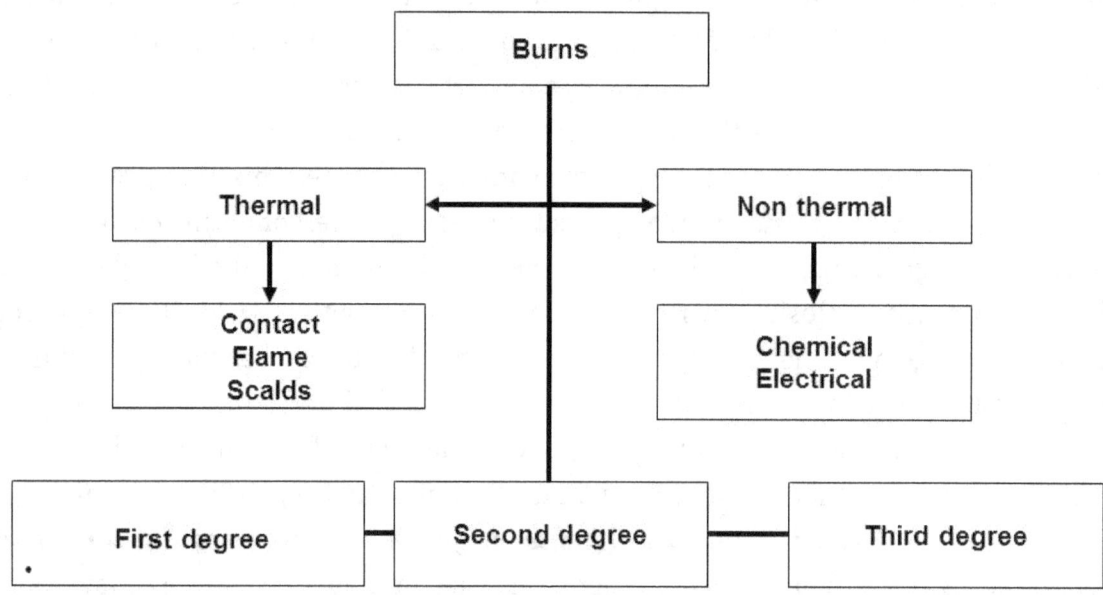

Fig. 6: Classification of Burns

The pathophysiological effects are dependent upon two factors: the extent of body surface area involved and the depth of the burn injury. The 1st degree burns are always superficial, affecting only the epidermis, and leave skin functions (pain and touch, water vapour and barrier functions) intact. They are partial thickness burns as they affect only the epidermis. They heal in about 4–5 days without scar formation.

Second degree burns are also partial thickness because though the epidermis and dermis are both involved, the subcutaneous layer is left intact. When the injury extends only partly to the dermis, the extent of the burn is said to be superficial, but when the injury envelopes the dermis fully, it is called the deep partial thickness burn **(Fig. 7).** In the superficial 2nd degree burn the pain sensation is intact while in the deep variety, it may be diminished or lost. As the protective function of the skin is lost in this degree of burn, evaporative water loss occurs readily from the area affected leading to rapid dehydration.

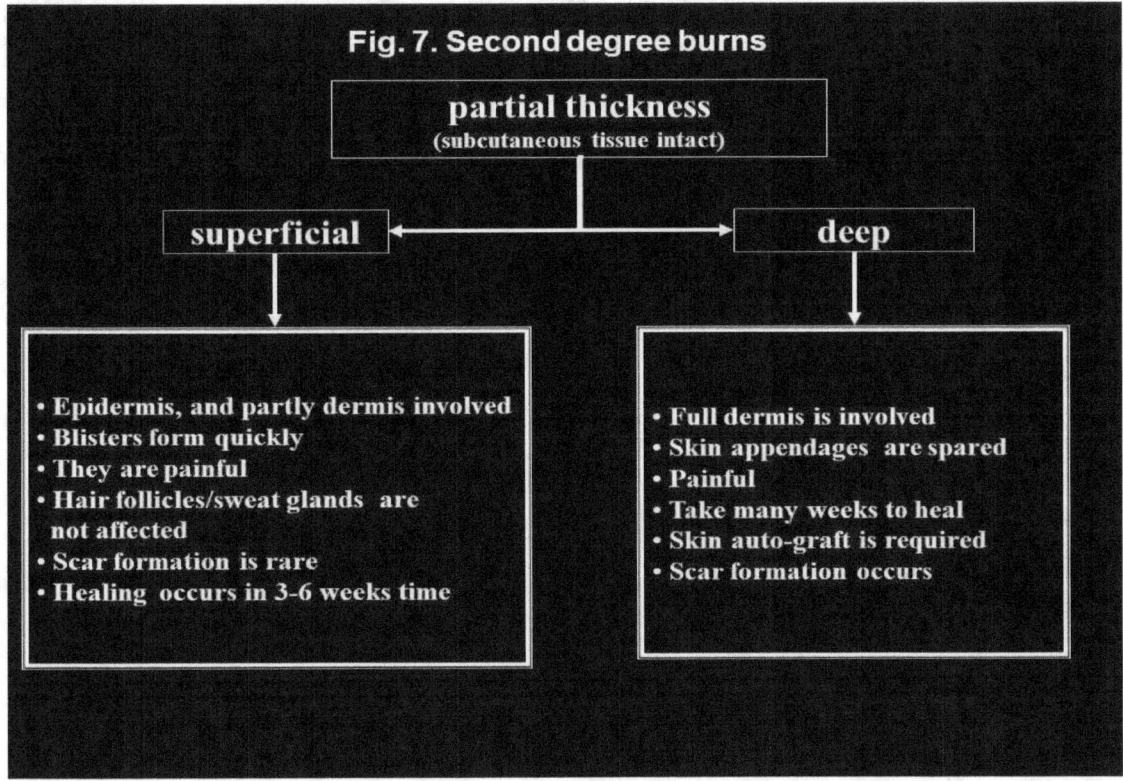

Fig. 7. Second degree burns

partial thickness
(subcutaneous tissue intact)

superficial ↔ **deep**

- Epidermis, and partly dermis involved
- Blisters form quickly
- They are painful
- Hair follicles/sweat glands are not affected
- Scar formation is rare
- Healing occurs in 3-6 weeks time

- Full dermis is involved
- Skin appendages are spared
- Painful
- Take many weeks to heal
- Skin auto-graft is required
- Scar formation occurs

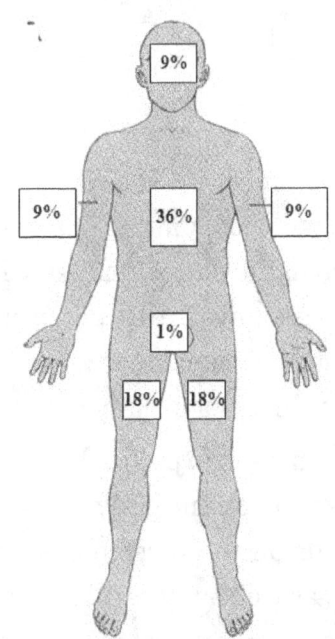

9%

9% 36% 9%

1%

18% 18%

Fig. 8: Wallace's Rule of 9%. The figures given are combined for both anterior and posterior surfaces.
Body surface area invoved: Major burns: > 20%; Extensive burns : > 40%

Third degree burns are *full thickness* because the epidermis, the dermis and the underlying subcutaneous tissue are all affected.

The severity of burn injury is dependent upon age of the patient, pre-existing disease, depth of the burn and the extent of the body surface area involved. The determination of the latter is done on the basis of "Rule of Nine" **(Fig. 8)**. A "Burn diagram" is used to depict the depth of the burn where the areas with 3rd degree burns are marked in red and those with 2nd degree with blue.

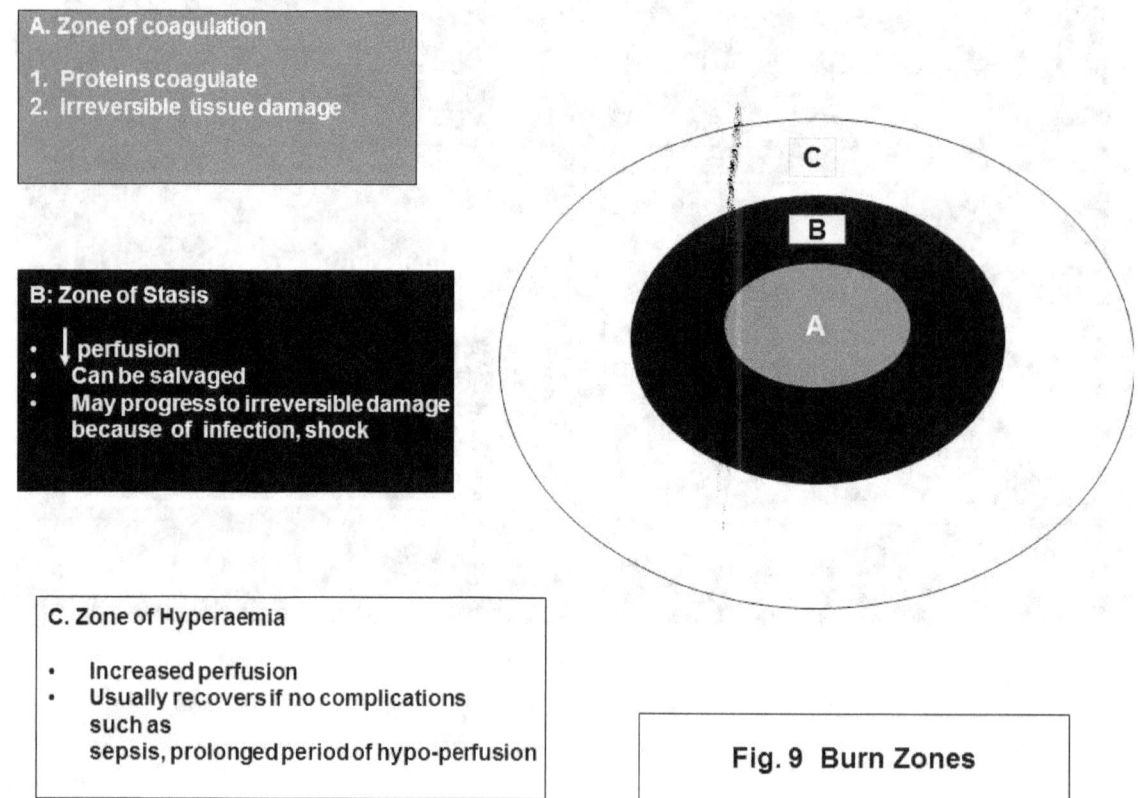

A. Zone of coagulation

1. Proteins coagulate
2. Irreversible tissue damage

B: Zone of Stasis

- ↓ perfusion
- Can be salvaged
- May progress to irreversible damage because of infection, shock

C. Zone of Hyperaemia

- Increased perfusion
- Usually recovers if no complications such as sepsis, prolonged period of hypo-perfusion

Fig. 9 Burn Zones

There are three distinct zones of burn injury **(Fig. 9)**.i. The central **zone of coagulation** in which there is severe irreversible tissue damage with coagulation of proteins. This is surrounded by ii. a **zone of stasis** in which there is decreased tissue perfusion. Delay in correction of hypovolemia, superimposed infection, and oedema, may convert this potentially salvageable area in to a zone of irreversible damage. iii. The outer most ring is the **zone of hyperaemia** where there is an increase in tissue perfusion. Tissues in this area usually recover.

Burns prevent normal skin function. Therefore, pathophysiological derangement involves water and electrolytes, body temperature control, and body surface lubrication. The magnitude of involvement of these functions depends on the surface area involved and the burn depth. The injury produces local tissue damage, as well as systemic effects which progress with time, and peak after about 5–7 days. The major organ systems involved are the cardiovascular, respiratory, renal and the gastrointestinal, and most of the damage done to these is by way of burn toxins produced in response to inflammatory mediators and the free O_2 radicals that are generated in large amounts **(Fig. 10)**. The extent of damage caused by the latter is determined by the extent to which natural mechanisms which scavenge free radicals are able to neutralize them.

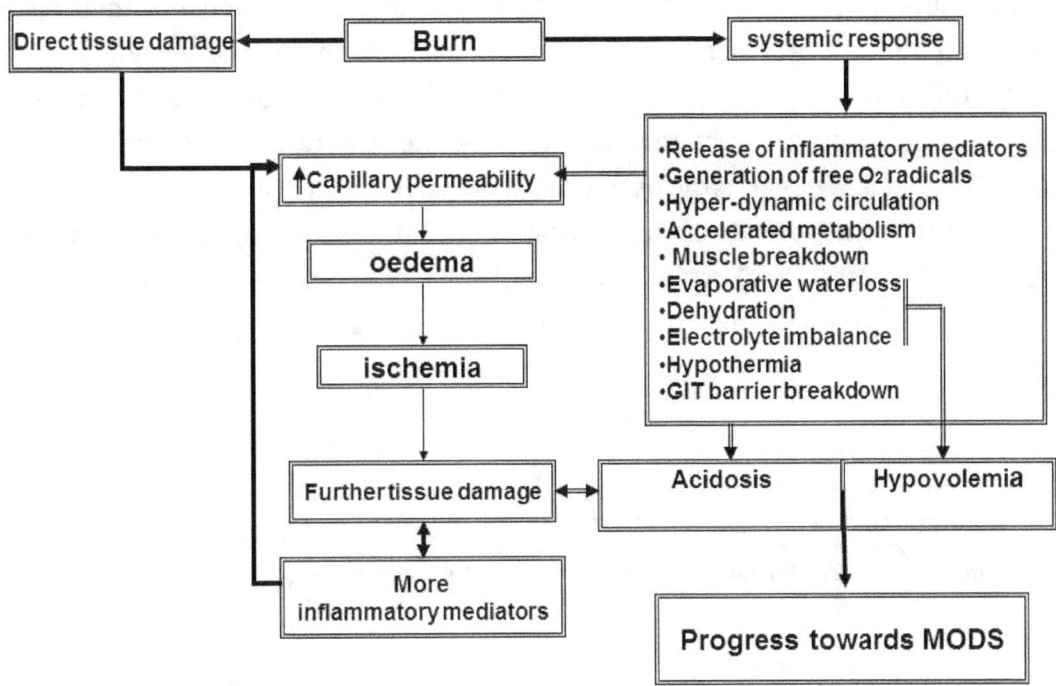

Fig. 10 Pathophysiological events because of burn injury

Systemic Pathophysiology in Burn Injury

The cardiovascular response to burns is triggered off by an immediate sympathetic activation brought about by the stressful event. This is rapidly followed by a response to the hypovolemia which develops because of two main factors: i. Evaporative water loss from the surface area damaged by the burn. This is the most potent source of loss of circulating fluid volume, and is also responsible for the initial state of hypothermia which may supervene. (Rapid infusions of cool fluids may also contribute to the hypothermia).More the surface area of burn, greater is the extent of evaporative fluid loss. ii. Release of inflammatory mediators which induce an increase in capillary permeability which then allows large amounts of fluid to escape out. The effect occurs not only locally, but also in other vascular beds distant to the area of injury. This is complicated by a state of myocardial depression as a result of release of TNF-α, one of the inflammatory mediators that is generated. The overall effect is a reduction in cardiac output and thus a reduction in tissue perfusion. (*Given the above, what physiological principle of treatment do you feel is indicated in a patient with burn injury?*)

Pathophysiology of the respiratory system in burn injury. The initial damage is done by the inhalation of hot fumes and smoke. Both the upper and lower airways may be affected. Appearance of laryngeal stridor a short while after exposure to hot fumes indicates upper airway injury **(How??)** Direct thermally induced tissue damage is followed by release of inflammatory mediators and generation of free O_2 radicals, particularly hydroxyl radicals. The latter have been held responsible for the lipid peroxidation that ensues. The pathophysiological effects of burns with inhalation injury are likely to result in pulmonary hypertension, increased airway resistance, and atelectasis which is a result of an increase in surface tension in the alveoli. Hypoxia therefore occurs, and is often worsened by inhaled smoke which contains carbon monoxide. The immediate effects are followed by the generalized systemic inflammatory reaction with the release of O_2 radicals which continues over a period of about a week **(Fig. 11)**. It is important to realize that direct inhalation of hot fumes is not mandatory for lung injury to take place in burn trauma. Systemic release of inflammatory mediators because of burn is implicated, and makes the respiratory system a potential target as a part of MODS which may complicate burn trauma.

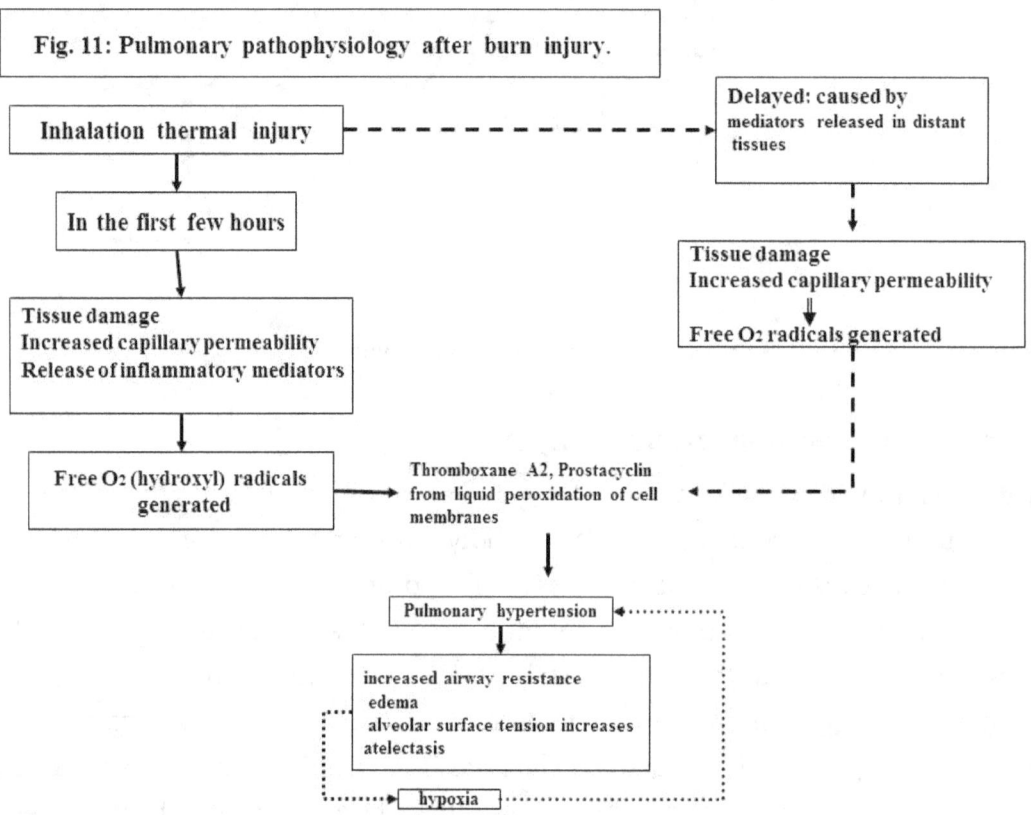

Fig. 11: Pulmonary pathophysiology after burn injury.

Gastrointestinal system involvement in burns adds to the development of MODS as complication of burn injury. Initially there is an intestinal ischemia as a result of a decrease in splanchnic blood flow. This may activate neutrophils which generate inflammatory substances and various enzymes which damage the intestinal mucosal barrier, and translocate intestinal bacteria which release Endotoxin. This substance activates macrophages and neutrophils which in turn help in the release of free radicals, thus setting up a vicious cycle of ongoing tissue damage and release of inflammatory mediators. Apart from the intestines, liver and stomach have been found to be targets of remote organ damage in burn injury cases. Arachidonic acid metabolites, the cyclo-oxygenases (COX-1 and COXC-2), have been implicated in small intestinal and liver injury in burns. It is being speculated that COX inhibitors may have a role to play in protection against inflammatory changes induced in the GIT of burn patients.

The kidneys are an obvious target for involvement in the pathophysiology created by burn injury. The immediate post injury problem is the possibility of setting in of acute renal failure (ARF). (*How does ARF occur in a burn patient?*)At a later stage, ARF is associated with tubular necrosis. It is a complication of burn sepsis and forms a part of MODS. The damage has been attributed to the release of free oxygen radicals with a concomitant decrease in renal glutathione which is the major anti-oxidant in the renal tissue. Renal failure has been known to occur more frequently in patients with inhalation burn injury.

Metabolic Problems in Burn Injury

A number of metabolic disturbances may result. Hypoalbuminemia occurs (*work out on your own as to why this is likely to happen*) Hyperkalemia often occurs because of extensive tissue damage, and metabolic acidosis because of the accompanying shock. Acute tubular necrosis may supervene because of massive breakdown of muscles (rhabdomyolysis), and hemolysis.

A hypermetabolic state sets in between the 4th and the 7th day of injury, and is a consequence of resting of metabolic set point in the hypothalamus indicated by release of stress hormones such as the catecholamines, cortisol and glucagon. The overall effect increases substrate availability because of hyperglycaemia, lipolysis and proteolysis. The insulin also increases in response to hyperglycaemia, but the latter overpowers the effect of insulin. The body temperature may not rise. The main triggers for the hypermetabolic state are pain, and release of inflammatory mediators. There is a direct correlation between the magnitude of injury and the degree of hypermetabolic state. (*Physiologically, what will be the disadvantages of such a metabolic state?*). This state may last as long as a year after the injury, and is often responsible for impaired wound healing, risk of repeated infections, and a continued loss in body tissues and weight.

Burn Injury and Hemolysis

Hemolysis is known to occur after burns. This has been attributed to direct thermal injury to the red blood cells and is seen as the presence of microsperocytes and schistocytes (RBC which have jagged, fractured margins) in peripheral smears. Such cells are found in peripheral smears after direct heating of blood. Apart it has been seen that life span of RBCs of patient with burns is reduced significantly, and contributes greatly to the post burn anaemia. Further, life of donor cells given as transfusion to burn patients is also shortened. However, if RBC of a burn patient are transfused in the early post burn period in to a healthy person, they do not undergo premature breakdown. This suggests that extracorporeal substances circulating in the blood of a patient with burns are responsible for the post burn anemia. Exact nature of these substances is as yet unknown.

Immuno-suppression in Burns

By far the most plausible reason as to why burn patients are susceptible to uncontrolled infection is the fact that immune mechanisms are compromised. The exact mechanisms as to why his happens is still unclear. A possible set of events is outlined in **Fig. 12.**

Loss of innate immunity. To begin with damage to the skin especially when extensive, removes the natural protective barrier, and exposes the underlying tissue to the environment. The invading organisms which enter the tissues are nurtured by the rich fluid and protein medium available to them as a result of tissue injury and increased capillary permeability. The resultant sepsis by itself may overwhelm the existing immune mechanisms.

Suppression of immune mechanisms. Release of various pro-inflammatory mediators is thought to act by i. Suppression of phagocytic activity of neutrophils and macrophages. Normally complements C3a and C5a are required for opsonisation of invading organisms, making them susceptible to phagocytic action by macrophages and neutrophils. Inflammatory mediators produced by burn injury reduce activation of C3 and C5, lowering the efficacy of phagocytoses by macrophages and neutrophils. ii. Induce a neutrophil dependent free oxygen radical injury; iii. Suppression of the proliferation and differentiation of T and B lymphocytes, and iv. Suppression of killer T cell activity. The likely mediators that are involved are Interleukin (IL) 1 and 6; Tumour Necrosis Factor-α, and Tumour Growth Factor-β. Though the latter is a chemo-attractant of monocytes, neutrophils, and fibroblasts, and helps to stimulate tissue repair, it simultaneously suppresses activity of immunity conferring cells.

Burn injury therefore is a complex entity which not only induces hypovolemic shock, which by itself may progress to MODS, but complicates the situation by producing systemic inflammation which may rapidlyprogress to MODS. The likelihood of MODS setting in increases rapidly in patients with major burns (>40% body surface area involved).

Fig. 12. Immune-suppression in burns

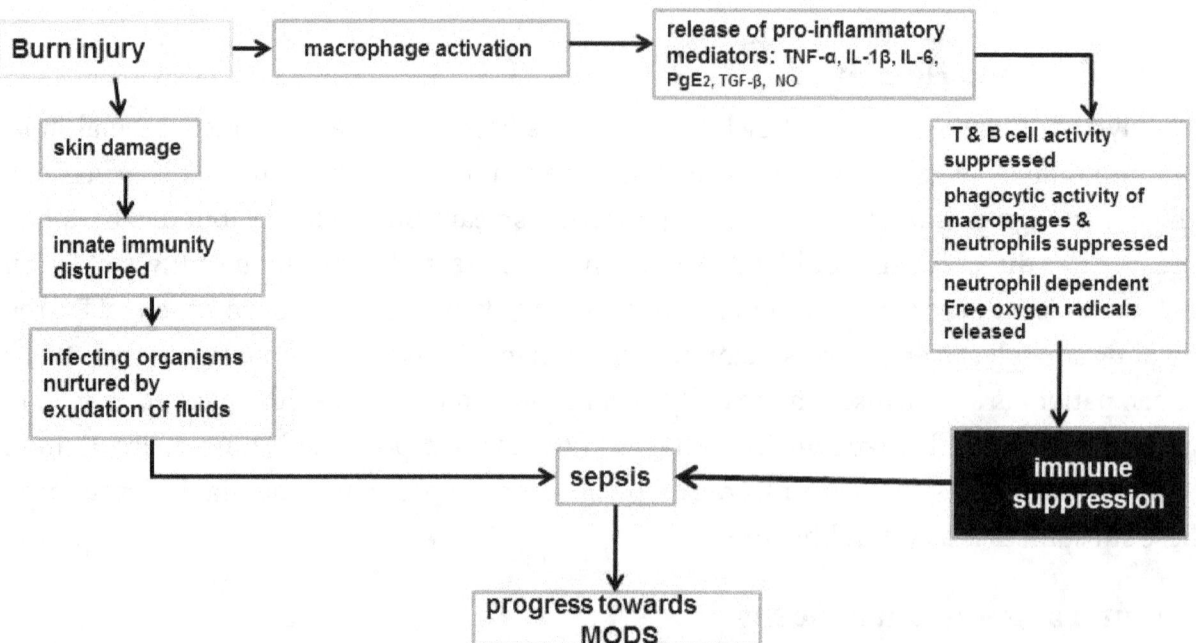

TNF-α: Tumour Necrosis Factor- α; IL : Interleukin; TGF: Tumour Growth Factor; NO: Nitric Oxide

In conclusion, the bottom line in any variety of shock, whether caused by hypovolemia or burns, is the development of MODS- the major cause of "late death" after hypovolemic shock/ burns/trauma/sepsis. As of now, the pathophysiology of injury-induced organ dysfunction is poorly characterized but has been linked to systemic inflammation. Subsequently, organ damage may also occur as a consequence of immune suppression. Multiple organ dysfunction syndrome (MODS) is the summation of responses to severe systemic injury, integrated at the cellular, organ, and host levels. An attempt at realizing this will only help in offering a better prognosis to patients who may be threatened by it in the course of their illness.

References

1. Best and Taylor's Physiological Basis of Medical Practice, 13th edn.; Ed. Best JB. William & Wilkins, London; 1990

2. Cakir B, Yegen B. Perspective in Medical Sciences: Systemic responses to burn injury. Turk. J Med. Sci. 2004; 34: 215–226.

3. Cobb PJ, Buchmann TG, Karl IE, Hotchkiss RS. Molecular Biology of Multiple Organ Dysfunction Syndrome: Injury, Adaptation, and Apoptosis. Surgical Infections. 2000, 1(3): 207–215.

4. Ganong WF. Review of Medical Physiology. 21st edition. Lange Medical Books/McGraw-Hill, London. 2003

5. Gauer OH, Henry JP. Circulatory basis of fluid volume control. Physiol. Rev.. 1963; 48: 423–481.

6. Goldsmith SR, Francis GS, Cowley AW, Cohn JN. Response of vasopressin and nor-epinephrine to lower body negative pressure in humans. Amer. J Physiol. (Heart and Circulatory Physiology). 1982; 243: H970-H973.

7. Hettiaratchy S, Dziewulski P. Pathophysiology and types of burns. British Med. J. 2004; 328: 1427–1429.

8. Holmes CL, Walley KR. The evaluation and management of shock. Clin. CVhest Med. 2003; 24: 775–789

9. Maier RV. Approach o the patient with shock. Chap. 253 in Harrison's Principles of Internal Medicine." Edited. Kasper DL, Braunwald E, Fauci AS, Hauser SL, Longo DL, Jameson JL 16th ed. MCGraw Hill, London. 2005; pp 1600– 1612.

10. Kolecki P, Menkhoff CR. Hypovolemic shock in eMed: Web MD. 2006; July

11. Kreimeir U. Pathophysiology of fluid imbalance. Critical care. 2000;(suppl. 2): S3–S7.

12. Loebl E C, Baxter C R, Curreri P W. The mechanism of erythrocyte destruction in the early post-burn period. Ann Surg. 1973;178: 681–686.

13. McCance KL, Huether SE. Pathophysiology: The Biologic Basis for Disease in Adults and Children. 4th edition. Mosby; St. Louis USA; 2002

14. Rea RF, Hamdan M, Clary MP, Randels MJ, Dayton PJ, Strauss RG. Comparison of sympathetic response to haemorrhage and lower body negative pressure in humans. J Appl. Physiol. 1991; 70: 1401–1405

15. Sanford A, Herndon D. Modulation of hypermetabolic response after trauma and burns, and therapy. Chap. 33 in Multiple Organ Failure: Pathophysiology and Prevention. Ed. Baue A, Faist U, Fry DE. Springer Verlag, New York, 2000: pp 322–329

CHAPTER - 6

Physiological Disturbances in Respiratory Disease

1. Normal designated functions of the respiratory system

   i. Ventilation ii. Diffusion iii. Perfusion; iv. carriage of respiratory gases v. tissue (cellular) respiration

2. Relevant functional Anatomy of the respiratory tract

   i. upper and lower airways; ii. dichotomous branching; division in to about 23 "families" iii. small airway; iv. smooth muscles in airways v. secretory cells; vi. innervation of airways: efferent and afferent

3. Common clinical symptoms and signs in respiratory patho-physiology:

   i. cough; ii. dyspnea; iii. cyanosis; iv. clubbing; v. adventitious sounds vi. chest pain

4. Pathophysiology of lung chest wall mechanics

   i. normal physiology;

   ii. Pathophysiology of resistance to airflow:

   a) sirway resistance

   b) tissue resistance

   iii. pathophysiological aspects of asthma, bronchiectasis & COPD.

   iv. physiological principles of investigations & treatment of obstructive disease

   v. deranged tissue resistance and ventilation

   a. restriction oflung movement

   b. restriction of chest wall movement

5. Pathophysiology of diffusion of respiratory gases:

   i. review of normal physiology

   ii. effectsof deranged diffusion

   iii. principles of investigation and treatment

6. Control of respiration

   Hypo and hyperventilation

   Respiratory failure

General Introduction

Normal Functions of the Respiratory System

The respiratory system contributes to the maintenance of constancy of the Internal Environment (Milieu Interior) by ensuring that the arterial blood oxygen and carbon dioxide (PaO_2; $PaCO_2$) tensions remain normal as far as possible under all situations. For this to happen, i. the system must trap fresh air with each breath **(ventilation)**; ii. exchange the respiratory gases across the alveolar-capillary membrane **(diffusion)**, iii. **carry** these gases from the lungs to the tissues and back in the blood which iv. **perfuses** the lungs, and finally v. put to work the respiratory enzymes at tissue level **(Fig. 1)**. If disease affects any of these processes of respiration or its **control,** physiology of the respiratory system will malfunction. The processes involved in respiration, ventilation and diffusion are usually the ones which are more commonly affected by disease, while perfusion problems may be considered to be less common. The former two are also easily measurable in a pulmonary function laboratory, and these measurements help in understanding the patho-physiology that supervenes when disease strikes. Cellular respiration will not be considered here.

Relevant Functional Anatomy of the Respiratory Tract

The respiratory tract is divided into upper (from the nose up to the larynx) and lower airway (trachea and below). We will deal only with the lower respiratory tract. The first part of the trachea lies outside the thorax, while the remaining part of the bronchial tree and the lungs are intra-thoracic. The latter are thus subjected to pulls and pushes of the thoracic cage. Disease in the thoracic cage will therefore affect expansion and deflation of the lungs.

The bronchi divide dichotomously, about 23 times in all **(Fig. 2)**. Maximum resistance to air flow is present in the large airways. The small airways (<2 mm internal diameter) and beyond offer low resistance to airflow because of the vast surface area involved. Disease process in this region often goes undetected because it produces few symptoms in the early phases. Therefore the small airways are often known as the "silent zone." The epithelium gradually changes from multi –layered pseudo-stratified to cuboidal to squamous at the alveolar level. There is no smooth muscle in the alveoli. Histologically, simple squamous, type I pneumocytes line the alveoli. They are very thin (about 0.2μm) and hence suitable for respiratory gas exchange. If they thicken, diffusion will be difficult. Type II pneumocytes are found singly or in small groups between the type I cells. They secrete surfactant which plays an important role in lung chest wall mechanics. Migratory macrophages may be seen outside the capillary in the septum or within the alveolus itself, and are involved in inflammatory reactions. There are dopamine containing cells in the neuro-epithelial body. Their possible function may be the local release of catecholamines. These may relax the smooth muscle of the airways during anoxic or hypoxic episodes and thus be of help in obstructive airways disease.

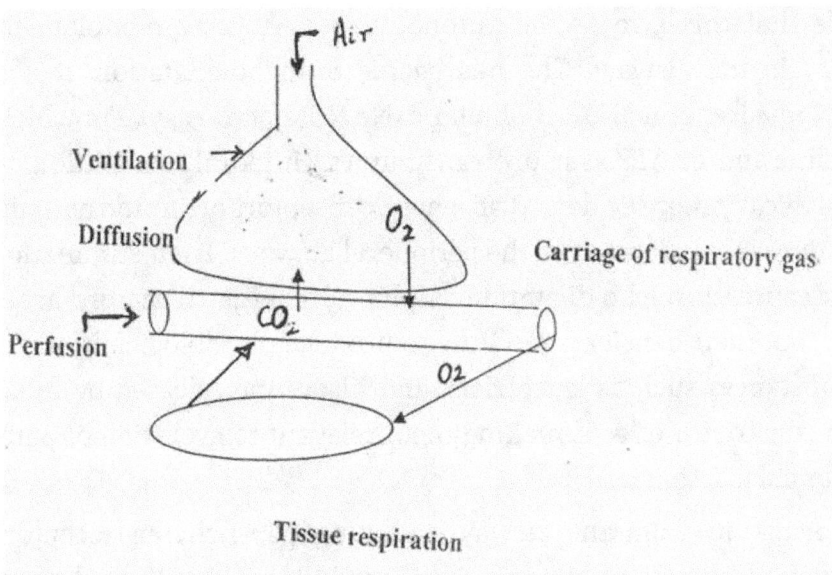

Fig.1

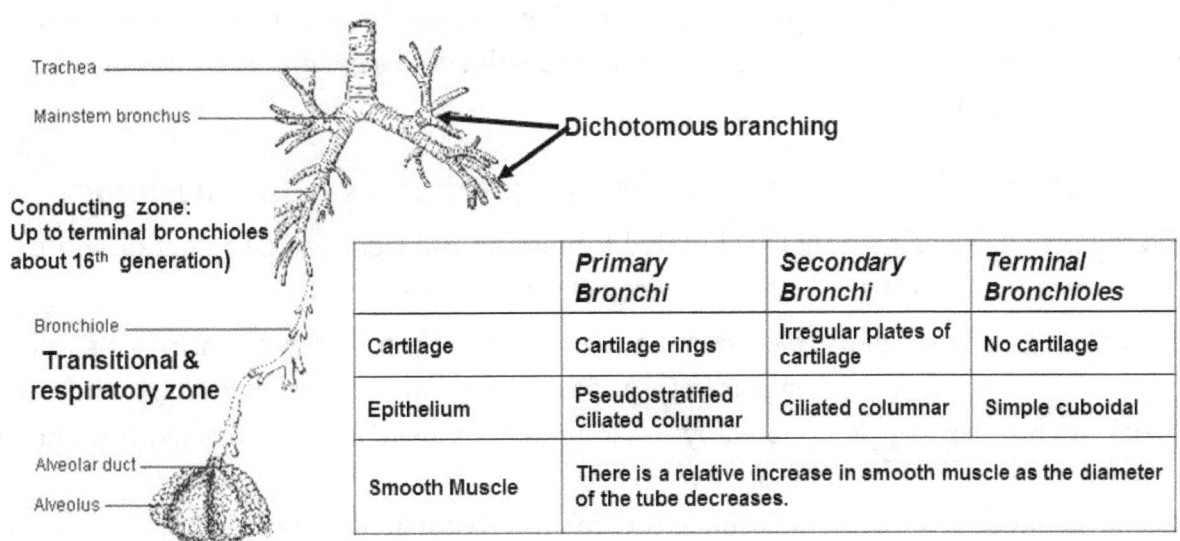

	Primary Bronchi	Secondary Bronchi	Terminal Bronchioles
Cartilage	Cartilage rings	Irregular plates of cartilage	No cartilage
Epithelium	Pseudostratified ciliated columnar	Ciliated columnar	Simple cuboidal
Smooth Muscle	There is a relative increase in smooth muscle as the diameter of the tube decreases.		

Fig. 2 Functional anatomy of the lower respiratory tract

Each tube keeps dividing in to two branches (dichotomous). Each division (generation) is called a "family" Small airways (internal diameter < 2 mm) start around the 15th - 17th generation, while the alveolar ducts to the sacs constitute on an average, about 23rd generation. The conducting zone is only a passage for air *(anatomical dead space)*, while respiratory gas exchange starts at the transitional zone.

The parasympathetic nervous system is the dominant efferent neuronal pathway in the control of airway smooth muscle tone. Stimulation of cholinergic nerves causes bronchoconstriction and mucus secretion, mainly in the central airway. Sympathetic nerves may control tracheo-bronchial blood vessels, **but no sympathetic innervation of human airway smooth muscle has been demonstrated.**

It is possible that this arm of the autonomic system may modulate parasympathetic ganglion activity in the airways. The neurogenic broncho-dilatation, if it occurs, may be mediated by a Non-adrenergic Non cholinergic (NANC) nervous system which may have VIP, adenosine, adenine and/or ATP as neurotransmitters. Humoral modulation of airway tone is however physiologically more evident. β-adrenergic receptors are found abundantly on human airway smooth muscle, particularly in the peripheral airways. Their stimulation by circulating catecholamines causes broncho-dilatation. Alpha adrenergic receptors are also present in human airways, but their density is too little to produce physiologically significant broncho-constriction. Substances such as leucotriens and histamine released by inflammation, have potent broncho-constrictor effects, making them relevant to evolution of pathophysiology of diseases like bronchial asthma.

Sensory information from the airway and lung parenchyma receptors (irritant and pulmonary stretch receptors) is conveyed to the medulla via myelinated nerves which travel in the vagus. Paintal's Juxta-pulmonary capillary receptors (J receptors) signal the increase in interstitial fluid pressure in the lung parenchyma and send their information to the brain via C fibres. More recently all these airway sensory have been clubbed together as bronchial C fibres. Stimulation of the latter has been associated with breathlessness during severe exercise andin patients of pulmonary edema.

Common Clinical Symptoms and Signs in Respiratory Pathophysiology

Deliberations on the commonly encountered symptoms and signs are given here in orderto enhance the understanding of Pathophysiology at a later stage.

1. **Cough**: This is one of the most common symptoms of respiratory disease. It is a reflex, mediated entirely by the vagus nerve, and is perhaps the only reflex that may be mimicked voluntarily. It is initiated by inflammatory (most common), mechanical and at times psychogenic stimulation of irritant receptors in the larynx and the large airways, most abundant at the bifurcation of the trachea. Distortion of the peripheral airway by lung fibrosis and collapse may induce the symptom. Sensory input via the5[th] and the 9th cranial nerves may also be involved. Cough can not be induced by irritation of the smaller airway and alveoli. A simplified sequence of events which may occur during reflex coughing is given in **Fig. 3.** In the so called "cough centre," impulses from the stimulated receptors are received in the Nucleus of the Tractus Solitarius, and the motor outflow goes to the respiratory muscles via the ventral respiratory group of neurons.. Voluntary coughing is thought to bypass this pathway. Compression of the airways by the increase in pleural pressure during the expiratory effort helps the reflexly induced broncho-constriction in increasing velocity of air flow during the explosive expiration. Cough is a protective mechanism which helps to clear blockages in the airways. Use of cough suppressing medication, particularly in young children and the elderly must therefore be used with caution. The reader may refer to text books of medicine for causes of cough.

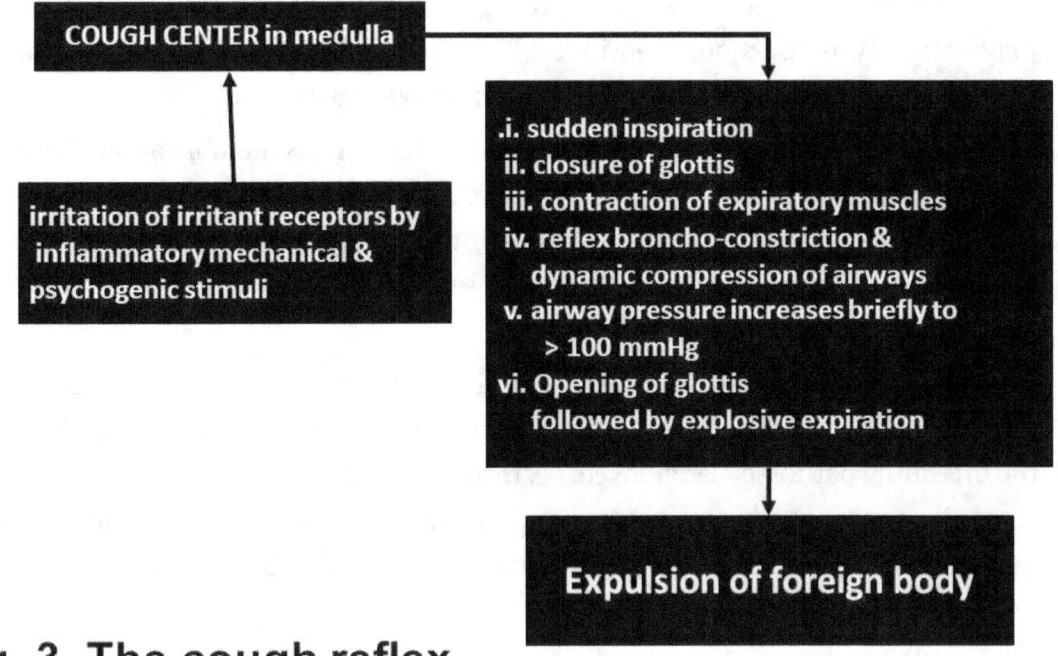

Fig. 3. The cough reflex

2. **Dyspnea**: This means an awareness of shortness of breath. Patients who suffer from it often describe it as "Not getting in enough air." It is usually reported subjectively by the patient (symptom). Also, the breathing effort and the distress of the patient may be observed by the physician as a sign. Normal individuals develop dyspnea at high level of exercise, but when this happens at an unacceptably low level of exertion for that individual, it indicates presence of pathology. The symptom is generally associated with diseases of the respiratory and cardiovascular systems, but may also be present in diseases of other organ systems.

> **Table I: Respiratory causes of dyspnea**
>
> 1. Obstruction to airflow
> 2. Resistance to expansion of the lung-chest wall system
> 3. Excessive respiratory drive
> 4. Increases sensitivity of respiratory centres to physiological stimuli

Pathophysiology and mechanisms of dyspnea are poorly understood. The major afferents which send excitatory inputs into the brain stem respiratory centres are i. J receptors, ii. pulmonary stretch and irritant receptors; iii. peripheral chemoreceptors (hypercapnia and severe hypoxia); iv. Central chemoreceptors (hypercapnia); and v. Respiratory muscle and chest wall afferents. The sense of respiratory effort is believed to arise from a signal transmitted from the motor cortex to the sensory cortex coincidently with the outgoing motor command to the respiratory muscles. The involvement of the emotional cortex is irrevocable as the "distress" felt by the sufferer constitutes an important part of the syndrome. Cortical association areas may also be involved. The basic pathophysiology in the respiratory system which may cause dyspnea is outlined in Table I. ***The reader may work out the contributing pathophysiological situations involved in the given processes.***

5. **Cyanosis** is the bluish discoloration in the skin and mucus membranes which occurs as a result of reduced Hb of 5gm% or more. In respiratory pathophysiology, ventilation /perfusion mismatch, severe hypoventilation, limitation of lung chest- wall movement may lead to cyanosis. The presence of this sign in a patient with respiratory disease usually indicates severe disease.

6. **Clubbing** (a typical thickening of base of finger/toe nail bed) may occur in chronic pulmonary disease (though it is an extra-pulmonary manifestation). It may be reversed if the offending pathology is removed. eg. resection of a bronchogenic carcinoma, lung abscess. Platelet micro-thrombi are present in clubbed fingers, and a platelet derived factor may be responsible for the peculiar thickening. Clubbing is also seen in cardiovascular and gastro-intestinal diseases

7. **Adventitious sounds.** Sometimes adventitious sounds may be superimposed on the normally heard breath sounds. These might be "crackles" which suggest gas bubbling through fluid accumulated in the alveoli as in pulmonary edema, bronchiectasis. These sounds are usually discontinuous. The continuous ones such as the "wheeze" and the "rhonchi" are produced by vibration of narrowing of airways because of bronchospasm, mucosal edema, or secretions in side the airway.

8. **Chest pain** Lung parenchyma and visceral pleura are insensitive to painful stimuli. Nociceptive receptors are present in the trachea, main bronchi and parietal pleura, and other structures in the chest wall. Some patients of pulmonary hypertension and pulmonary artery stenosis may complain of retro-sternal chest pain. Exact mechanism for this is not known. Pain of pulmonary embolism is usually pleuritic in character.

Other symptoms/signs such as hemoptysis have been omitted as they are more clinical indications of lung disease rather than pointers of physiological dysfunction.

Pathophysiology of Lung Chest Wall Mechanics

Normal physiology

Unhampered mechanical movements of the lung and the chest wall must occur in unison if adequate ventilation is to take place. The lung chest wall system is basically a "balloon in a box" arrangement. At the end of a quiet expiration (FRC), the chest wall is coiled and ready to spring outwards. This force is neutralized by the inward collapsing force of the elastic recoil of lungs, and hence no flow of air occurs. During inspiration, the contraction of the inspiratory muscles expands the chest wall, pulling at the lungs attached to it by the parietal pleura, thus sucking in fresh air. During expiration, the opposite happens. With inspiratory muscles now inactive, the elastic recoil of the lungs deflates the lungs until FRC is reached. Normal quiet expiration is thus passive. A forced expiration would transmit the expiratory force to the pleural space, increasing its pressure to positive from the usual negative value. This is added on to the elastic recoil pressure of the deflating lungs. *(At this stage, the reader should refer to the lung- chest wall compliance diagram in any textbook of physiology)*. Airflow to the outside continues until the positive pleural pressure applied to the outer surfaces of the small airway during forceful expiration exceeds that within these airways and closes them. The volume thus left behind is the residual volume. If peripheral airways are damaged, they tend to close earlier than usual. Traction applied by lung parenchyma to the small airway helps to keep them open. This is more effective during inspiration. This helps to some extent in widening of even inflamed airways on inspiration.

An air column which is basically a fluid, as it enters or leaves the lungs, must: i negotiate the airway. While doing so, the airway offer resistance to it as it flows in and out **(airway resistance)**. ii. the tissues of the lung and the chest-wall must be moved if the air column is to be allowed to ventilate the lungs. The tissues being visco-elastic (lungs) and rigid (parts of the chest wall), tend to resist this movement. This constitutes the **tissue resistance** which too must be overcome if air flow is to occur through the lungs. During each breath work has to be done to overcome both these resistances. Under normal conditions, only the airway resistance may alter to some extent as airway diameter is physiologically variable. In the absence of disease, tissue resistance remains constant. It may however increase in normal people with aging. As a corollary to the above, if disease makes movement of the lung and chest wall difficult, the work of breathing will increase, and the air column movement in and out of the lungs will become difficult. This may compromise ventilation, and hence the oxygenation of arterial blood. **Fig 4** highlights this concept.

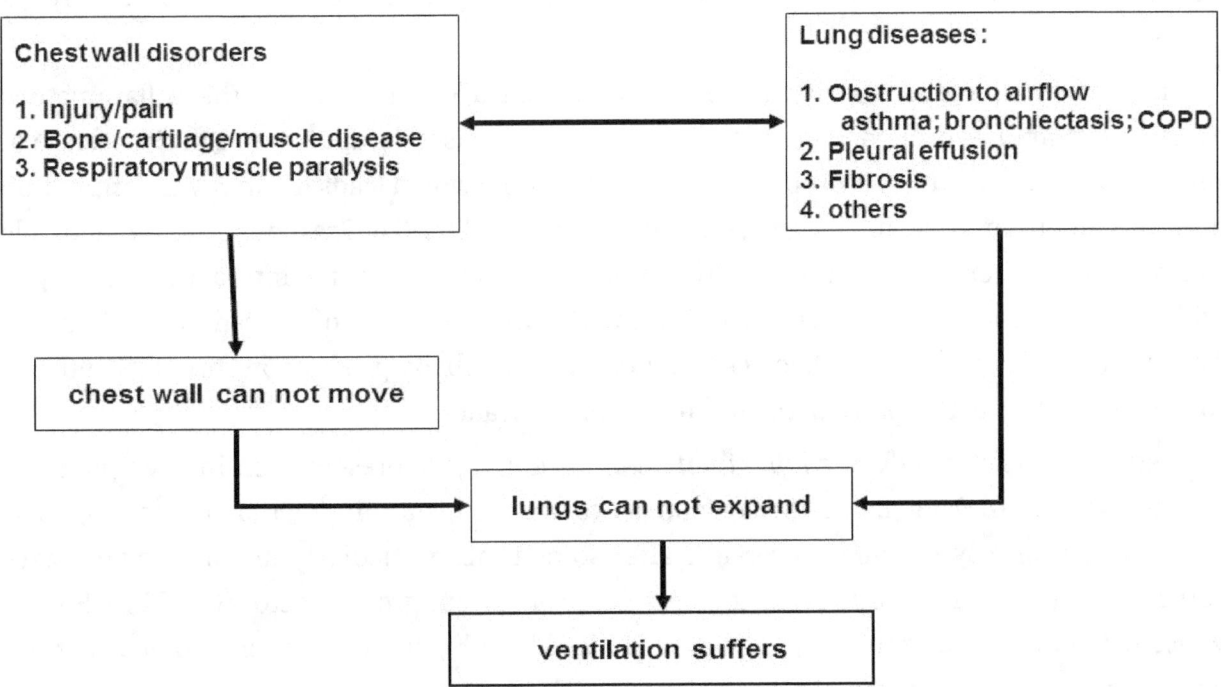

Fig. 4. Alterations in lung-chest wall mechanics leading to inadequate ventilation

Pathophysiology of Resistance to Air Flow

Airway resistance

The airways are a series of tubes which offer a certain amount of resistance to the air column as it moves through them. The relation is mathematically described by the equation $R = 8\eta l / \Pi r^4$ where $8/\Pi$ is a constant; η is viscosity of the air we breathe (does not usually change); l is the over all anatomical length of the airways (again does not alter at any given point of time); and r is the radius of the airway. It is noticeable that radius alters resistance by the inverse power of 4 ($1/r^4$). This means that resistance to airflow is markedly affected by small changes in the airway diameter. Physiologically, the diameter of the airway is under nervous control, and is constantly being regulated to suit the demand of oxygen (air) required. Disease of the airway often results in their narrowing because of i. actively induced bronchoconstriction; ii. accumulation of secretions, particularly mucus; iii. inflammation induced edema of the mucosal lining. Under all or any of these conditions, resistance to air flow will increase (remember the relation $1/r^4$). Disease may affect both the small and the large airway (bronchial asthma); mainly the large airway (bronchiectasis), or mainly the small airway (chronic bronchitis with emphysema). It is fortuitous that airway diameter can be manipulated pharmacologically by the use of bronchodilator agents, and anti-inflammatory medications are available in order to reduce airway edema. All of these conditions produce **"airflow limitation."** Pathophysiology of three typical disorders which generate this problem is now discussed.

A. Bronchial asthma

By definition, Bronchial Asthma is a recurrent, episodic, widespread, variable airway obstruction of both large and small airway, brought about by inflammation of the airways. It is often reversible.

The airway epithelium is both the target of, and a contributor to, the inflammatory cascade. It amplifies bronchoconstriction and promotes vasodilatation through the release of inflammatory compounds. A probable sequence of events which leads to airway inflammation and subsequent increase in airway resistance, is given in **Fig. 5.** IgE response has a major role to play in atopic (extrinsic) asthma. In the appropriate genetic setting, interaction of antigen with a naïve T cell in the presence of IL-4 leads to the differentiation of the cell to a TH2 subset. This in turn induces theB lymphocytes to switch their antibody production from IgG and IgM to IgE which then participates in the inflammatory cascade.

A state of *persistent subacute inflammation of the airways* is present even in asymptomatic asthma patients, making them hyper-responsive to inhaled histamine/methacholine challenge. The obstruction may become irreversible after some time, particularly in patients who have Intrinsic asthma. Inflammation may induce release of various growth factors (EGF, PDGF, IGF1) which promote smooth muscle hyperplasia, followed by subepithelial fibrosis. Other structural changes include hypertrophy of mucus glands, angiogenesis and changes to the extra cellular matrix. These features constitute the newer concept of **"airway remodeling"** in asthma. The

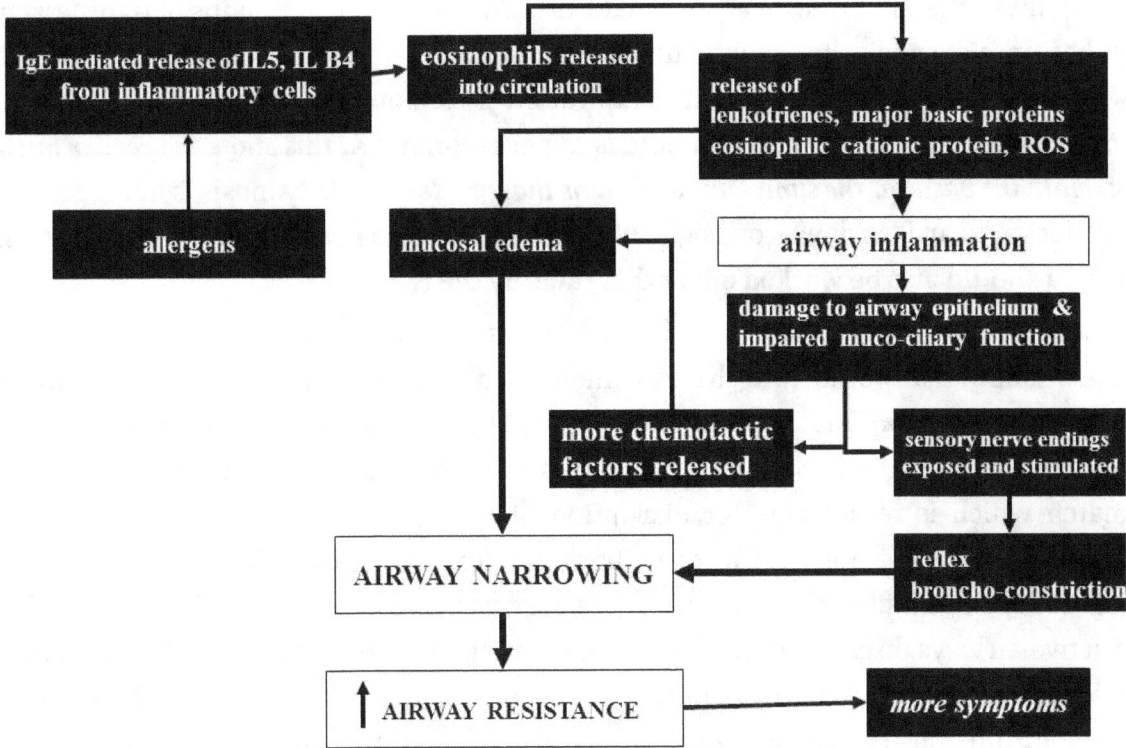

Fig. 5: Pathophysiology of increased airway resistance in Asthma

substances released by the inflammatory process may directly stimulate smooth muscles to contract, may stimulate sensory receptors which form the afferent path of the vago-vagal bronchoconstrictor reflex; and alter capillary permeability to produce local edema. Smooth muscle spasm often constricts veins which become tortuous. This adds to the mechanical blockade of the airway lumen.

The increase in airway resistance decreases forced expiratory volumes and flow rates, causes hyperinflation of the lungs, and increased work of breathing. The cardinal feature of airway obstruction is demonstrated on spirometry as a reduction in FEV_1/FVC ratioto <75% (normal is about 83%). With increase in the narrowing/closure of airways in various parts of the lung, the FVC and FEV_1reduce further, the latter more so than the former. A mismatch of ventilation and perfusion in the lungs takes place, leading to a fall in PaO_2, causing a hypoxia induced increase in ventilation with a lowering of $PaCO_2$, and an alkaline pH. (An element of anxiety because of the distress felt is likely to contribute to the tachypnea, as also the sympathetic excitation which may occur). If the obstruction goes unchecked, the work of breathing will keep increasing, and cause fatigue of respiratory muscles, and worsening of hypoxia. **After a certain stage, a rapid increase in $PaCO_2$ with the pH turning acidic heralds onset of respiratory failure**. Enthusiastic treatment of what may appear to be a routine attack of bronchial asthma is therefore mandatory at all times. In severe asthma, FEV1 or the Peak Expiratory Flow Rate (PEFR) may reduce to <40% of predicted. Residual volume frequently approaches 400% of normal, while functional residual capacity doubles. The lungs are therefore in a state of hyperinflation.

Clinically, in the beginning of an asthmatic attack, the patient complains of tightness in the chest; becomes dyspneic, has cough, initially non–productive, later productive; the accessory muscles of respiration are active, and expiration is prolonged with both inspiratory and expiratory wheezing. The percussion note is hyper-resonant. ***At this stage the reader must try and explain the basis for the symptoms and signs that are described.*** Cyanosis is not usually seen, but if it does appear, it's a dangerous sign *(why??)*. **The physiological principles of management of asthma should also be worked out at this stage by the reader**.

B. Bronchiectasis is a dilatation of bronchi/large airways. This happens as a result of complications of many other lung diseases (mention of which is beyond the scope of this book). Repeated inflammation leads to production of purulent sputum which results in inflammatory stenosis/blockage of the peripheral airway. It is the latter that causes ventilation/perfusion mismatch which is responsible for disruption of lung physiology. If the disease starts in childhood, it is often complicated by disturbances in lung growth and development. Peripheral airway collapse, and fibrosis of the exchange surfaces occurs. Therefore, not only is this an obstructive airways disease, but also a restrictive lung disease with serious diffusion defects. Functionally a chronic airflow limitation with ventilation perfusion mismatch going into a phase of chronic obstructive pulmonary disease (COPD) is the end result. Arterial blood gas picture will show hypoxia with progressive hypercapnia and respiratory acidosis. With destruction of lung parenchyma including the blood vessels, right heart failure supervenes.

The usual symptoms are productive cough with purulent expectoration, repeated lung infections; hemoptysis; progressive breathlessness, clubbing; cyanosis and right heart failure.

Chronic Obstructive Pulmonary Disease (COPD). The hall mark of the condition is a **chronic airflow limitation** which may be only partially reversible. The archetypal COPD is chronic bronchitis with emphysema. It is predominantly a small airway (the so called silent zone) disease. Repeated infections produce airway inflammation accentuated by tobacco smoking and air pollution. The former is considered to be a major risk factor. There is production of thick, tenacious mucus and defective muco-ciliary clearance, thickening of the small airway by inflammation, and hypertrophy of smooth muscle. Inflammatory cells active in the peripheral airways and alveolar ducts may release proteolytic enzymes which destroy elastic tissue in them. These factors further accentuate airway fibrosis and collapse. The surfactant secretion is reduced and this increases tendency of alveolar collapse. The radial traction which normally helps to keep airways patent, is thus lost Alpha$_1$ Antitrypsin deficiency in some cigarette smokers increases their susceptibility to developing emphysema which is a part of the syndrome. Airway hyper-responsiveness is often present as in asthma. The narrowing/blockage of small airway causes air trapping [increase in both, the residual volume, and its ratio to the TLC (normal ratio is about 20%)]. There is thus a state of hyperinflation which is initially helpful because it tends to increase lung recoil pressure. But with progressive disease, it produces mechanical limitation in diaphragmatic movements because of flattening of the muscle fibres. Increasing inspiratory effort is thus added on to the already existing expiratory difficulty. Superimposed on this is the gradual destruction of gas exchanging tissue. COPD thus develops into a combined obstructive

> **Table II**. Arterial blood gas and pH values in a normal adult
>   i. PO$_2$ approximately 98 mmHg (range 85- 100 mmHg or 11.3 -13.3 kPa)
>   ii. PCO$_2$ 40 mmHg (range 35- 45 mmHg; 4.6-6.0 kPa)
>   iii. pH 7.40 (range 7.38-7.43)

and restrictive lung disease. Because of loss of elasticity, the hyper-inflated lungs become highly compliant. This results in a left ward shift of the lung compliance curve. Simply put, this means that the inflated lungs find it difficult to deflate. The overall situation develops in to a state of HYPOVENTILATION. The PaO$_2$ at rest may remain normal until the FEV1 falls to about 50% of predicted, while the PaCO$_2$ starts to build up at a later stage (FEV1 about 25% predicted). The hypoxia is a consequence of ventilation/perfusion mismatch. As more and more lung exchange surface is destroyed, pulmonary hypertension that occurs is complicated by right ventricular failure (Cor-pulmonale). When PaO$_2$ drops to <60 mmHg, the ventilatory drive is maintained by stimulation of peripheral chemoreceptors. At this stage, too enthusiastic a treatment with supplemental oxygen must be curtailed as this will remove the hypoxia drive.

COPD patients most often have productive cough, and gradually worsening exertional dyspnea, which finally becomes dyspnea at rest with cyanosis and symptoms of heart failure. The symptoms set in insidiously and patients seek medical help many months/years after the disease process has begun (remember small airway are the silent zone). *The reader should review the above information and derive the various clinical findings which a patient with advanced COPD is likely to have.*

Physiological basis of investigation and treatment of obstructive airway disease

Normal, adult arterial blood gas levels (ABGs) are given in Table II. (If the normal alveolar oxygen tension is about 104 mmHg, why does the PaO2 reduce to 95–98 mmHg?). In the elderly, blood oxygen tension reduces as per the relationship: PaO2 = 109 – (0.43 x age in yr). Blood carbon dioxide is unaffected by increase in age. Logically, the normalcy of respiratory function should be best assessed by measurement of ABGs. However the procedure is relatively risky as it involves arterial puncture if reliable data is to be obtained, and expensive. Indirect, non-invasive measurement of hemoglobin saturation by oxygen can be made more easily using oxy-hemoglobinometer applied to the finger or the ear lobe. Normally the saturation recorded should be between 95–98%. (At this stage the reader should refer to the standard oxy-hemoglobin dissociation curve). This parameter is helpful in the physiological evaluation of patients suspected to have early COPD, or restrictive lung disease such as interstitial fibrosis. Such patients who have normal ABGs at rest, become hypoxic during exercise as indicated by a drop in the HbO$_2$saturation%.

Measurement of ventilation is easy, reliable and cost effective. The tests also give an overall insight into the state of the lung chest wall mechanics. The gold standard for diagnosing airway obstruction is a lowering of the FEV$_1$ /FVC ratio to <75%. These "dynamic" parameters may be measured by using a spirometer. Computerized equipment integrates the forced expiratory effort into flow volume loops in which air flow is plotted against volume. The middle part of the expiratory curve (FEF $_{25-75\%}$) and the last part (towards RV) the FEF $_{75\%}$ (which means that 75% of the FVC has been breathed out) together indicate airflow through the small airways Typical spirometry patterns in various conditions with air flow limitation are given in **Fig. 6**. *(If obstructive disease is a COPD, the TLC is likely to be higher than normal: why?).*

Fig.6 . Typical spirometry results

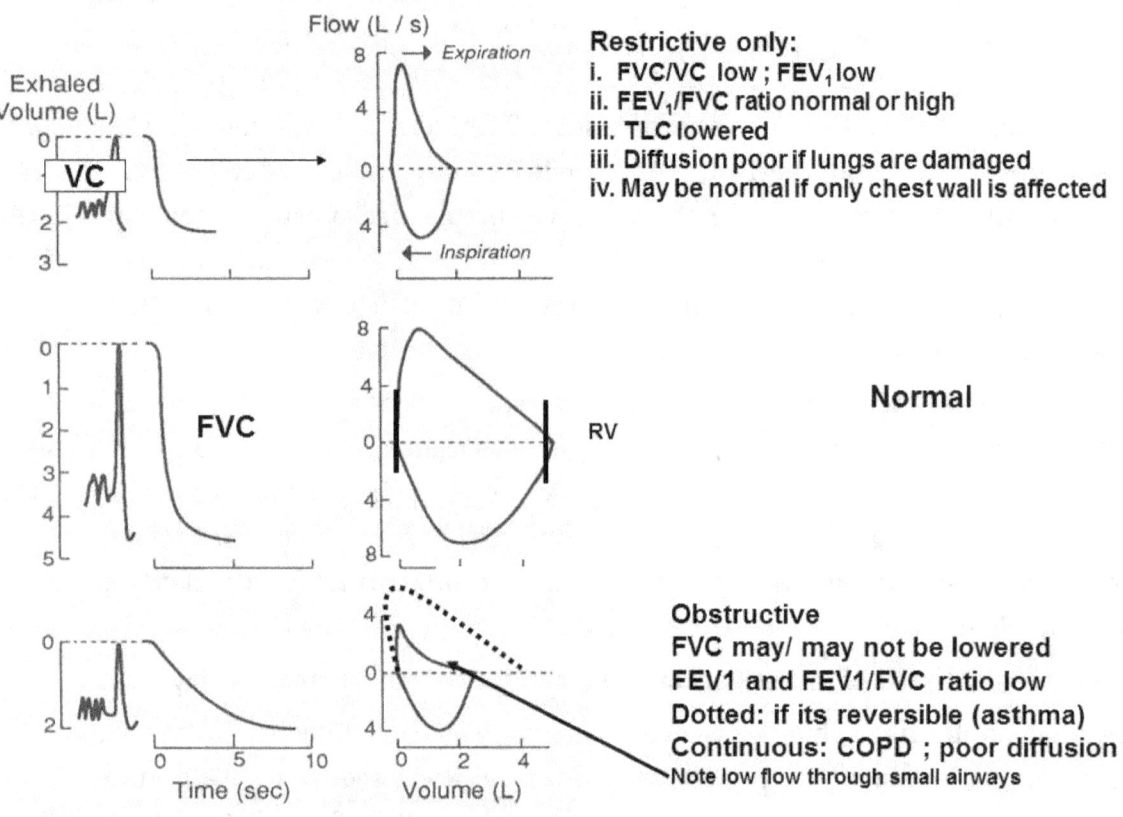

Physiological principles in the treatment of obstructive airway disease must logically encompass: 1. airway dilatation 2. reduction of inflammatory process in the airway lumen. Bronchodilatation may be brought about actively by using beta$_2$ adrenergic receptor stimulants which are abundantly distributed in the smaller bronchi and peripheral airways. Alternatively, anti-cholinergic agents will prevent vagally induced bronchoconstriction mainly in the larger airways.

Apart from eliminating bronchoconstriction, the inflammatory process in the airways needs to be subjugated. Corticosteroids are by far the most powerful and effective anti inflammatory agents. LeukotrienesLTC$_4$, LTD$_4$ and LTE$_4$ are involved in the inflammatory process, and drugs which block their receptors on mast cells or those which directly inhibit leukotrienes may be used. Mast cell membrane stabilizers help by preventing release of Interleukins 4, 5 and Tumor Necrosis Factor, all potent inflammatory substances. Use of anti-histamine drugs should help in theory but are not very effective. Blocking just the histamine effect is inadequate because many other potent inflammatory substances go unchecked. The principles outlined above are more effective in treatment of acute airflow limitation disorders like asthma. In patients with chronic airflow limitation such as bronchiectasis, and the classical COPD, these principles of treatment are more effective during acute exacerbations of airway blockage. Supplemental oxygen to overcome hypoxia is often resorted to. In COPD patients with resting Hb-oxygen saturation<90%, with pulmonary hypertension and corpulmonale, this form of oxygen treatment at night is known to reduce mortality and improve quality of life. When blockage of the airways becomes difficult to manage with drug treatment, mechanical ventilation may be used to enhance oxygenation. Viscosity of the air we breath in affects resistance to flow. Normally this remains constant as we breath atmospheric air which has a fixed composition. Use of gas mixture containing Helium and Oxygen helps to reduce work of breathing in patients who have beleaguered breathing. Deep sea divers use such gas mixtures. The disadvantage is the distortion of speech which low density gas mixture produces.

Increased Tissue Resistance and Its Effects on Ventilation: Restriction of Lung-Chest Wall Movement

At lung level

The surfactant lining of the alveoli is required for facilitating lung expansion. In its absence/ deficiency, the lung compliance reduces, and the alveoli tend to collapse. More than normal inspiratory effort is required to inflate the lungs. The compliance curve shifts to the right (stiff lungs) which means more than the usual inspiratory pressure has to be generated for inflation to occur. Deficiency of surfactant secretion by type II alveolar cells occurs in the Respiratory Distress Syndrome of the new born because of poor maturation of type II pneumocytes. There is also pulmonary edema. In the Acute Respiratory Distress Syndrome and in patients of chronic airway inflammation, alveolar-capillary membrane and thus the Type II pneumocytes are damaged. Important physiological disturbances are arterial hypoxia and tachypnea/ dyspnea. Arterial hypoxia happens because the gas exchange surface is reduced, while the tachypnea/dyspnea probably occurs because of a) stimulation of peripheral chemoreceptors by the hypoxia; b) presence of fluid in the interstitial spaces which stimulates the J receptors; and c) stimulation of lung irritant receptors by distortion of collapsed airways.

Loss of elasticity of lungs (emphysema) hyper-inflates the lungs, and with it the chest wall, both of which remain fixed in that position (compliance of the lungs in this case shifts to the left). The end result is the inability of the lungs to deflate, restricting the lung chest wall movement. This has already been mentioned in the part on chronic airflow limitation. Thus a patient of chronic airflow limitation also suffers from restriction of lung chest wall movement.

Fibrosis of the airways, and the exchange surfaces of the lungs, which causes severe diffusion defects (see below),prevent the lungs from expanding. The compliance curve of the lungs will shift to the right and the lungs will become stiff.

Fluid or air (pneumothorax) collection in the pleural cavity will cause a lung collapse on the affected side, and restrict lung and therefore chest wall expansion. In the extreme situation, hypoxia and tachypnea/breathlessness are likely. Penetrating injury of the chest wall produces its physiological derangement because of the accompanying lung collapse. *(Why do such patients usually have lowered blood CO_2 tension?)*

The cardinal clinical feature of restrictive lung disease is exertional dyspnea which progresses to dyspnea at rest. The ABGs will show hypoxia at all times with raised carbon dioxide and acidic pH in severe disease. In some situations like traumatic lung collapse and pleural effusion, the increase in ventilation will lead to hypocapnia. Dyspnea and its probable causes have been mentioned above.

Restriction of Chest Wall Movement

Injury (eg. fracture ribs); inflammation in various structures in the chest wall cause pain induced restriction of movement which may be severe enough to induce hypoxia and breathlessness.

Defects in the spinal column: Patients of kyphoscoliosis, and kyphosis often have fixed a chest wall. Ankylosing spondylitis may cause severe dyspnea and hypoxia.

Neuromuscular junction disorders: A typical example is a patient with Myasthenia gravis who may initially feel exertional dyspnea which may progress on to respiratory muscle paresis/paralysis. Treatment will rapidly restore the deranged respiratory physiology to normal.

Diseases of the nervous system causing paralysis of respiratory muscles; The list includes depression of respiratory centre, bulbar polio myelitis, injury to the spinal cord polyneuropathy (viz. Guillain Barre syndrome). These may cause a paralysis/paresis of the respiratory muscles and thus limit chest wall expansion.

Physiological derangements in chest wall restriction. Whenever chest wall movement restriction takes place, arterial hypoxia with a rise in blood CO_2 is the usual physiological derangement, with dyspnea as the related symptom. In the early stages, the altered physiology may manifest itself only during exercise. With advancing disease, the work of breathing will increase progressively.

Physiological Basis of Investigations in Patients of Restrictive Lung Disease

Spirometry. Classically, spirometry will indicate shrunken lung volumes with a normal or raised FEV_1/FVC ratio. A normal lung diffusion capacity rules out involvement of lung parenchyma.

ABGs Resting values for ABGs may be normal to begin with, and may show hypoxia only during exercise when O_2 demand goes up. With progressive disease, the PaO_2 starts to decline. $PaCO_2$ may be variable eg. it may be normal or low when there is interstitial fibrosis (early stage) as the gas has a high coefficient of diffusion, while in COPD, it may be normal initially, followed by retention as pathophysiology deteriorates. Retention of CO_2 occurs in the terminal phases of all restrictive as well as obstructive lung disease.

Pathophysiology of Diffusion of Respiratory Gases

Normal diffusion. The respiratory gases diffuse across the alveolar-capillary membrane **(Fig 7).**

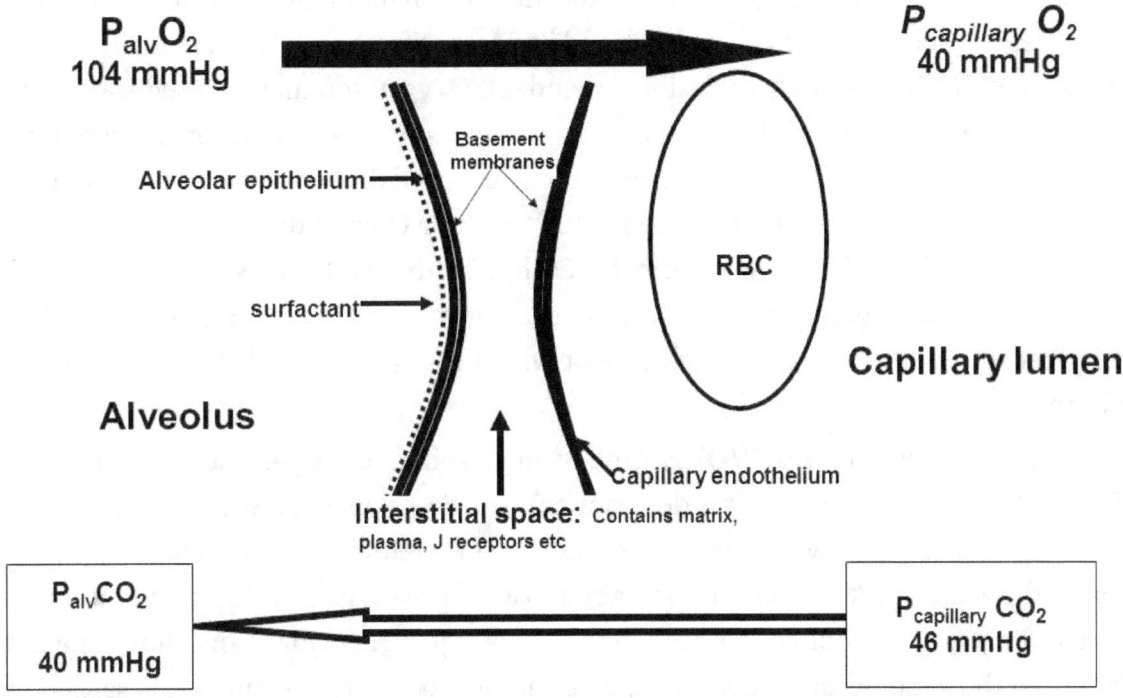

Fig. 7. The alveolar capillary membrane extends from the surfactant layer to the capillary endothelium. The membrane thickness is about 0.1 μ to 0.5 μ.

Oxygen moves from the alveoli into the capillaries and CO_2 in the opposite direction. In some areas, the basement membranes of the alveolar epithelium and the capillary endothelium are nearly fused, and the gaseous exchange occurs most efficiently here. In other places the basement membranes are separated by an interstitial matrix which consists of collagen and proteoglycans. Plasma infiltrates in between the matrix. It is this constituency of the interstitium that gives it its variable thickness. J receptors are located in the interstitial spaces. Factors that determine diffusion are:1. tension gradients of the respiratory gases. For example, at high altitude, and in patients with obstructive airways disease, the alveolar O_2 tension reduces, thus limiting the diffusion of this gas. This leads to hypoxia. **(*Suggest a method of overcoming this problem and improve blood oxygenation in a patient of hypoxic hypoxia?*)** 2. Thickness of the respiratory membrane. This is normally about 0.1 μ to 0.5 μ. If the membrane thickens to beyond 1 μ, diffusion is affected. This is the basis of hypoxia in pulmonary fibrosis where the membrane thickens. 3. surface area of the membrane that is available. This is self explanatory. The reason for the poor diffusion in COPD is because large areas of the membrane are damaged.4. Coefficient of diffusion of gas. Notice that the tension gradient for diffusion of CO_2 is only about 6 mmHg as compared with the 60 mmHg required for Oxygen **(Fig. 7)**. This is because the diffusion coefficient of CO_2 is very high-about 20 times that of O_2. The corollary to this is that even a slight increase in the thickness of the membrane will increase the diffusion distance and limit the passage of O_2 across the membrane, but will not affect CO_2. This explains why CO_2 is not retained in blood during the early phases of diseases which affect respiratory membrane thickness.

Diffusion capacity of the lungs is measured using carbon monoxide. The advantage of doing so is that this gas does not dissolve in blood, and its passage from the alveoli in to the RBC where it combines with Hb, is entirely diffusion dependent. Oxygen too may be used for measuring diffusion capacity, but the method is complicated by the fact that pulmonary capillary PO_2 has to be taken into account. This is a difficult and unreliable measurement. The diffusing coefficient for CO is less than that of oxygen (0.81 vs 1 for O_2) and diffusion capacity for O_2 is obtained by multiplying diffusing capacity for CO by a factor of 1.23 (1/0.81). The normal value is about 20 ml/min/mmHg. It is virtually impossible to get an accurate estimate of CO_2 diffusing capacity because of the extremely high rate at which this gas diffuses across the respiratory membrane.

Ventilation perfusion ratio (V/Q). Normally all alveoli must be perfused with blood giving a V/Q of 1. However functionally, this does not happen. Some alveoli are over ventilated/under perfused giving aratio > 1 while others are under ventilated/overperfused with a ratio of <1. Normally, the average V/Q works out to be about 0.8 at an average alveolar ventilation of about 4 l/min and a pulmonary blood flow of 5 l/min. This is just right for maintaining the normal ABGs. When the ratio becomes deranged, a ventilation/perfusion mismatch is said to occur. The end result is arterial hypoxia. When lung disease is responsible for this mismatch, it may be corrected by breathing 100% oxygen **(explain why).** If the hypoxia cannot be corrected by breathing 100% O_2 a pathological vascular shunt more than the normal anatomic shunt of about 2% is present.

Physiological Effect of Deranged Diffusion, and Basis for Its Investigation and Treatment

The disease that showcases the pathophysiology of defective diffusion is Interstitial lung disease with diffuse parenchymal fibrosis. The etiology of development tissue fibrosis is not well known. The main physiological issue is hypoxia which is a consequence of the ventilation/perfusion mismatch that ensues. Arterial hypoxia stimulates peripheral chemoreceptors which in turn cause an increase in ventilation with CO_2 washout, and a respiratory alkalosis. Retention of carbon dioxide is a development in the late stage of the disease.

Patients of COPD also present with progressive lowering of diffusing capacity. Here too the contributing pathophysiology is the loss of functional respiratory membrane because of fibrosis around the small airways, as also the break up of alveolar septa because of the accompanying emphysema.

Other pathophysiological events which result in inappropriate diffusion are collection of fluid in alveoli (pulmonary edema) and pneumonias which cause inflammation and edema on the alveolar tissues. The pathophysiology involved in the causation of hypoxia is the increase in the diffusion distance. ***(Why could drowning result in a diffusion abnormality?)***

Measurement of diffusion capacity for CO (DLCO) forms physiological basis of investigation of diffusion defects. This has to be combined with spirometry which must include measurement of lung volumes and capacities. ABGs may be used with discretion. *(In a patient with low DLCO, how will spirometry and ABGs help to distinguish between COPD and Interstitial lung disease such a interstitial fibrosis?)*

The physiological basis of treatment of a diffusion abnormality is to increase the concentration of oxygen in the breathing mixture so that the PO_2 rises and is able to negotiate the thickened membrane barrier. This is usually resorted to when the resting arterial oxygen tension drops to <55 mmHg. It is essential to titrate the percentage of oxygen in the breathing mixture to ensure that the PO_2 does not increase to beyond 55–60 mmHg as this will remove the hypoxic drive that keeps respiration going.

Carriage of Respiratory Gases

After diffusion, O_2 is carried in blood in two forms: in the dissolved state, and as oxy-hemoglobin. Normally, 0.3ml O_2 dissolves in 100 ml of plasma at a pressure of 100 mmHg, which is the pressure at which O_2 is delivered across the respiratory membrane when breathing atmospheric air at sea level. From this it is deduced that if the concentration of oxygen in the breathing mixture is increased, the tension of oxygen will increase, and hence the gas in solution will also increase. For example, if 100% O_2 is breathed at sea level, the PaO_2 will increase to about 673 mmHg and this in turn will increase the dissolved gas to about 2.02 ml/100ml. This forms the basis of *Hyperbaric oxygen therapy* (where 100% O_2 is delivered at 2–3 times atmospheric pressure). Conversely, if O_2 tension in blood falls below 100 mmHg (as when exposed to high altitude, or in patients with hypoventilation at sea level), the amount of the gas in solution will decrease and hypoxia is the result. **(Anemia is usually not seen in patients of chronic hypoventilation such as COPD??)**. Most of the oxygen in blood is carried as oxy-hemoglobin (1 gm Hb carries 1.34 ml of O_2). But it is the PaO_2 that controls this relationship which is described by the Oxy-hemoglobin dissociation curve (please look up this curve in a standard text book of Physiology). For normal functioning, Hb saturation by O_2 must not drop below 90% (when oxygen tension is 60 mmHg). In anemia, this curve is normal as the oxygen tension is normal. But in hypoxic hypoxia, the Hb saturation by O_2 is reduced. Hence even though tissue hypoxia occurs in both situations, the mechanism differs.

Control of respiration. Normally ventilation is controlled reflexively. In lower animals, the Hering Breuer's reflex inhibits inspiration during tidal volume breathing. But in human beings, this reflex becomes operative only after the tidal volume exceeds about 1000 ml. Routinely the central chemoreceptors responding to blood CO_2, regulate respiration. Therefore, an increase in blood CO_2 increases respiration, while the opposite happens if hypocapnia occurs. A schematic representation of control of respiration is offered in **Fig 8**. Some of the inputs into the medullary centres are excitatory while others are inhibitory. During voluntary breath hold or hyperventilation, higher centres by pass the medulla to directly control respiratory muscle activity (dotted line in Fig 9). Hypoxia is a direct depressor of the medullary respiratory neurons. However, hypoxia stimulates the peripheral chemoreceptors which in turn stimulate the inspiratory neurons to maintain ventilation. Peripheral chemoreceptors are also stimulated by hypercapnia and acidic pH but this role is weaker than that of arterial hypoxia.

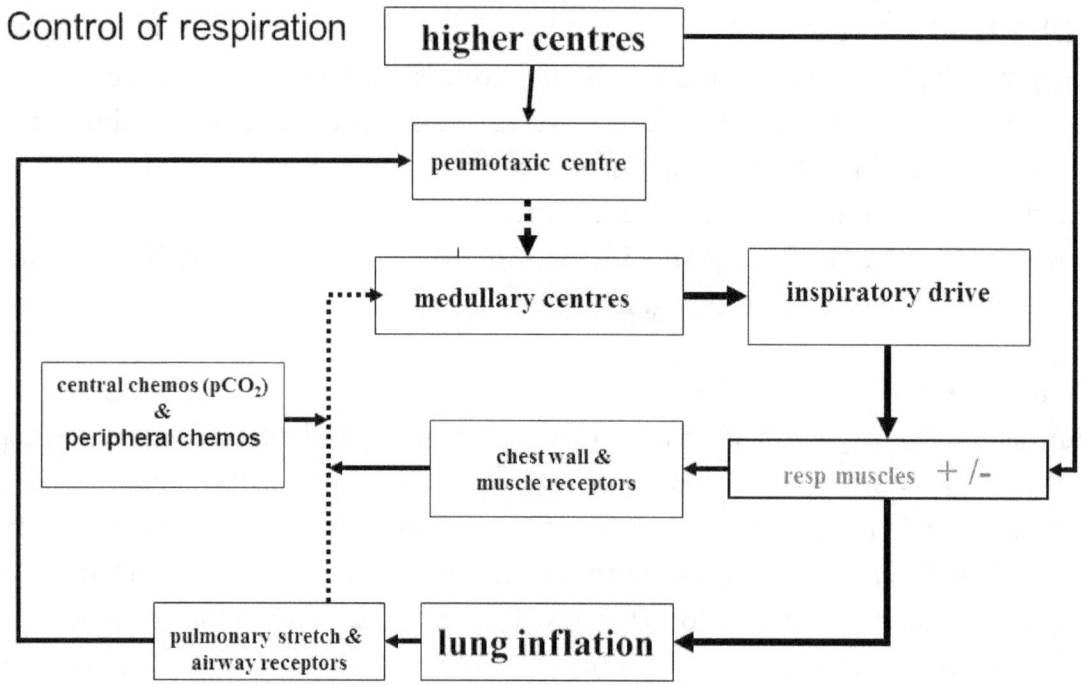

Fig. 8 Control of respiration. The dotted lines indicate inhibitory influence.

Hypo and Hyperventilation

Ventilation is adjusted to keep pace with metabolic demand. If it falls below that which is demanded, ALVEOLAR HYPOVENTILATION occurs, while if it is in excess of metabolic demand, ALVEOLAR HYPERVENTILATION is the result.

Alveolar hypoventilation may occur in a variety of circumstances from depression of the respiratory neurons, to various causes which disrupt lung-chest wall movement **(Fig 4)**. The arterial blood picture of a patient with hypoventilation is a low PO_2, high PCO_2, and an acidic pH (respiratory acidosis). The classical example of a disease which results in this syndrome is advanced COPD. (**Usually, patients with chronic hypoventilation such as COPD do not suffer from anemia?**)

Breath holding is an acute form of hypoventilation with a fall in blood oxygen, and an increase in blood CO_2. Both the factors are powerful stimulants of respiratory centres. At about a PaO_2 of 55 mmHg and a $paCO_2$ of 50 mmHg, an irresistible urge to take a breath finally ends the breath hold. This is the "break point." The symptoms felt by a normal person at break point is the classic sensation of dyspnea and discomfort in a patient of COPD. (Readers should try out this manoueuvre and realize what a patient of COPD is constantly suffering). Breath hold by strangulation results in asphyxia. Cyanosis is an important sign of asphyxiation.

Ventilation in excess of metabolic demand is alveolar hyperventilation. There is a general misconception that any increase in ventilation means hyperventilation. That is not so. Classically, a normal or slightly raised arterial oxygen tension with a $PaCO_2$ of <35 mmHg and a pH >7.45 indicates hyperventilation. The PaO_2 may be low in certain situations of hyperventilation such as exposure to high altitude (physiological response) or early interstitial lung disease. The metabolic disorder under the circumstances is respiratory alkalosis. Hypocapnia induces a vasoconstriction, particularly in cerebral and coronary beds. The former is responsible for the symptoms of blurring of vision, faintness which may progress on to a vaso-vagal syncope. Patients of epilepsy may precipitate a seizure during hyperventilation. Because of this, hyperventilation is used as a provocative test while recording EEG of patients suspected to be suffering from epilepsy. Ischemic cardiac pain may occur because of coronary vasoconstriction in susceptible patients. Respiratory alkalosis secondary to hyperventilation may rapidly induce a reduction of blood ionic calcium. This increases irritability of peripheral nerves resulting in tingling sensations in the finger tips, nose and lips, and may precipitate tetany. Interested readers may like to experience physiological effects of voluntary hyperventilation. From Fig 8 note that the cortex bypasses the medullary centres to exercises direct control over respiratory muscles during voluntary hyperventilation, or to that matter breath hold. This method of hyperventilation can lower the $PaCO_2$ to almost 50% of the normal value. *Please note that the experiment must be done in the presence of a medical teacher after ensuring that the volunteer subject is not a patient of seizures.*

The most common cause of hyperventilation is anxiety. Physiologically, exposure to high altitude (hypoxic hypoxia), increase in body temperature induce hyperventilation. Hyperventilation which occurs in aircrew while in flight could be a serious flight safety hazard as it may hasten sudden loss of consciousness especially during high G manouevres. During exercise hyperventilation as per definition occurs only when a person is nearing his/her maximum exercise limit.

Respiratory Failure

Whenever the respiratory function is impaired enough to cause failure in other organ systems and endanger life, respiratory failure is said to have set in. The failure is called type I when the PaO_2 is <60mmHg (hypoxemic). Here the lung-chest wall movements are limited by disease, but the respiratory drive is intact. The end result is dyspnea, hypoxia and at least in the early stages, normal or low CO_2 with V/Q mismatch as the primary pathophysiology. Onset of hypercapnia with acidosis heralds a dangerous situation. In type II failure, the hypoxia is accompanied by hypercapnia ($PaCO_2$ > 50 mmHg) and the respiratory drive may be blunted. Type I failure is the more common variety (**refer Fig. 4**). Clinico-physiological diagnosis of respiratory failure has to be made on the basis of ABGs. Physiological basis of management revolves around supplemental oxygen to reduce the V/Q mismatch in type I failure. Assisted ventilation is indicated when there is reduced respiratory drive either in type I or type II failure.

References

1. Beasley R, Page C, Lichtenstein L. Airway remodelling in asthma. Clinical & Experimental Allergy Reviews. 2002; 4: 109–116

2. Best and Taylor's Physiological Basis of Medical Practice, 13th edn.; Ed. Best JB. William & Wilkins, London; 1990

3. Crapo R. Current concepts in pulmonary function testing. New Eng. J Medicine. 1994;331: 25–30.

4. Fishman AP. Pulmonary Diseases and Disorders. Vol.1; 2nd edition; McGraw –Hill, London;1988

5. Ganong WF. Review of Medical Physiology. 21st edition. Lange Medical Books/McGraw-Hill, London. 2003

6. Guyton AC and Hall JE. Textbook of Medical Physiology 11th edition. Elsevier, Philadelphia; 2006.

7. Manning HL, Schwartzstein RM. Pathophysiology of dyspnea. New Eng. J Medcine. 1995; 333: 1747–1553

8. McCance KL, Huether SE. Pathophysiology: The Biologic Basis for Disease in Adults and Children. 4th edition. Mosby; St. Louis USA; 2002

9. McFadden ER Jr. Asthmain "Harrison's Principles of Internal Medicine." Edited. Kasper DL, BraunwaldE, Fauci AS, Hauser SL, Longo DL, Jameson JL.16th ed. MCGraw Hill, London. 2005; pp 1508–1516.

10. Murray and Nadel's Textbook of Respiratory Medicine; ed. Mason RJ, Broaddus VC, Murray JF, Nadel J A. Vols. 1 & 2,4th edition; Elsevier Saunders, Phil. 2005.

11. Paintal AS. Vagal sensory receptors and their reflex effects. Physiol. Rev. 1973; 53: 59–227

12. Ravi K, Kappagoda C. Responses of pulmonary C fiber and rapidly adapting receptor afferents to pulmonary congestion and oedema in dogs. Can. J. Physiol. & Pharmacol. 1992; 70: 68–76.

13. Riley JJ, Siverman EK, Shapiro SD. Chronic obstructive pulmonary disease. in "Harrison's Principles of Internal Medicine." Edited. Kasper DL, Braunwald E, Fauci AS, Hauser SL, Longo DL, Jameson JL.16th ed. MCGraw Hill, London. 2005. pp 1547–1554

14. Saldhana MJ. Pathology of Pulmonary Disease. JB Lippincott Co. Phil. 1994

15. Text Book of Medicine. Ed. Souhami RL, Moxham J. 2nd edition. Churchill Livingstone, London; 1994

16. Understanding Medical Physiology: A Text Book for Medical Students. Ed. Bijlani RL 2nd edition; Jaypee Brothers Medical Publishers, New Delhi. 1997

17. Vincent H.J. van der Velden, Hulsmann AR. Autonomic Innervation of

18. Human Airways: Structure, Function, and Pathophysiology in Asthma. *Neuro Immuno Modulation* 1999; 6:145–159

CHAPTER - 7

Pathophysiology of Some Gastro-intestinal Disorders

Plan

1. General introduction
2. Physiological anatomy of the GIT
3. General principles of GIT function
4. Commonly encountered GIT symptoms
5. Pathophysiology of some motility disorders

i. Thepharynx & the esophagus

– Dysphagia

– GERD

ii. The stomach

Gastroparesis

– Normal stomach emptying

– Pathophysiology of Gastroparesis

– Co-relation of pathophysiology & symptoms

– Physiological basis of investigations

– Principles of management

Pyloric obstruction

– Normal function of the pyloric sphincter

– Effects of obstruction of the pylorus

iii. Intestinal Pathophysiology

– Physiology of intestinal motility

– Diarrhea

– Intestinal obstruction

6. Pathophysiology of GIT secretions

i. General aspects of GIT secretions

ii. Natural protection of stomach mucosa

iii. Peptic Ulcer Disease

iv. Pancreatitis

v. Some disorders of absorptive function

General Introduction

The GIT may be likened to a conveyor belt which traps food stuff (SWALLOWING) and pushes it along the belt (MOTILITY) only in one direction- towards the anal exit (Aboral movements).Along the way, there are many "factories" and "treatment rooms" which add SECRETIONS to the material that is traveling past them on this belt, while other agencies

ABSORB whatever is necessary by body Physiology. Finally, the unwanted remnants are evacuated as feces.

The physiological anatomy of the whole of the gastro-intestinal tract (GIT) remains virtually the same throughout its course. It will be useful if the reader reviews this **(Fig 1),** as also the minor variations in various parts of the GIT which facilitate its designated purpose. There are however a few **general principles of GIT function** which need to be reviewed before dealing with its Pathophysiology.

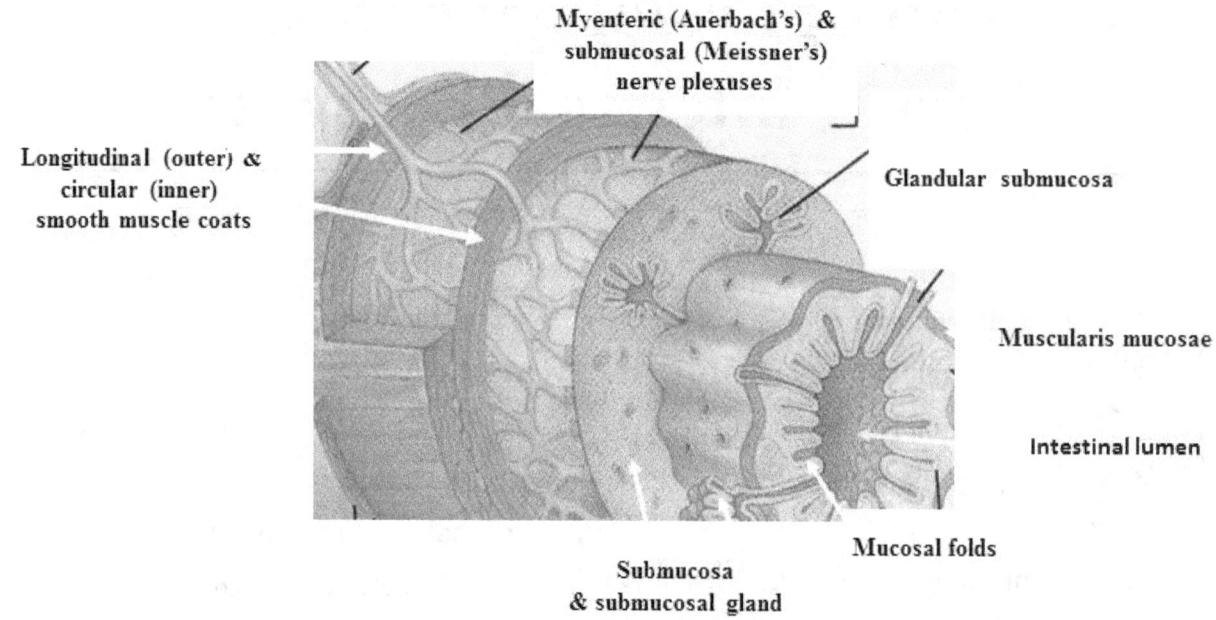

Fig. 1. General architecture of the gastro intestinal tract.

The neural control of the GIT is exercised by two agencies- the autonomic nervous system, and an independent enteric nervous system. The former modulates the activity of the GIT and is not essential for it, while the enteric nervous system may control the GIT entirely on its own. It extends from the esophagus to the anus, and is constituted by the myenteric plexus (Auerbach's plexus) which lies in between the outer longitudinal and the inner circular smooth muscle layers, and an inner Meissner's plexus or the submucosal plexus which lies in the submucosa. The former is associated with the control of movements of the GIT, while the latter influences secretions and the local blood flow. The sympathetic arm of the autonomic nervous system almost invariably inhibits the activity of the GIT while the parasympathetic arm excites it *(Use of which kind of medication could be guided by this type of innervation?)*. Mechanical or chemical irritation and distension in the GIT is sensed by epithelial receptors and transmitted to centres in the spinal cord as also the brain stem. Majority of these sensory fibres are multi- modal: they respond to more than one type of stimulus. Perhaps this explains why GIT afferents constitute only about 2–10% of all afferent fibres that reach the spinal cord. The process of nociception in the GIT is as yet not clearly understood. Probably, there are no specific nociceptors as in the skin. The pain felt in the internal organs may be signaled by excessive stimulation of the existent sensory receptors. There is also a possibility that there are some receptors which respond with the sensation of pain only to high intensities of mechanical stimulus. More recently, considerable interest has been generated by an endocannaniboid system in the GIT. Physiological functions such as relaxation of the lower esophageal sphincter, inhibition of gastric acid secretion, and intestinal motility and secretions have been attributed to CB1 receptors of this system, while the CB2 receptors may be modulating intestinal inflammation and visceral sensitivity and pain. They may hold a promise in therapeutics as they are not associated with psychotropic effects which occur when the CB1 receptors are stimulated.

There is a continuous, undulating change in the resting membrane potential of the GIT smooth muscle, the intensity varying from 5–15 mvolts, and the frequency from 3–12/ minute. These are probably generated by the Interstitial cells of Cajal which act as the pacemakers, and constitute the basic electrical rhythm (BER) in the GIT. The slow waves are responsible for contractions of the smooth muscle only in the stomach, while in other parts of the GIT, the mechanical events follow generation of sudden electrical activity (the spike potentials) which are superimposed on the resting slow waves. Acetylcholine is the main excitatory neurotransmitter which induces the spikes by calcium influx while epinephrine is responsible for influx of K^+ which inhibits the action potentials. During fasting, the BER is replaced by the migrating motor complexes (MMCs) which originate in the stomach and travel down into the ileum. They disappear as soon as food enters the GIT.

Table I: GIT Hormones

1. Gastrin
2. Gastrin Release Peptide (GRP)
3. Cholecystokinin(CCK)
4. Secretin
5. Motilin
6. Gastric Inhibitory Peptide (GIP)
7. Somatostatin
8. Ghrelin

Gastrointestinal control is exercised with the help of a number of reflexes **(Fig. 2)**. A number of hormones are secreted in the GIT. Ones that are more relevant to this write up are listed in **Table I**. Of these, Gastrin and Glucagon, Somatostatin and GRP are released mainly in the stomach, and to an extent in the duodenum, while Secretin, CCK, and GIP are concentrated in the small intestines. These agents help to modulate mainly the secretory activity, and to some extent the motility. The GIT hormones are a part of the neuro-endocrine cell system which is present in various parts of the body, inside and outside the nervous system. They belong to a family of cells called the Amine Precursor Uptake and Decarboxylase (APUD) cells. Some of these will be recalled later while discussing pathophysiology.

integrated entirely by the
enteric nervous system in the gut wall:

i.   GIT secretions
ii. peristalsis
iii. mixing movements
iv. local inhibitory reflexes

GIT
↓
sympathetic prevertebral ganglia
↓
GIT → gastrocolic reflex
          enterogastric reflexes
          colono-ileal reflexes

GIT → spinal cord/brain stem →GIT

i.   gastro/duodenal via the vagus for
     secreto-motor activity
ii. nociceptive inhibition
iii. defecation reflex

Fig. 2: GIT reflexes

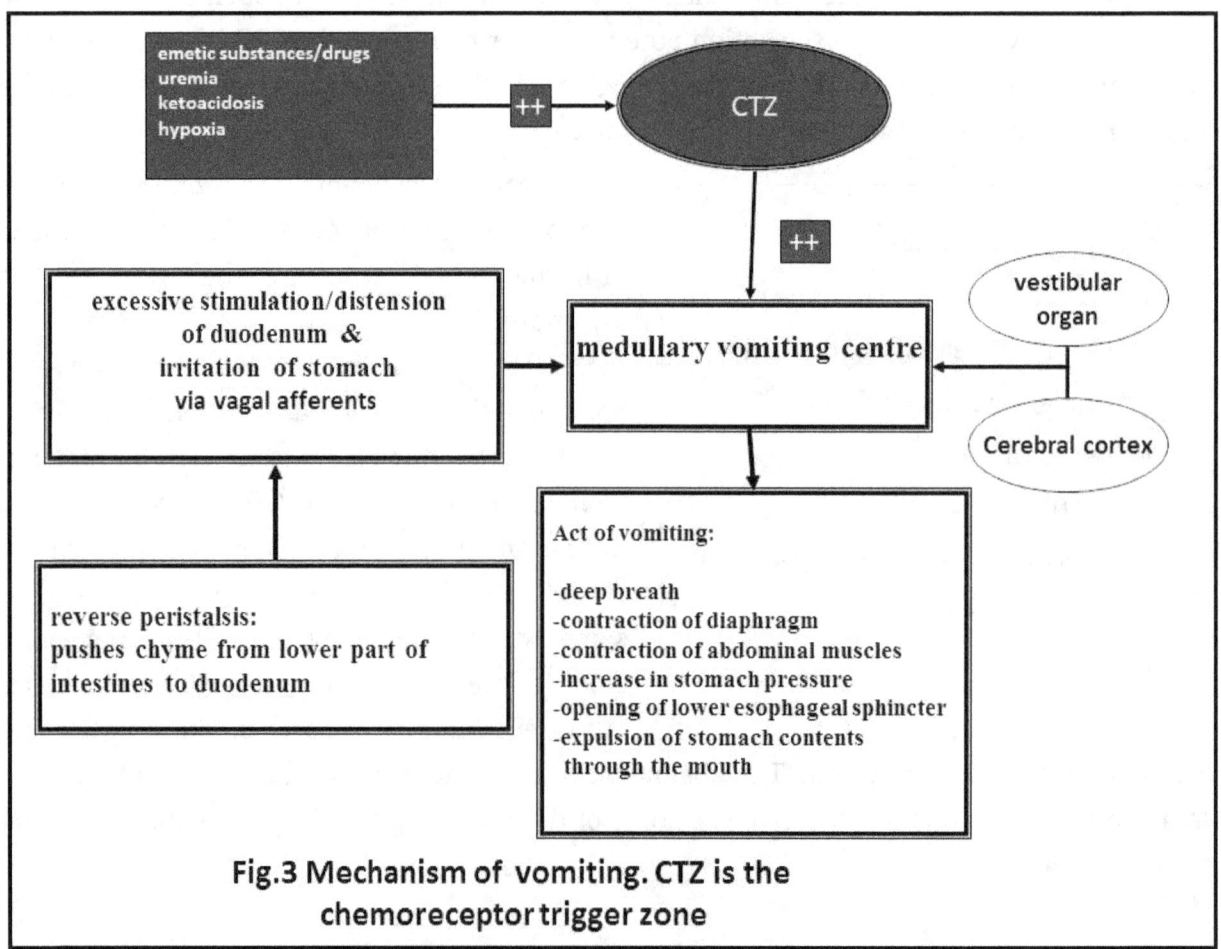

Fig.3 Mechanism of vomiting. CTZ is the
chemoreceptor trigger zone

GIT Symptoms and Signs

Some symptoms pertaining to the GIT are given in **Table II,** while the more common ones are discussed below.

Table II: Symptoms associated with GIT pathophysiology

1. Anorexia
2. Abdominal pain
3. Nausea and vomiting
4. Weight loss
5. Dyspepsia
6. Diarrhea
7. Bleeding in the GIT
8. Constipation
9. Jaundice

Anorexia (loss of appetite) is the absence of desire to eat even when physiological stimuli which generate hunger may be present. This is a non-specific symptom which may accompany nausea, pain in the abdomen, and diarrhea. All of these may be considered teleologically as protective in nature. Even if the overall incidence of anorexia in the general population may only be about 4%, it is much more frequent in patients having upper GIT disease, with an incidence of up to about 40–50% in dyspepsia related conditions. Persistent anorexia accompanied by weight loss could bean indication of serious underlying pathology.

Nausea, retching and vomiting may occur together as symptoms of GIT disease. **Nausea** is a conscious sensation which may occur independently of vomiting. **Vomiting** is a forceful evacuation of gastric and intestinal contents via the mouth, while **retching** is the forceful, rhythmic respiratory muscle activity that precedes vomiting. The basic pathways which operate are given in **Fig. 3** and pertain mainly to GIT afferents which stimulate the vomiting. Serotonin, acetylcholine, histamine and substance P are the major neurotransmitters associated with the stimulation of GIT receptors. Other areas associated with stimulation of the vomiting centre are the Chemoreceptor Trigger Zone (CTZ), the vestibular organ and the cerebral cortex. CTZ is a bilateral area in the brain stem which lies under the floor of the fourth ventricle. It is stimulated by toxic substances –some of which may have been ingested in the GIT. Other stimulants are uremia, ketoacidosis and hypoxia. Serotonin and Dopamine are the neurotransmitters associated with stimulation of the CTZ. These act via 5HT3 and D2 receptors. On being stimulated, the CTZ in turn stimulates the vomiting centre. In the absence of an active vomiting centre, the CTZ is ineffective. Many anti-emetics act on this area to inhibit it.

The onset of retching and vomiting is accompanied by a set of physiological events ie; i. inhibition of the ongoing motility of the proximal part of the small intestines which is accompanied by a reduction/loss of the spontaneous electrical rhythm *(what will be the end result of such a phenomenon?).* ii. This is followed by a single large amplitude retrograde giant contraction (RGC) which pushes the small intestinal contents into the stomach. This contraction can occur only when the normal electrical activity of the gut is interrupted. It is possible that the RGC is initiated in a specific part of the small gut which is especially innervated for the purpose. iii. Relaxation of the stomach to accommodate the backflow of chyme. This relaxation of the stomach antrum occurs prior to the generation of the RGC so that the reversed chyme is easily accommodated. The esophagus plays a part by getting shortened. This accompanied by the dilatation of the cardiac sphincter helps to deliver the chyme in to the esophagus.

Abdominal pain is by far a most distressing symptom of GIT disease.

Dyspepsia may occur in up to 80% of the population. It is a vague symptom which is often described as "Indigestion." More often than not, it is functional in origin, and no definite cause may be attributed to it. Peptic ulcer disease (PUD), and gastro-esophageal reflux disease (GERD) however may present with dyspepsia, often accompanied by the classical triad of GIT symptoms; pain, nausea, and vomiting.

GIT bleeding is often an indicator of serious GIT pathophysiology. It may occur as i. hematemesis (bloody vomiting); ii. Melena (black stools) iii. Hematochezia or fresh bleeding from lower GIT and iv. Occult or hidden bleeding which occurs as a slow, chronic blood loss from the GIT.

   i.   Hematemesis is vomiting of blood mixed with gastric secretions. The pH is acidic, and the colour black because of acid hematin which is formed when the blood comes in contact with the acid. It is foul smelling. Its occurrence is a sign of danger. The better known causes are a. bleeding esophageal varices; b. erosion of a blood vessel in the stomach or the duodenum because of peptic ulcer. c. gastric cancer. (***How do you differentiate between hematemesis and hemoptysis?***). A large volume of bleeding particularly from the esophageal varices, can result in hemorrhagic shock and acute renal failure. Accumulation of blood in the gut may stimulate peristalsis and cause diarrhea. The digestion of blood proteins may cause an increase in blood urea nitrogen. Hematocrit changes are not good indicators of large GIT bleeds because there is a lag period between the bleed and its reflection in the hematocrit.

   ii. Melena is light bleeding in the GIT which gets digested by gut enzymes. The result is tarry black foul smelling stools. The presence of melena indicates that blood has been present in the GIT for about 12–14 hours. Bleeding in the proximal part of the GIT is more associated with melena, and this thus is an important indicator of a bleeding peptic ulcer.

   iii. Hematochezia is fresh bleeding from the lower part of the GIT. Occasionally, if an upper GI tract bleed is very large and fast, some of it may present as hematochezia, but this is mostly mixed with acidified blood. Large intestinal polyps, internal hemorrhoids, and, in the elderly, cancer in the ano-rectal region may present as hematochezia.

   iv. Occult bleeding is a slow innocuous GIT bleeding which manifests as iron deficiency anemia because of chronic blood loss. It may be detected only by special techniques as it is not obviously seen in the stools. Hook worm infestation is a common cause.

In **diarrhea,** large volumes of fluid may be lost rapidly, and this may result in hypo-volemic shock and acute renal failure. Young children and the elderly are particularly susceptible. Normally, the ascending and transverse colon act as a reservoir while the descending colon is the conduit for passage of stools to the outside. Normal defecation follows a series of high amplitude propagated contractions. When these contractions increase in frequency, diarrhea may be the result. Diarrhea is one of the commonest symptoms of GIT disease.

Constipation is another common gastro-intestinal symptom. Because of the wide range of "normal" pattern of bowel movements, it is difficult to define it precisely. For example, Indians are used to a daily bowel movement while in the west, defecating about 3 times a week is considered to be normal, and is reported as 'Incomplete' defecation. Some call it a sense of not being "satisfied" after defecation. There are a number of causes for constipation. Some of the more common ones are i. physical inactivity; ii. consumption of low roughage diet; iii. dehydration; iv. poor dietary intake; v. pain in the perineum/anorectal region; vi. use of antispasmodic medication; vii. autonomic neuropathy (eg in diabetes mellitus). In the elderly, constipation of recent onset, reflecting a change in bowel habits, may bean indicator of cancer in the rectum. Such a history is of particular relevance to the elderly.

Disturbed Motor Functions of the Pharynx and the Esophagus

Disturbances in swallowing: Dysphagia

Table III: Factors associated with occurrence of GERD	
A. Lowered tone of LESB.	*Increased intragastri pressure/volume*
1. Idiopathic	1. Increase in gastric contents
2. Smoking	2. Gravitational:
3. Pregnancy	i.bending; recumbency
4. Anti-cholinergic drugs	3. Hiatus hernia
5. Smooth muscle relaxants	4. Pregnancy
	5. Ascites
6. Incompetence of diaphragmatic crural muscle	6. Gastroparesis

The motor process of swallowing is initiated by the swallowing centre in the medulla in response to afferent input from the sensory receptors in the mouth and the pharynx, particularly around the pharyngeal opening. The basic concept of the swallowing reflex, and the sequence of events that follows the initiation of swallowing is diagrammatically represented in **Fig. 4**. A disruption of this sequence leads to oral or pharyngeal reflux of the bolus, or entry of the bolus in to the respiratory tract (Oro-pharyngeal dysphagia). For patients in coma, such a disruption may pose a serious threat to life **(Explain why?)**. Other non neurological causes of dysphagia are i. acute inflammatory conditions in the oro-pharynx- septic tonsillitis in children. ii. severe iron deficiency anemia (IDA) and iii. pharyngeal pouches. The IDA when severe which may lead to the formation of upper esophageal webs just below the crico-pharyngeus muscle. They act as mechanical obstruction to passage of a bolus and are present in about 10% of iron deficiency anemia patients, mostly females. The dysphagia disappears on successful treatment, though the webs may persist in some cases. This raises some doubts as to the exact relationship of these webs to the IDA, and makes the clinical significance of this syndrome (Plummer Vinson syndrome) doubtful.

Dysfunction of the lower esophageal sphincter: Gastro-Esophageal reflux Disease (GERD)

A brief review of the physiological anatomy of the lower esophageal sphincter (LES) is required at this point of time. The LES is constituted by an intrinsic sphincter, the crural fibres of the diaphragm, surrounded by an external sphincter. The oblique fibres of the stomach act like a valve to close off the gastro-esophageal junction in order to prevent regurgitation of stomach contents into the esophagus whenever the gastric pressure increases suddenly. In addition, the angle at which the esophagus enters the stomach (Angle of His) constitutes an important part of the anti reflux barrier (ARB) and prevents duodenal bile, enzymes, and stomach acid from traveling back into the esophagus by creating a high pressure zone. Acetylcholine and substance P are the major excitatory neuro transmitters which act by producing an influx of Ca^{++} which in turn activates the calmodulin myosin light chain kinase to cause the contraction. All this ensures that the esophageal pressure recorded in this zone will be higher than that in the stomach. Nitric oxide (NO) is the main inhibitory neurotransmitter in the esophagus and is activated while swallowing, thus producing a fall in the pressure. Vasoactive Intestinal Peptide functions as an intermediate in enhancing electrical spike–induced augmentation of calcium influx and NO synthesis by nNOS. More recently, endocannaniboids have been associated with LES relaxation. The secondary peristaltic wave of the esophagus is the main regulator of the opening of the LES. This allows the bolus to enter the stomach. The LES closes once the bolus has passed through, restoring the +ve pressure gradient between the esophagus and the stomach.

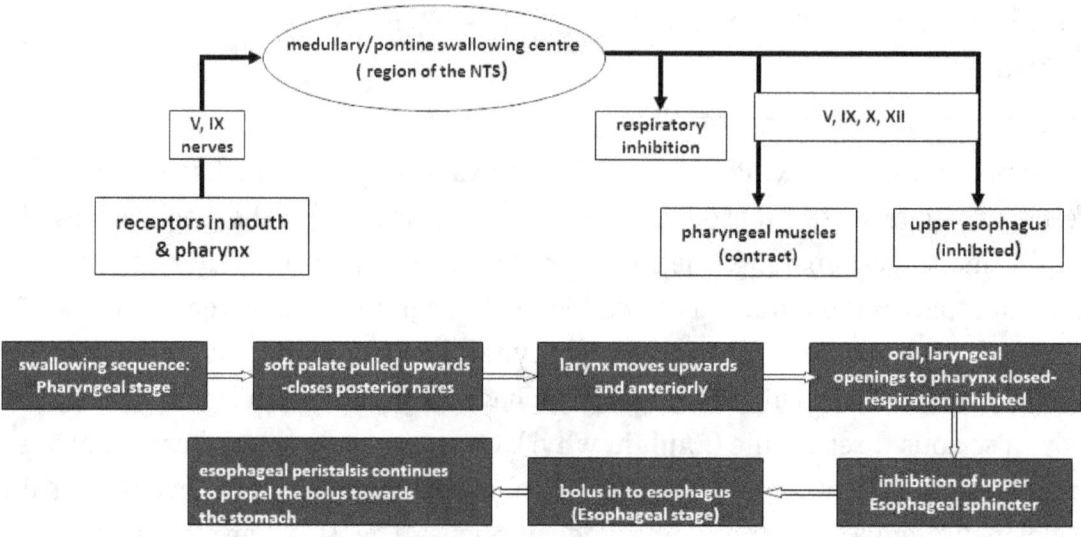

Fig. 4: The swallowing reflex. It is bilaterally represented, and the areas are interconnected. Cortical connections though present, are not required to effect swallowing.

A relatively common disorder of esophageal motility is the gastro-esophageal reflux. This may occur occasionally in normal people, when there is a transient inefficiency of the LES. This is usually easily cleared by esophageal peristalsis and gravity. The acid material which enters the pharynx and the mouth is easily neutralized by the saliva. *(Explain why drinking excessive tea/ coffee often produces a transient gastro-esophageal reflux).* But if the incompetence of the LES is sustained, GERD is the result. Here, the LES sphincter tone is reduced/lost, the angle of His is obliterated, thus eliminating the pressure gradient between the esophagus and the stomach. This allows a reflux of stomach contents (acid, bile and enzymes) into the more sensitive lower one third of the esophagus **(Fig. 5)** to produce esophageal irritation which causes heart burn and retro-sternal discomfort which "rolls" up and down the chest, at times to the side of the neck and the angle of the jaw.

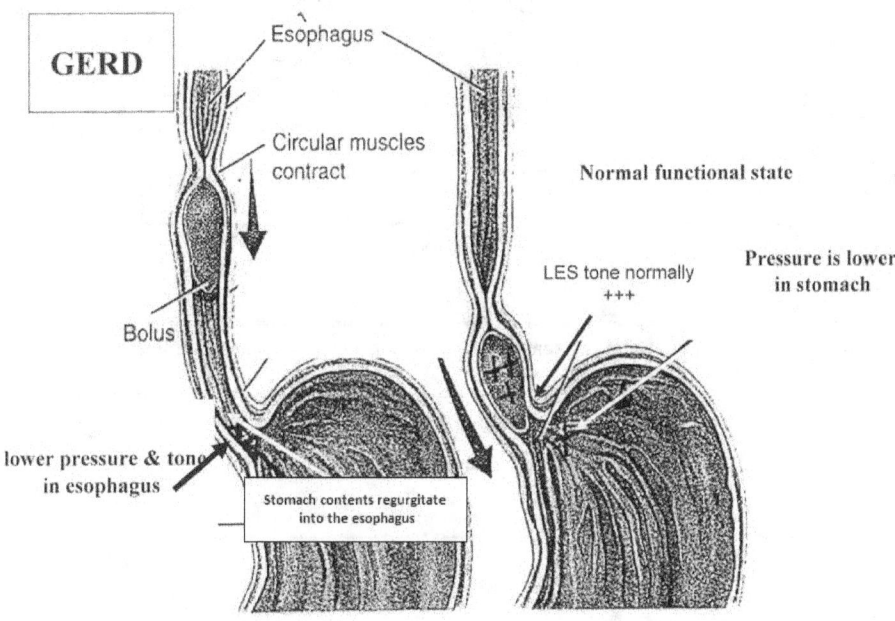

Fig. 5: Gastroesophageal Reflux Disease (GERD)

Epigastric pain often accompanies this irritation. All these symptoms may mimic ischemic heart disease pain. This happens because the afferent nociceptive fibres travel up the same first four thoracic segments which also innervate the heart. The possible pathways for transmission and perception of esophageal sensations at various levels are shown in **Fig. 6.** The gastric contents may get effortlessly regurgitated into the mouth and may track backwards in to the laryngeal opening causing bouts of cough and bronchospasm. The acid reflux is likely to inflame the esophageal lining, and may at times erode the mucosal blood vessels. It is the former that is responsible for the retrosternal burning, epigastric pain and discomfort. In the long run, this may lead to a fibrosis of the lower esophagus resulting in a stricture, and present clinically as a progressive dysphagia. Sometimes, the chronic inflammation may undergo a malignant change.

Diagnosis of the condition may be made on the basis of the clinical history which is typical in most cases. Investigations are done to i. Document esophageal mucosal injury and assess pathophysiological damage by barium swallow and direct visualization of the affected part by esophagoscopy and ii. to quantify the reflux phenomenon. This is done by long term (24 hour) monitoring of esophageal pH **(Fig. 7).** If the pH remains below4 for a period which is > 8% of the total recording time, a diagnosis of GERD is confirmed. Routine and Exercise Stress ECG test may be required to rule out ischemic heart disease.

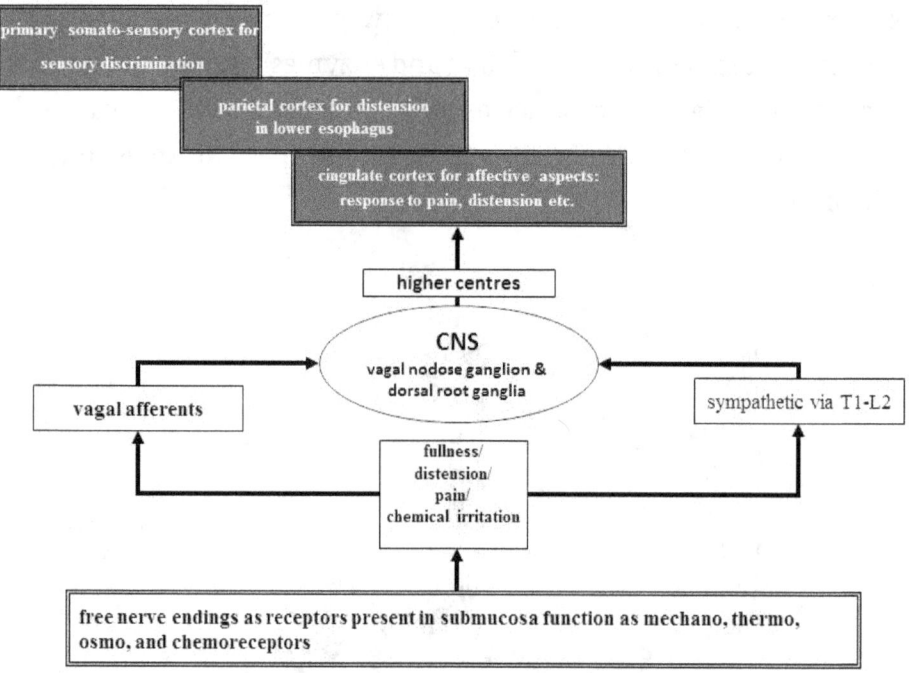

Fig. 6 Esophageal sensations

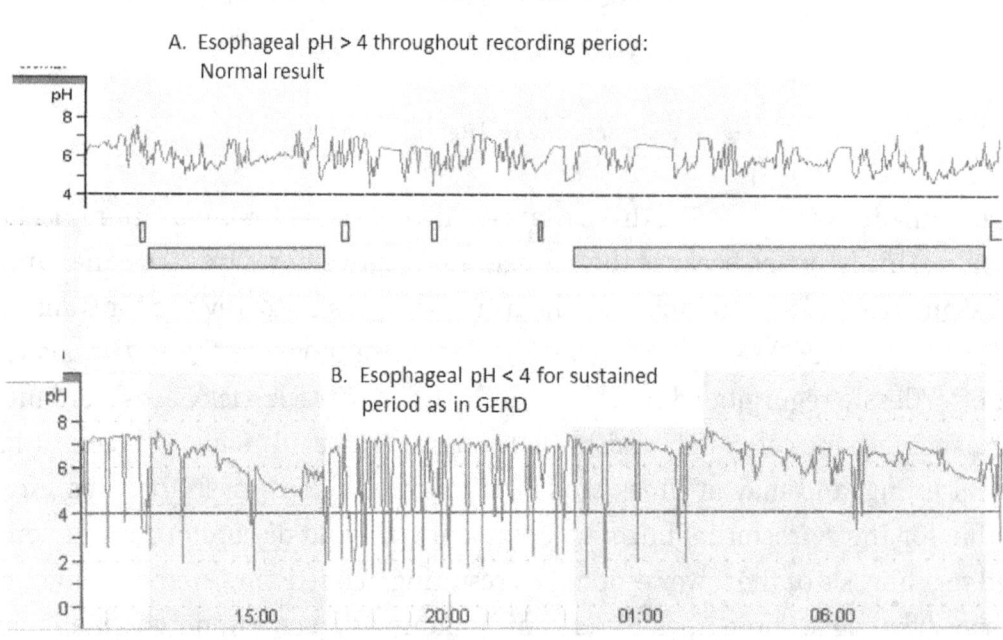

Fig. 7. A typical recording of long term esophageal pH recording. Fig. courtesy Depart. Of Clinical Physiology SQU Hospital and HoD PhysiologySQU Muscat.

In esophageal achlasia, the LES fails to relax while swallowing. This is because there is a degeneration of the inhibitory neuronal cell bodies which synthesize NO synthase, and VIP. The condition is usually idiopathic, but may occur as a complication when cancer of the lower third of the esophagus invades the enteric plexus of the area. Dysphagia is the predominant symptom, and regurgitation of accumulated food and saliva occurs with increasing severity. Weight loss occurs because of caloric and nutritional deficiency. In achalasia secondary to cancer, the dysphagia and weight loss occur rapidly. The esophageal pressure when recorded may be normal or high at the LES, but fails to reduce, or reduces less while swallowing.

Disturbances in the Motility of the Stomach

Gastroparesis

Normal gastric emptying is a complex phenomenon in which there is an integration of tonic contractions of the fundus, phasic contractions of the antrum, and the inhibitory influences of pyloric and duodenal contractions. These complex activities are initiated by specialized pacemaker cells –the Interstitial cells of Cajal (ICC). There are two types of ICC in humans. The first type is present in the circular and longitudinal muscle layers from the fundus to the antrum. Their loss is associated with low basal gastric tone and increased compliance. The second variety of ICC is found in the myenteric plexus of the body and the antrum, and is responsible for the generation of slow waves which occur at about 3/minute. A number of neurotransmitters are involved in initiating, maintaining and regulating gastric contractility, namely noradrenaline, acetylcholine, 5-hydroxy tryptamine (5-HT) and nitric oxide (NO). The latter has an important function as a non-adrenergic, non-cholinergic (NANC) neurotransmitter in the gut including the stomach and pylorus. It is synthesized by neuronal NO synthase (nNOS) in the myenteric plexus and causes relaxation of smooth muscle in the proximal stomach and appears to inhibit gastric motility. When the stomach contains food, the intensity of the contractions may result in a pressure generation of 50–70 cmH_2O while the usual mixing movements generate only about 8–12 cm H2O pressure. In between meals, Migrating Motor Complexes (MMCs) are generated about every 90 minutes, and travel at about 5 cm/sec in an empty stomach. Motilin is the neurotransmitter associated with their generation, and drugs which increase motilin activity have been used therapeutically to promote gastric motility. These compounds are referred to as motilides because of their affinity for binding to motilin receptors and inducing both fasting and postprandial gastric and small bowel motility. **Fig. 8.**

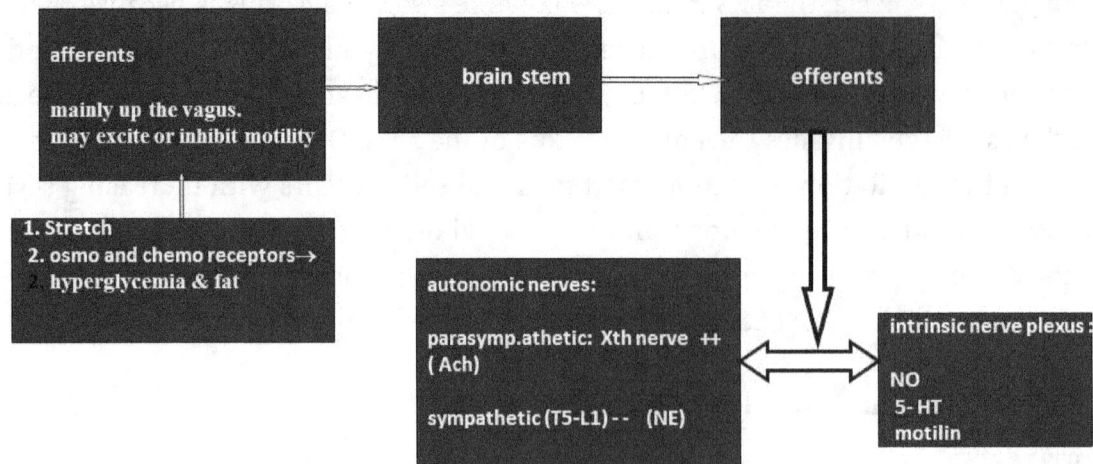

Fig. 8. Afferent and efferent mechanisms which affect gastric motility

Metabolic factors such as rapid changes in blood glucose levels, affect gastrointestinal motility reversibly. Gastric emptying is slowed during hyperglycemia and accelerated during hypoglycemia. Blood glucose fluctuations even within the normal postprandial range also influence gut function and may be important in the regulation of gut motility and sensation in healthy individuals. The neuro-endocrine axes may influence gastrointestinal function.

Gastroparesis is impaired gastric motility with delayed gastric emptying *in the absence of a mechanical outlet obstruction.* More recently the term has been expanded to include situations such alterations of gastric tone, myoelectric rhythms, antral contractions, pyloric relaxation, duodenal receptivity, and an overall in coordination of these events. The normal electrical rhythm is replaced with bradygastrias, tachygastrias, and mixed or non-specific dysarrhythmia.

Histology of gastrectomy specimens from patients of Gastroparesis have shown the presence of abnormal submucous and myenteric plexuses with a 60–100% reduction in the density of the ICC both in the myenteric plexuses and within the circular and longitudinal muscle layers of the corpus, antrum and pylorus. Thus, disordered gastrointestinal motility is likely to involve autonomic neuropathy, enteric neuropathy involving excitatory and inhibitory nerves, ICC, sudden fluctuations in blood glucose, and psychosomatic factors. In diabetic mice, defects in nNOS expression, pyloric function and gastric emptying may be reversed by insulin treatment. However even if these findings add to our understanding of the pathophysiology of gastroparesis, and may suggest development of novel therapeutic methods, their relevance to the human condition requires further elaboration.

Symptoms of Gastroparesis and their correlation to pathophysiology

The pathophysiology usually presents in a patient with long standing diabetes mellitus (50% of the cases) with recurrent nausea, vomiting, easy fatigability, post prandial fullness, early satiety, abdominal bloating and discomfort, often accompanied by burning epigastric pain, loss of appetite and constipation. The symptoms may occur after gastric surgery. In about 33% of the cases, there is no detectable cause (Idiopathic).The symptoms are nonspecific, and may suggest the presence of conditions not associated with non-motility such as peptic ulcer disease, mechanical obstruction, or non-ulcer dyspepsia.

Nausea and vomiting may be explained by over-distension of the stomach and its irritation **(Fig. 3)** with consequent stimulation of the vomiting centre. A constant feeling of satiety with a loss of appetite may also be attributed to this. The summative effect of both the effects is a loss in weight and malnutrition. The hypothalamic-pituitary-adrenal axis may also influence gastrointestinal function. Corticotrophin releasing hormone has been shown to decrease gastric motility. This may also play a role in activating the appetite suppressing pathway. The occurrence of vague epigastric pain and tenderness are is difficult to explain.

Investigations that are used in the diagnostic procedure include gastric emptying studies, intraluminal pressure measurements ,and recording of myoelectrical activity of the stomach.

Nuclear scintigraphy with barium as a contrast medium is considered the gold standard for diagnosing and quantifying delayed gastric emptying. Normally half of the stomach contents are expected to be cleared in about ninety minutes. The additional advantage of using barium is that it helps to rule out peptic ulcer. Recording of myoelectrical activity of the stomach may be done by using externally applied electrodes (**External electrogastrography (EGG)**. In gastroparesis, the normal electrical rhythm is replaced with bradygastrias, tachygastrias, non-specific dysrhythmias. Its main drawback is its inability to monitor accurately the contractile activity of the stomach. Gastric manometry measures pressures generated in the antrum, the pylorus and the duodenum. Endoscopy is done basically to rule out the presence of pathology such as gastric ulcer and mechanical obstruction of the gastric outlet.

The symptoms of gastroparesis cannot always be correlated with abnormal test results. Poor concordance between the severity of symptoms and degree of motility impairment in diabetics has been noted. Delayed gastric emptying is present in up to 50% of all patients with diabetes mellitus. On the other hand diabetes may exist in the absence of symptoms.

The physiological principles of management of gastroparesis centres around arresting of vomiting, promotion of gastric motility by using drugs (motilinides), control of blood sugar as this condition is often found in long standing diabetics with poor glycemic control, and supportive measures to improve weight. Attempts have been made at using gastric pacing by applying electrodes on the serosal surface of the stomach. However this is presently in an experimental stage.

Pyloric obstruction

The passage of stomach contents into the duodenum is controlled by the pyloric sphincter which is a collection of thick circular muscle at the duodenal end of the stomach. The pylorus though is kept slightly open most of the time so that a thin stream of fluid keeps trickling into the duodenum. The overall state of opening is controlled by nervous and humoral signals from the stomach and the duodenum. Gastric contractility increases the rate of stomach emptying while duodenal factors tend to inhibit it. The latter is logical as the duodenum by inhibiting the emptying, helps its own cause by protecting itself against a sudden influx of the acid chyme. (At this stage, the reader should review the function of the pyloric pump, and the role of the pyloric sphincter in the control of stomach emptying). An obstruction at this level will produce an accumulation of stomach contents which in turn will stretch the stomach and cause symptoms of nausea vomiting, and early satiety. This mechanism of stomach distension is different to the distension caused by gastroparesis described earlier, though clinically, the presentation and effects of both pathologies may be similar. Sometimes, a peptic ulcer may be responsible for a mechanical obstruction of the stomach outlet. This may happen on a transient basis because of development of local edema and inflammation, and would be resolved when the ulcer is treated. If the ulcer progresses to fibrosis, a permanent mechanical obstruction of the outlet may occur. However, with the medical treatment currently available for peptic ulcer, such a development is rare.

Pathophysiology of Intestinal Motility

Here the discussion will be restricted to pathophysiologic consequences of increased intestinal motility and intestinal obstruction.

Physiology of Intestinal Movements

Typically, the small intestines demonstrate mixing and propulsive movements. Functionally, both these occur together to propel the chyme towards the ileo-cecal valve for entry in to the large intestines. The mixing movements results in segmentation of the intestines. This helps in mixing the chyme with the secretions of the intestines so that the digestive juices can have their desired effect. The slow propulsive movements (peristalsis) are required to push the chyme towards the large intestines. These movements occur slowly, the aim being to give the maximum time possible for the digestive juices to have their action, and to distribute the chyme over the intestinal surface for the same purpose. Usually, the transit of chyme from the pyloric end of the stomach to the ileo-cecal valve takes about 3–5 hours. The rate of these movements is about 3–4 times/minute under normal circumstances. (*What do you think is the clinical relevance of this effect?*). The muscularis mucosa also keep folding and unfolding upon itself (like a Mexican wave at the cricket stadium!!), the objective being to expose the chyme to the intestinal enzymes which are secreted at the tips of the villi. The basic principles of the generation of the slow waves and the spikes which translate themselves into contractions has already been mentioned earlier.

Pathophysiology of Diarrhea

Diarrhea is said to have occurred when a person passes large amounts of fluids, electrolytes, mucus and undigested food material in the feces. This may at times be accompanied by blood. The intestinal contractions occur rapidly and strongly, and are accompanied by colicky pain. The condition, like vomiting, is meant as a protective step by the body to get rid of unwanted substances. However, on the other side of the coin is the fact that it may lead to rapid dehydration, especially in children and the elderly, and what begins as a protective phenomenon may lead rapidly to a fatality.

Clinically, diarrheas may be acute or chronic. The former last for less than2 weeks. Most of the acute types are caused by infectious agents. Chronic diarrheas are longer lasting. In adults, passage of stools more than 200gm is considered as diarrhea. (The reader may refer to any standard text book of Medicine for further details).Pathophysiologically, diarrhea may be Secretory, Osmotic, Motility induced, or Inflammatory.

Most secretory diarrheas are caused by irritation produced by bacterial enterotoxins. These induce secretion of large amounts of water, Cl^- and $HCO3^-$. There is also deranged net absorption of Na^+. The sum total effect is rapid dehydration which may progress to hypovolemic shock, acidosis and hyponatremia. The acidosis is the result of loss of bicarbonate. The lowering of pH stimulates the peripheral chemoreceptors to increase ventilation with the aim of inducing a counter active respiratory alkalosis. The end result is a typical pattern of breathing called Kussmaul respiration. In children, diarrhea and dehydration may be accompanied by hypoglycemia. This probably happens because ofreduced caloric intake becauseof the general morbidity induced by the diarrhea itself. Inadequate gluconeogenesis may also be involved. (*What is the corollary to this?*).

When a non-absorbable substance pulls out water from the intestines, it increases the volume of the feces, and the end result is an Osmotic diarrhea. The best known of these is the lactase deficiency diarrhea. If the intestines have been resected, the shortened length does not allow adequate time for absorptive processes to occur. This in turn increases the volume of the feces, resulting in a diarrhea categorized as a motility type of diarrhea. As most of these diarrhea categories result in loss of large volumes of fluid, they are known as large volume diarrheas.

Inflammatory bowel disease (IBD), typically represented by Crohn Disease and Ulcerative Colitis, is a potent cause of bloody diarrhea in the 15–30 age group, and later in the 60 + yrs age group. Its prevalence in Asian populations is however much lower than that found in westerners.

Intestinal bacteria normally found in the GIT release antigens which enter the intestinal mucosa. The antigens are presented to the regulatory lymphocytes by the local antigen presenting cells (often the mucosal epithelial cells) and are thus neutralized. Secretion of factors such interleukin 10 and transforming growth factor $-\beta$ by regulatory T cells help to prevent the inflammatory process from dominating. At the same time, the process of apoptosis of the mucosal T cells is kept ongoing. The innate immune system seems to be activated by an intracellular protein of the regulatory lymphocytes which combines with remnants of the cell wall of the digested bacteria. When this natural process fails to deliver, inflammatory bowel disease is the end result. The tight junctions of epithelial cells of the intestinal mucosa usually form an effective barrier which prevents fecal antigens from entering the mucosal cells. It has also been postulated that excessive secretion of inflammatory cytokines such as Tumour

Necrosis Factor, and Interleukins 1 & 6 secreted by the mucosal epithelium may be responsible for the inflammatory reaction.

(Whatever the cause of diarrhea/dysentery, what will the major clinico-physiological implications?

Oral rehydration solution (ORS)

One of the most effective methods of preventing the pathophysiologic effects of rapid fluid loss as a result of diarrhea, especially in children and the elderly, is the use of oral rehydration solution (ORS). At this stage the student may recall that glucose and Na^+ are absorbed together via a symport mechanism involving a Sodium-dependent Glucose Transporter (SGLT). The amount of glucose carried by the transporter is dependent on the concentration of Na^+ available in the luminal fluid. Water is dragged along with the Na^+. The standard ORS constituents as per WHO recommendations are: Na^+ 90mmol/l with K (20), Cl⁻ 80, Citrate 10 (as a source of HCO_3^-) and glucose 110 mmol/l. For an effective home made formulation, 1 teaspoon of table salt with 8 teaspoonfuls of sugar may be added to 1 litre of boiled water. The K^+ supplementation is done by addition of coconut water or orange juice. On the Indian subcontinent in areas where rice is the staple diet, about 10 teaspoonfuls of water of freshly cooked rice may replace the sugar. ORS is the ideal first aid for patients with diarrhea.

Intestinal Obstruction

Intestinal obstruction may develop with increased peristaltic activity which is brought into play in order to overcome a mechanical obstruction (Dynamic) or one in which there is no peristalsis (Adynamic). The reader may refer to any standard text book of Surgery for the various causes of intestinal obstruction. Briefly however, about 40–50% of the dynamic variety is caused by adhesions, while intestinal inflammation (15%) and carcinoma (15%) constitute the other major reasons. Whatever the cause, in the dynamic variety the part of the intestine proximal to the obstruction dilates while the distal part continues to have the normal physiological function until such time that it empties itself of the contents. Following this, it ceases further peristalsis but remains in a state of contraction. At this point of time the situation of "absolute constipation" is reached. Clinically this is diagnostic of obstruction.

Initially, an attempt is made to overcome the obstruction by increased peristalsis which may be heard (borborgymi), and felt or seen clinically. A possible sequence of events which compounds the effects of obstruction in the intestine proximal to the obstruction is outlined in **Fig. 9.** The distension of the bowel is a result of accumulating gas and fluids. The source of the gas is by swallowing, and metabolism of intestinal organisms. Most of the gas is the nitrogen and some hydrogen sulphide (H_2S). The fluids are the secretions of the gut which tend to accumulate as their absorption by the intestinal epithelium is adversely affected. A plain radiograph of the abdomen in the sitting posture may easily demonstrate this fluid accumulation (Fluid levels). The progressive distension leads to a decrease in peristalsis and later to a cessation of this activity.

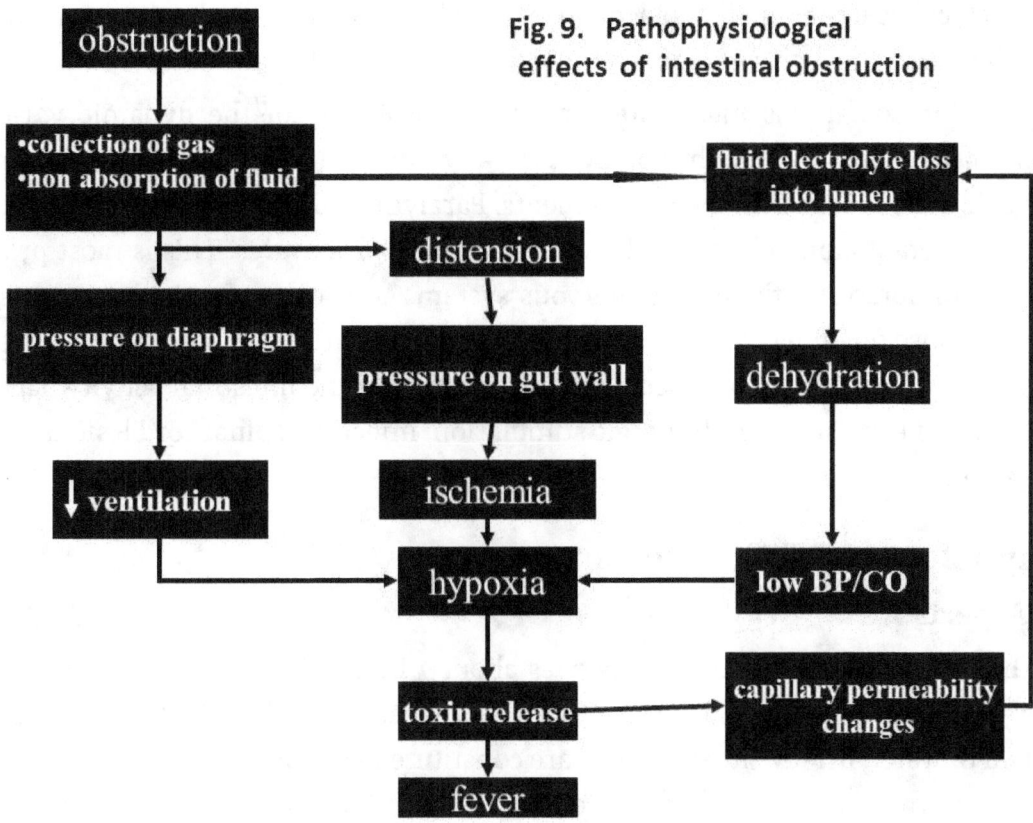

Fig. 9. Pathophysiological effects of intestinal obstruction

The distension of the intestines may restrict the movements of the diaphragm, resulting in some degree of hypoventilation and contributes to the overall hypoxia. It may be severe enough to restrict intestinal blood flow and in this manner induce local tissue ischemia. The end result is release of local toxins which in turn initiate an inflammatory reaction. A vicious cycle then sets in, perpetuating more capillary permeability changes, and enhances further fluid loss.

The loss of fluid and electrolytes into the intestines sets up the stage for dehydration. This is enhanced by vomiting and reducedoral intake. There is therefore a grave danger of dehydration and shock setting in rapidly. When this happens, the lowering of blood pressure and cardiac output further accelerates the overall tissue hypoxia.

The distension of the intestines causes abdominal pain, but not tenderness. When the latter appears, it heralds severe gut ischemia because of compromised arterial blood flow (Strangulation). This may be followed by infarction, after which the weakened gut wall may perforate easily. If fever sets in, it suggests that there is inflammation. This may be a consequence of gut ischemia which releases inflammatory mediators, or perforation which may follow the ischemia. In both situations, the ensuing pathophysiology is the harbinger of this important sign.

(From the above, derive the cardinal clinical features of intestinal obstruction).

If the pathophysiology, and the symptoms and signs that intestinal obstruction sets up are reviewed, the physiological principles of management may be easily discerned. These should be to i. prevent/limit the distension of the intestine proximal to the obstruction by sucking out the fluid which collects therein; ii. replace the lost fluid and correct the electrolyte imbalance. iii. remove the mechanical obstruction.

The discussion so far has highlighted the pathophysiology of the dynamic variety of obstruction. In general, the pathophysiology of the **adynamic** variety is also similar, the major difference being the absence of bowel movements. Paralytic ileus typifies the Adynamic type. The so called "silent abdomen" usually follows lapratomy procedures. This is most probably because of an inhibition of the enteric nervous system. Activation of opioid receptors may be involved because opiod blockers are of some help in mitigating the obstruction. Irritation of the peritoneum is known to produce a reflex inhibition of the intestinal activity. This may happen because of excessive noradrenergic stimulation, infection, spinal cord lesions, uremia and hypokalemia.

Pathophysiology of GIT Secretions

General Aspects

The adult human gastrointestinal tract secretes about 7 litres of fluids in a day. That almost all of it is reabsorbed as it travels down the GIT is evidenced by the fact the feces contain only about 200 ml of water. Briefly, the secretions are constituted by mucus, water and electrolytes, and digestive enzymes. The pH of these secretions is determined by two major considerations: environment required for optimal activity of the various digestive enzymes, and protection of the lining of the intestinal epithelium. The control of these secretions is influenced by the mechanical stimulation of the mucosa (particularly effective for mucus secretion), the enteric nervous system (activated by local tactile, chemical stimuli and distension of the lumen), the autonomic nervous system, and a number of GIT hormones. The parasympathetic autonomic system stimulates the secretions. The sympathetic part may also cause some stimulation, but at the same time, it may vasoconstrict the local blood supply, thereby restraining the stimulatory effect. In this write up, pathophysiology of only the stomach and the pancreas will be dealt with.

Physiology of Secretions of the Stomach

The stomach secretes about1. 5 litres of gastric juice/ per day, with a pH between 1–3.5. The secretions come from the mucus glands which line the epithelial surface, the gastric (Oxyntic) glands which secrete the acid, intrinsic factor, pepsinogen as well as mucus. They also contain enterochromaffin cells and endocrine cells. These glands are located in the body and the fundus of the stomach. The pyloric glands are located in the antral part, and secrete mainly mucus and gastrin. The student at this stage must review the mechanism of acid secretion by the stomach, and reiterate the importance of acetylcholine, histamine and gastric receptors on the parietal cells. The location of the proton pump (H^+K^+ ATPase) at the tip of the luminal end of the parietal cells must also be recalled.

At this point of time, it is critical to review the process of **natural protection of the gastro-duodenal mucosa.** The trio of endogenously produced factors which constantly attack the gastric /duodenal mucosa are i. HCL; ii. pepsinogen and pepsin and iii. bile salts. A number of exogenous factors also try to undermine the continuity of the gastric mucosa eg bacterial infections, alcohol, medications etc. The defence of the mucosa against these assaults is three tiered; i. above the epithelium; ii. the epithelium itself and iii. below the epithelium. The most prominent amongst these is the surface layer of mucus and bicarbonate. The former consists of glycoprotein mucin and phospholipid emulsion. It is secreted by the mucosal cells. Its presence prevents the entry of pepsin and ions because of the hydrophobic layer that this combination forms over the mucosal surface. The bicarbonate is also secreted by the mucosal cells, and is under influence of prostaglandins E and I, Ca^{++}, and the pH of the surface layer. Even if the quantum of HCO_3^- secreted is not enough to neutralize the bulk of the acid being secreted, it is sufficient to react with the little acid that infiltrates in to the mucus layer. Cholinergic activity which on one hand promotes acid secretion by the parietal cells, simultaneously stimulates bicarbonate secretion. The gastric epithelial cells by themselves offer resistance to invading material by virtue of the tight junctions between them. Further more, small breaks in the continuity of the cell barrier can be plugged by a process of RESTITUTION where adjoining gastric epithelial cells move into these breaks to seal the gaps. A number of growth factors are involved in the process. If the break is larger, restitution may not be enough to offer further protection. In this situation, epithelial cell regeneration is required. Prostaglandins play a major role in this phenomenon along with secretion of growth factors. This is supported by laying down of new microcirculation in the area. It is the latter that is responsible for offering the third level of barrier protection by providing sufficient nutrition to the gastric epithelial cells, as also providing the HCO_3^-. The genesis of peptic ulcer disease lies therefore in the damage to the naturally protective barrier.

There are two basic reasons for the mucosal damage in peptic ulcer disease (PUD). i. Excessive secretion of acid which can break through the existing mucosal defense and ii. the damage to the natural barrier which then becomes ineffective in protection against even normal amount of acid secretion.

Gastric hypersecretion may be attributed to i. an increase in the number of parietal cells; ii. hyperplasia of G cells; iii. Imbalance between secretion of gastrin which stimulates acid secretion, and the somatostatin which inhibits it; iv. gastrinoma in Zollinger-Ellison syndrome. Cholinergic hypersensitivity and parasympathetic overactivity has been involved, not only in increased acid secretion, but also in stimulation of pepsin release which is an important co-factor in the development of mucosal injury. Age-related decline in the synthesis of PGs, smoking, alcohol, consumption of non-steroidal anti-inflammatory medications (aspirin in particular) have all been recognized as factors which promote damage to the gastric mucosa. However, since 1982, *Helicobacter pylori* infection has been recognized to be by far the most important causative factor of PUD. Up to 95% of PUD in the duodenum has been attributable to this bacterial infection. A high percentage of gastric ulcers too have been associated with H. pylori infection.

The acid pH in the stomach is normally an innate barrier which denatures proteins of bacteria which may otherwise infect the human being. H *pylori* usually colonizes itself in the mucus layer of the stomach in the antral region and protects itself by producing an enzyme, probably urease, which creates a low acid zone in its immediate environment by generation of ammonia. This may by itself be damaging to the gastric epithelium. The damaged cells then release a variety of cytokines (IL 8, IL1 and Tumour Necrosis Factor) which in turn attract inflammatory cells such as neutrophils, monocytes and lymphocytes. The monocytes in turn release inflammatory mediators, function as antigen presenting cells and activate T cells. The inflammatory mediators released then may directly excite the antral G cells to induce greater acid production. They also inhibit somatostatin. As a result, excess acid is produced. Hence the basic mechanism by which H pylori induces ulcer formation is by production of excess acid which then reaches in to the duodenum to damage its epithelium. H pylori has also been linked to gastric cancer which is secondary to a chronic gastritis and hypochlorhydria which may be caused by the infection. Detection of H pylori may be done by the urea breath test and the presence of the bacterial antigen in stool.

Most other factors which have been associated with causation of PUD appear to do so by damaging the mucosal lining rather than production of excess acid. Smoking, for example, is an important risk factor. The exact mechanism by which it is responsible for ulcer is not known. Various possibilities include depletion of bicarbonate generation. Consumption of non steroidal anti-inflammatory drugs (NSAIDs) on the other hand is known to reduce the generation of the protective PGs.

Symptoms and Signs of PUD vis-à-vis the Pathophysiology

Epigastric pain is by far the most prominent symptom of PUD. It is referred pain, but the exact mechanism of its etiology is as yet uncertain. A distinguishing feature is its relationship to meals. In most patients of duodenal ulcers, it occurs a few hours after a meal, and is relieved by eating, while in gastric ulcer patients, it may be brought on by a meal. It may be caused by irritation of chemoreceptors in the stomach/duodenum by the acid, pepsin or bile salts. Increased motility with spasm of smooth muscles may be involved. If the pain persists after food and antacids, and becomes continuous and boring in nature, radiating to the back, it signals the emergence of a penetrating ulcer. Pain accompanied by nausea and vomiting of undigested food indicates the development of pyloric obstruction. Distension of the stomach and irritation of its mucosal lining by the acid probably explains these symptoms. Currently, this complication of PUD is rarely seen because of the diagnostic measures, and effective medications available in the early stages. Sudden severe epigastric pain with a feeling of "give way" in the abdomen" suggests a perforation. The acid contents may spill in to the peritoneal cavity, or may involve organs which lie posterior to the duodenum eg the pancreas. Erosion of gastric/duodenal blood vessels by the progressive ulcer is likely to cause catastrophic hematemesis. Of concern is the fact that such an episode may occur without any previous warning. Low grade bleeding presents as tarry black stools. This is often accompanied by an iron deficiency anemia for obvious reasons.

Diarrhoea and steatorrhoea may occur in patients of PUD. The former is caused by the large volume of gastric and duodenal secretions in response to the irritation of the mucosa. These in turn increase intestinal motility. Pancreatic enzymes which function ideally in alkaline conditions are inactivated in the presence of excess acid. Bile salts are unable to form micelles in this altered pH environment, and this disturbs lipid absorption. Steatorrhoea is the result. Also mucosal damage reduces the surface area of absorption. Weight loss in patients of PUD is attributable to the pain, nausea, vomiting, accompanying loss of appetite, and malabsorption of nutrients.

Principles of management of PUD revolve around i. reduction of acid secretion and ii. promotion of cyto-protection.

Pathophysiology of Pancreatitis

The exocrine pancreas secretes about 1 litre of fluid into the small intestine. This is a mixture of water and electrolytes (mainly $NaHCO_3$), and enzymes at a highly alkaline pH (8–8.5). The functional unit of the pancreas is the acinar cell which secretes the enzymes, while the sodium bicarbonate comes from the ductules which emerge from these cells. The latter have a columnar epithelium and secrete water and electrolytes rich in $NaHCO_3$. The acinar cells contain zymogen granules which house inactive pancreatic pro-enzymes. When the pancreas is stimulated, the granules are offloaded into the ducts to make their way to the duodenum. Here the enteropeptidase acts upon the trypsinogen to form the active enzyme trypsin. This in turn unleashes a cascade which transforms the digestive pro-enzymes in to their active form. The acinar cells also contain granules which store enzymes which canact on the precursors of pancreatic digestive enzymes to activate them. The corollary to this is that if by any chance the lysosomal membranes are disrupted, the enzymes they contain will enter the zymogen granules and activate the digestive enzymes, the most important event being the conversion of the pro-enzyme trypsinogen to the active trypsin. If that happens, auto-digestion of the pancreatic tissue will occur, resulting in pancreatitis which is a manifestation of acinar cell injury. The extent of injury may range from a local edema to extensive necrosis which may result in multi organ damage and death.

Normally, there is a multi-tiered protection mechanism which prevents auto-digestion of the acinar cells. Foremost amongst these is the synthesis of the enzymes in an inactive form, their packaging in a capsulated granule, and their transport in these granules to the duodenum where they are activated. Even if the digestive enzymes and the hydrolytic enzymes are manufactured in the common Golgi apparatus, they remain separated via an unknown mechanism. But if this mechanism fails, then even at this stage, the hydrolytic enzymes (Cathepsin B in particular) may prevail upon the proenzyme trypsinogen and convert it in to the proteolytic trypsin. Some trypsin is spontaneously formed in the acinar cells, but is promptly denatured by the availability of a specific trypsin inhibitor, and other enzymes such as mesotrypsin and enzyme Y. Apart from this trypsin also has the property of "self destruct" to some extent. If these naturally occurring protective mechanisms are disturbed, auto-digestion will occur. Some of the factors that have been implicated in the malfunction of these mechanisms are alcohol, regurgitation of biliary and pancreatic secretions because of obstruction of the common bile duct, ischemia, and viral infection. Activation of trypsin initiates an inflammatory response which starts with the recruitment of neutrophils and macrophages, and mast cells which release inflammatory mediators such as Tumor Necrosis Factor$_\alpha$, as also excessive amounts of free O_2 radicals. Apart from the various pro-inflammatory factors, there are also some protective elements which are released simultaneously, Complement Factor 5a being one of them. The multi organ involvement is determined by the severity of the disease which gives an upper hand to the inflammatory factors, and the up-regulation of their receptors in the lungs, the cardiovascular system and the kidneys. Acute Respiratory Distress Syndrome is a major complication of this involvement.

Chronic pancreatitis. Etio-pathogenesis of chronic pancreatitis is complex and ill understood. The most important factor seems to be alcohol consumption. It has been postulated that a protein, lithostathine, which is a by product of proteolytic activity of the small amounts of trypsin that is being formed in the acinar cells, prevents the precipitation of Calcium carbonate. Alcohol seems to inhibit the formation of this protein, thereby allowing the formation of $CaCO_3$ stones in the ductules/ducts. More recent evidence suggests that perforin secreted by activated inflammatory cells may be involved. A number of genes have been linked to the condition. One of them, the tryptase gene is induces the production of this enzyme, particularly in mast cells which are found abundantly in pancreatic tissue harbouring the disease. It helps in the proliferation of fibroblasts which are ultimately responsible for laying down the fibrous tissue which is the hall mark of the disease. Smoking, apart from enhancing proteolytic activity may also do damage by release of free O_2 radicals. The possible sequence of events is outlined in **Fig.10**

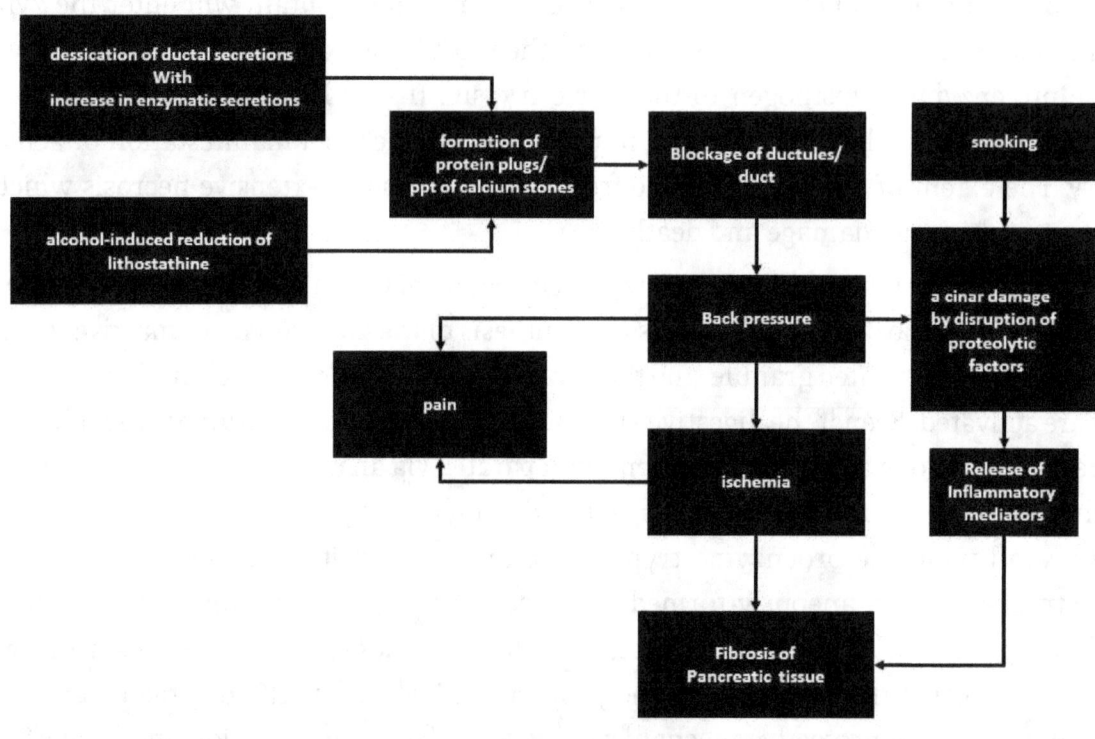

Fig. 10: Pathogenesis of chronic pancreatitis
(based on- Friess et al 2003, Khokhar & Seidner 2004)

Clinical symptoms and signs as related to the pathophysiology of pancreatitis.

Epigastric pain is the cardinal symptom of this condition, and may vary in severity. The etiology is thought to be multifactorial. Neurotransmitters, such as substance P and Calcitonin Gene-Related Peptide (CGRP) which are released may irritate the pancreatic and peri-pancreatic nerves. There is also the possibility that distension of the ducts by the rapidly collecting fluid and edema causes the pain. The observations that relief of pain cannot be attained by a single mode of treatment in all patients suggest the diversity of etiological mechanisms.

Nausea and vomiting are the other major features and are attributable to the copious secretions brought about by the inflammation.

Involvement of distant organs is indicated by cardio-respiratory symptoms and signs. Of grave pathophysiologic concern is the development of ARDS.

Long term effects of pancreatitis will present as malabsorption of the proximate principles (carbohydrates, fats and proteins), with weight loss and malnutrition. (*What may cause this malabsorption?*)

Principles of management are dictated by the pathophysiology, and include relief of pain, maintenance of fluid and electrolyte balance, management of multi organ involvement in case such problems develop.

Absorptive Dysfunction in the GIT

About nine litres of fluid (about 2 litres of which is ingested) are presented to the GIT every 24 hours. Of this, up to 98% is reabsorbed. Most of this (approximately 7.5 litres) are reabsorbed in the small intestines, mainly the jejunum and the ileum, while about 1.3 litres are reabsorbed in the proximal colon. The stomach absorbs only a little water.

In the small intestines and the proximal colon, Na^+ diffuses easily in to the lining cells across the concentration gradient. Once inside the enterocytes, it is pumped in to the extracellular fluid by the Na^+/K^+-ATPAse on the basolateral surface of the cell. In the small intestines, glucose and amino acids are absorbed along with Na^+ as secondary active transport. This attribute is used successfully for applying the oral rehydration solution therapy in patients with diarrhea, and is especially useful in children as mentioned earlier in the diarrhea section. Cl^- is secreted in to the lumen via protein channels activated by 3',5' cAMP. Cholera bacillus releases an exotoxin which enters the intestinal mucosa to generate cAMP. Large amounts of the ion then leaks out, taking water with it. At the same time, the Na^+ carrier is rendered relatively ineffective, allowing large amounts of water and electrolytes to accumulate inside the lumen, precipitating a severe diarrhea which may terminate fatally if not treated in time.

All water soluble vitamins with the exception of folate and vitaminB_{12} are dependent on the availability of a Na^+ carrier. The absorption of fat soluble vitamins is hindered by obstructive jaundice and pancreatitis. Almost all the vitamins are absorbed in the upper intestine with the exception of B_{12}, which is absorbed in the ileum

In brief, Malabsorption is the interference with absorption of nutritional elements by the small intestines. This may happen because of; 1. maldigestion which is the failure to digest nutrients because of enzyme deficiency, absence of bile salts etc.;2. mal-absorption which is the failure to take in the digested elements because of disruption of intestinal cell membrane because of disease, intestinal resection, blood supply deficiency in region of the GIT concerned with the digestive process. DIGESTIVE problems may arise because of i. pancreatic insufficiency; ii. unsuitable pH in intestines in which the digestive enzymes cannot function effectively; and iii. absence of a specific intestinal enzyme; eg. Intestinal lactase.

Clinically, the above mentioned pathophysiology is reflected as symptoms and signs of weight loss, fatty stools (steatorrhea), gaseous distension of the abdomen, osmotic diarrhea and anemia.

References

1. Akheel A, Rattansingh A, Furtado SD. Current perspectives on the management of gastroparesis. J Post Grad. Med.2005; 51: 54–60

2. Bailey and Love's Short Practice of Surgery. Edt. Williams NS, Bulstrode C, O'Connell P. 25th edn. Hodder Arnold, London. 2008; pp 1188–1203

3. Bamias G, Nyce MR, De la Rue SA, Cominelli F. New concepts in the Pathophysiology of Inflammatory Bowel Disease. Ann. Intern. Med.2005; 143: 895–904.

4. Best and Taylor's Physiological Basis of Medical Practice, 13th edn.; Ed. West JB. William & Wilkins, London; 1990.

5. Conklin JL, Christensen J. Motor functions of the pharynx and the esophagus in Physiology of the Gastrointestinal Tract. Vol. 1; 3rd Edn. Edited. Johnson LR. Raven Press, New York, 1994; pp 903–928

6. Davidson's Principles and Practice of Medicine. Edt. Haslett C, Chilvers ER, Boon NA & Colledge NR. 19th Edn. Churchill Livingstone., London. 2002

7. di Sebastiano P, di Mola PF, Büchler MW, Friess H. Pathogenesis of pain in chronic pancreatitis. Dig.Dis. 2004; 22: 267–272

8. Farthing MJ. Helicobacter pylori.: an overview. British med. Bull. 1998; 54: 1–6.

9. Friess H, Kleef J, Buchler M. Molecular pathology of chronic pancreatitis. J. Gastrointestinal Surgery. 2003; 7: 943–945.

10. Ganong WF. Review of Medical Physiology. 21st edition. Lange Medical Books/McGraw-Hill, London. 2003.

11. Gray's Anatomy: The Anatomical Basis of Medical Practice; 39th Edn. Ed. Standring S; Elsevier, Churchill Livingstone, London; 2005.

12. Grundy D, Reid K. The physiology of nausea and vomiting in The Physiology of the Gastrointestinal Tract. Vol. 1; 3rd Edn. Edited. Johnson LR. Raven Press, New York, 1994; pp 879–901

13. Guyton AC, Hall JE. Textbook of Medical Physiology. 11th edition. Elsevier, Philadelphia; 2006.

14. Harrison's Principles of Internal Medicine Edt. Longo D., Fauci A, Ksper D, Hauser S., Jameson J, Joseph Loscalzo J, 18th ed. MCGraw Hill, London. 2011.

15. Khokhar AS, Seidner DL. The pathophysiology of pancreatitis. Nutrition in Clinical Practice. 2004; 19: 5–15.

16. Liu J, Kahrilas P J,. Pharyngeal and esophageal diverticula, rings, and webs. *GI Motility Online.* 2006. 16th May. doi:10.1038/gimo41;

17. Luman W. Helicobacter pylori: causation and treatment. J. R Coll Phsicians Edinb. 2005: 49: 35–45.

18. McCance KL, Huether SE. Pathophysiology: The Biologic Basis for Disease in Adults and Children. 5th edition. Mosby; St. Louis USA; 2006

19. McCallum RW, Brown RL Diabetic and Nondiabetic Gastroparesis. Current Treatment Options in Gastroenterology 1998, 1:1–7

20. O'Donovan D, Feinle-Bisset C, Jones K, Horowitz M. Idiopathic and Diabetic Gastroparesis. Curr Treat Options Gastroenterol. 2003; 4: 299–309.

21. Porth C. Essentials of Pathophysiology: Concepts of Altered Health States. 2nd edn; Lippincott Williams and Wilkins; Philadelphia; 2007

22. Robbins & Cotran. Pathologic Basis of Disease. Edited. Kumar V, Abbas AK, Fausto N. 7th edn. Elsevier Saunders, China. 2005.

23. Scharschmidt B. Peptic ulcer disease: Pathophysiology and current management. West J Medicine. 1987; 146:724–733.

24. Sengupta JN, Gebhart GF. Gastrointestinal afferent fibres and sensation. In The Physiology of the Gastrointestinal Tract. Vol. 1; 3rd Edn. Edited. Johnson LR. Raven Press, New York, 1994; pp 483–519.

25. Spiller RC. ABC of the upper gastro-intestinal tract. Anorexia, nausea, vomiting and pain. British Med. J. 2001; 323: 1354–1357

26. Wright KL, Duncan M, Sharkey KA. Cannabinoid CB2 receptors in the gastrointestinal tract: a regulatory system in states of inflammation. British J Pharmacol. 2008; 153: 263–270.

CHAPTER - 8

Pathophysiology of Obesity and Weight Loss

Plan

1. General introduction: Obesity
2. Obesity amongst Asian Indians
3. Criteria for diagnosing obesity
4. Pathophysiology of obesity and pathogenesis of some obesity induced disorders
5. Pathophysiology of weight loss
   i. Protein calorie malnutritionby starvation
   ii. Cachexia
   iii. Anorexia nervosa and Bulimia
   iv. Dehydration
   v. Sarcopenia
   vi. Weight loss in the elderly

Table I: Methods for measuring obesity

1. Quetelet's Body mass Index (BMI) = Wt in Kg/ht m^2
2. Waist circumference
3. Waist to hip ratio (W/H)
4. Measurement of skin fold thickness
5. Estimation of Body fat %

General Introduction: Pathophysiology of Obesity

Obesity has been defined by Kane and Vinay Kumar as a "Disorder of energy balance in which the food derived energy chronically exceeds energy expenditure and the excess calories are stored as triglycerides in adipose tissue." Alternatively, the WHO defines it as "A condition of excess fat accumulation in the body to the extent that health and well being are adversely affected." It has been called a "State of excess adipose tissue mass" by Flier and Maratos-Flier.

Obesity has been associated with considerable morbidity and mortality the world over. It has been shown that as compared with the western population, **Asian Indians** are more susceptible to cardiovascular manifestations of obesity and its "Sister affliction," the Metabolic syndrome. It is possible that genetic factors, early adverse life events, and progressive life style changes, particularly with urbanization, contribute to this problem. More recently this was emphasized in a "Indo-US Healthcare Summit" in 2007, and the challenges and opportunities to curb the evolution of the problem were discussed critically, emphasizing the serious epidemic proportions of the problem in India. Alarmingly, obesity and hypertriglyceridemia with Insulin resistance seems to be surfacing at an early age amongst Indians irrespective of whether they live in India or abroad. Even uncomplicated obesity has been associated with cardiac dysfunction in Asian Indians.

A number of methods have been suggested for the **measurement and diagnosis of obesity,** some of which are given in **Table I.** Weight measurement may appear to be the most obvious but in many instances, the weight of lean muscular individuals may in fact be higher than the predicted. Measurement of body mass index (BMI) (Quetelet's Index) has since been used as measure of obesity because the index corresponds fairly well with body fat content. The index is however influenced by ethnicity, age and sex. Most surveys suggest that the BMI between 18.5 to 24.9 is desirable. If it ranges between 25 and 29.9, it qualifies the individual as "overweight." Epidemiological surveys have indicated that the incidence of metabolic disorders and other morbidities like cancer start to surface in people who have a BMI of 25 or more. Those with a BMI between 30 (obese i) and 35 (obese ii) are considered frankly obese, and beyond 40 (obese iii) obesity is said to have attained a morbid proportion. It has been suggested that waist circumference (WC) may express obesity related disorders better than the BMI alone. This may be combined with the hip circumference and expressed a ratio (W/H) which helps to identify the deposition of fat as central (visceral). Those women and men who have W/H ratio of >0.9, and >1 respectively fall into the category of obesity with central distribution The BMI may be combinedwith the waist to hip circumference ratio. Central obesity which means fat deposition in the abdominal tissues such as the mesentery and around the viscera, has been associated with a higher risk of metabolic and other over nutrition related diseases as compared with fat deposition in subcutaneous tissues inthe lower limbs and buttocks. The relationship between BMI and body fat percentage may be skewed in populations. For example, BMI in westerners may be higher than in some Asiatic populations, and yet their body fat % may be lower than the Asians. It is therefore difficult to fix a definite cut off point for diagnosis of obesity for a given population. Another factor that limits apportioning of a cut off factor is the difficulty in obtaining reliable body composition data.

At this point of time, a brief insight into the current understanding of factors which influence body feeding behaviour and hence weight gain or loss, is in place. The process is as yet an ill understood complex interaction of the brain-gut axis. The main player involved in the maintenance of energy expenditure and intake, **leptin,** is a secretion of the fat cells. It is chemically a cytokine, and a product of the adipocyte obesity gene. Receptors for leptin are present in the gastro intestinal tract (GIT) as well as the arcuate nucleus of the hypothalamus. Leptin crosses the blood brain barrier to generate two types of activities. 1. It occupies receptors on hypothalamic neurons which produce **Pro-Opio-Melano-Cortin** (POMC neurons). These neurons release **anorexigenic** substances such as α Melanocyte Stimulating Hormone (α MSH) and **Cocaine and Amphetamine related Transcript (CART).** These in turn may stimulate the release of Thyrotrophin Release Hormone and Corticotrophin Release Hormone (TRH & CRH) through second order hypothalamic neurons to increase energy expenditure and decrease appetite (the so called catabolic arm). At the same time that this activity is taking place, leptin may also act by reducing food intake by suppressing the feeding centre in the hypothalamic paraventricular nucleus. It inhibits the hypothalamic neurons which produce **Neuropeptide Y (NPY)** and **Agouti related protein (Ag RP)** which in turn are believed to activate 2nd order neurons in the hypothalamus responsible for the release of **Orexins A and B, and a melanin concentrating hormone (MCH).** These substances may excite the feeding centre. Leptin therefore is an important factor that fine tunes the balance between energy intake and energy expenditure on a long term basis. A slightly different version of the same hypothesis based on Konturek et al is illustrated in **Fig 1**. This mechanism is centred around the feeding promoting substances NPY and AgRP which may either be stimulated or inhibited. On feeding, neuro-endocrine cells in the GIT release hormones and peptides such as Cholecystokinin (CCK), Peptides YY (PYY) and Oxyntomodulin (OXN). These in turn may act via gut afferents or directly on the arcuate neurons of the hypothalamus to inhibit the release of NPY and AgRP. Distension of the upper GIT, particularly the stomach, also excites stretch receptors which in turn signal satiety. This too may happen through the action of CCK. In the fasting state on the other hand, the gut mucosa releases orexigenic substances such as Ghrelin, Orexins A and B and **cannabinoid receptor agonists** which stimulate the feeding centre by releasing NPY and AgRP. Ghrelin is probably associated with the release of growth hormone during the fasting phase. A balance therefore must be struck between the two sets of agents to regulate food intake. A disturbance in the equation is likely to result in obesity or weight loss.

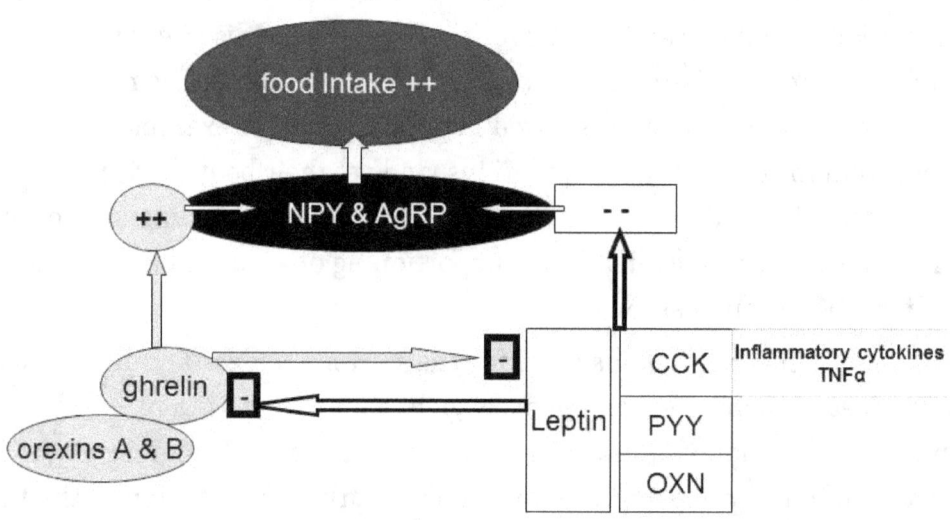

Fig. 1. Interactions of neuro-endocrine secretions for controlling food intake

Human beings have a predetermined number of fat cells at infancy, and when obesity occurs, these fat cells may either hypertrophy or undergo hyperplasia. It appears that adult obesity is more often associated with hypertrophy of fat cells. What then causes obesity? This is a multi-factorial problem. Role of genetic influences cannot be ignored. A genetic defect involving the leptin receptor causes severe obesity. More commonly, it is believed that a large group of genes acting in consort with various environmental factors is more likely to be involved. The interacting factors may be listed as: 1. Food intake and physical inactivity 2.Failure of satiety-causing mechanisms 3. Alterations in hormonal control of fat deposition. The role of leptin, the product of the obesity gene is the centre point of all these permutations, especially where long-term control of feeding behaviour is concerned. As mentioned above, there is a continuous adjustment in the decrease or increase of leptin, dictated by the fat content of adipocytes. For its action, leptin must be produced in adequate amounts, cross the blood brain barrier, and act on the associated hypothalamic neurons. A disruption anywhere in this circuitry may result in obesity. Suffice to say that this is at most a simplistic approach to the pathogenesis of obesity. In reality however, leptin and its interactions in the hypothalamus are extremely complex and not yet fully understood. There is some experimental evidence to support some of the possibilities. It has been found that blood levels of leptin are high in obese individuals. This is contrary to what is expected. The possible explanation is that the passage of leptin across the blood brain barrier is inhibited, probably because of defective transport mechanism across the endothelium in the barrier. In support of this hypothesis is the finding that in obese individuals, high blood leptin level is associated with low concentration of this hormone in the CSF. Peptide YY is a 36 amino acid peptide that is released rapidly with each meal, and is thought to inhibit feeding behaviour. It therefore plays a role in short term inhibition of energy intake. It has been found that in the obese, its activity is curtailed, and this may also contribute to the overall pathogenesis of obesity.

In summary, the pathogenesis of obesity is a complex combination of genetic factors of which leptin is a major contender in combination with multiple environmental factors. The human as a social animal is influenced by behavioural patterns which may govern his/her feeding behaviour. These may over rule the existent physiologic mechanisms which control food intake. Similarly, the response of the taste buds to a meal may overwhelm these mechanisms, as could stress. Injury may affect the satiety centre in the hypothalamus, and promote uncontrolled feeding behaviour. Lately, sleep deprivation has been linked to rapid weight gain. An interesting observation that the incidence of obesity in various populations was growing hand in hand with partial sleep deprivation led to a number of studies which have tried to correlate these relatively diverse life style events. The possible mechanisms suggested are 1. more awake time results in more calorie consumption. 2. partial sleep loss produces fatigue which in turn may reduce overall energy output. 3. sleep deprivation may increase appetite and with it caloric intake which is not dissipated enough by activity.

Pathogenesis of Obesity Induced Disorders

Obesity has been associated with a number of disorders **(Table II)**, of which the more important ones have been discussed below. The genesis of these disorders is complex.

Based on Redinger 2007 and Poirier et al 2006, **Fig 2** has been constructed as an attempt to explain the pathogenesis of what happens when excess fat accumulation occurs. Sympathetic activity is enhanced by excess fat accumulation. The mechanism by which this occurs is not well understood, but various possibilities have been Suggested.

Table II: Obesity related disorders

1. Cardiovascular

    i. Hypertension,

    ii. Ischemic heart disease

    iii. Congestive heart failure

    iv. Ischemic stroke

    v. Venous thrombosis

2. Respiratory

    i. Hypoventilation syndrome

    ii. Obstructive airway disease

3. Endocrine & Metabolic

    i. Insulin resistant Diabetes Mellitus & the Metabolic syndrome (Syndrome X) and Dyslipidemia

    ii. Disruption in reproductive physiology

4. Gall balder diseases; gall stones

5. Degenerative joint disease

6. Mobility disturbances

Some of these include baroreflex disturbance, excessive leptin level, and secretion of inflammatory cytokines. The resultant lipolysis produces large amounts of free fatty acids (FFAs) which in turn inhibit lipogenesis, promote insulin receptor dysfunction giving rise to hyper-insulinemia on the rebound. Rapid break down of lipids generates oxidative stress which in turn may be responsible enhanced secretion of inflammatory cytokines by the fat cells.

The brunt of the adverse effects of obesity is taken by the cardiovascular system. The effects, more often than not, occur as a part of the so called Syndrome X (Metabolic syndrome).The capillary density of fatty tissue is high, but not as high as some of the other tissues. The cells are placed in close proximity to the blood vessels which have a relatively high permeability. This relationship allows easy transfer of material to and from the adipose cells, and may assist these cells to function efficiently as endocrine cells. The vascularization promotes collection of tissue fluid which is not easily added to the circulating blood, and hence perhaps does not add significantly to an increase in the circulatory load of the myocardium. Even then, cardiac output is high in obese individuals. This has been attributed to the increased demand generated by the additional obese tissue, and partly to an increase in the lean body mass which occurs concomitantly with increase in fat deposition. It is the stroke volume that contributes to the increase in cardiac output. In the long run this adds to the work load of the myocardium.

Fat cells release a number of modulators which affect cardiovascular function. Renin, Angiotensinogen, and Angiotensin II seem to exist in fat cells. Angio II may be responsible for the hypertension seen commonly in the obese. In morbid obesity, cords of fat cells may get deposited amongst the myocardial cells and precipitate their degeneration. The propensity to thrombus formation is promoted by the inhibition of the plasminogen activator (**Fig 2**). This adds to the problems created by atherosclerotic lesions secondary to disregulation of Insulin activity. The pathophysiology may progress onto a variety of cardiovascular co-morbidities such as hypertension, ischemic heart disease, and heart failure.

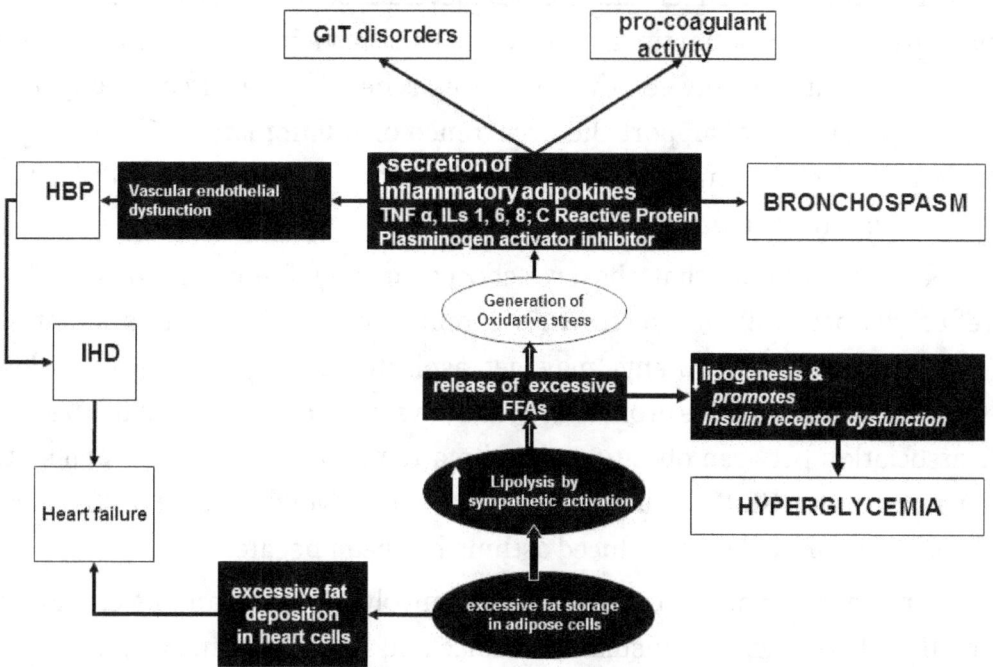

Fig.2. Likely pathogenesis of development of co-morbidities in obesity

Relatively less attention has been paid to the obesity induced changes in the respiratory system. The usual finding is a reduction in the functional residual capacity (FRC) without an alteration in the Residual Volume (RV). As a consequence, the expiratory reserve volume reduces. In some instances there is a reduction in the total lung capacity (TLC) and the vital capacity (VC), the former occurring only in morbid obesity. The probable reason is the mechanical disadvantage suffered by the chest wall because of the excess fat accumulation. If weight is reduced, the affected volumes tend to revert to baseline.

In some individuals with morbid obesity (BMI > 40kg/m^2), the mechanical load becomes so high that the FRC is reduced severely, particularly in the supine posture. Such individuals breath at low lung volumes and their small airways in the lung bases tend to remain closed. The overall pathophysiology is a disturbed ventilation perfusion ratio which translates into hypoxia, hypercapnia and polycythemia. **(*Explain the polycythemia?*)**. The persistent severe hypoxia may finally progress to pulmonary hypertension and right heart failure.

Evidence has been accumulating that obesity is associated obstruction to airflow as shown by a lower FEV1/FVC ratio and lower mid expiratory airflow (FEF$_{25-75\%}$) in young obese subjects. Asymptomatic obese individuals show a greater response to bronchodilator aerosol suggesting that sub clinical airway obstruction exists in them. The relationship is reversed with some loss of weight. It is now clear that obesity and asthma often co-exist and that the asthma follows the onset of obesity, there by suggesting that the pathogenesis of such asthma is triggered off by the obesity. The exact mechanism by which this happens is as yet undetermined. Various possibilities have been considered. Chronic airway inflammation may reduce airway diameter in response to the release of cytokines such asIL6 and TNFα by the fat cells. It has also been postulated that some of these inflammatory markers may be released by macrophages which lie amongst the fat cells. To their credit, the adipocytes also synthesize and release anti-inflammatory adipokines such as leptin, adiponectin, resistin, visfatin and IL10 to name a few. It is possible that an imbalance between the two varieties decides the outcome. Experimental evidence has been generated to support the occurrence of inflammation. Genetically leptin deficient rats are grossly obese. Such animals have been shown to have a heightened airway sensitivity to methacholine, and ozone.

There have been suggestions that the bronchospasm may follow the onset of gastro-esophageal reflux disease (GERD) which often complicates obesity. In this situation, the regurgitated gastro-esophageal contents may get aspirated in to the airway and set up inflammatory changes which induce bronchospasm. However, more recently it has been shown that the close association between obesity and asthma remained unaffected when statistical adjustments were made for GERD. In a similar manner the involvement of obstructive sleep apnea as a causative factor of obesity induced asthma has been negated.

Metabolic syndrome is a complex condition which involves obesity and its cardiovascular, endocrine (type II Diabetes mellitus: Insulin resistance and hyper Insulinemia) and metabolic (dyslipidemia) consequences accompanied by a consistent pro-inflammatory state. The DM follows the development of Insulin resistance. The exact mechanism which is responsible for the Insulin resistance is not fully comprehended. Adiponectin and resistin, two of the adipokines produced by fat cells have been linked to this phenomenon. Adiponectin expression by fat cells may be lower in the obese, and this may be responsible the Insulin resistance. Similarly, high leptin and TNFα levels may reduce tyrosine kinase activity in insulin receptors and decrease the efficacy of the signal transduction by the receptor, thus rendering the receptor ineffective. It has been suggested that macro and micro nutrient status during fetal life may be linked to the development of obesity and insulin resistance, and may be classified as one of early life adverse effects. The excessive breakdown of fats because of an increase in sympathetic drive **(Fig 2)** increases the free fatty acid (FFA) levels and inhibits lipoprotein lipase activity. The latter reduces the reformation of triglycerides in the fat cells, and allows the circulation of excess amounts of the FFAs. These enter the liver where they are redirected to form increased amounts of VLDL, while the HDL formation is relatively reduced. The VLDL then is responsible for the atherosclerosis.

Obesity has been associated with dysfunction of the reproductive system in both males and females. Males who are obese (central obesity in most instances) are known to have low sperm counts, poor semen quality, infertility and erectile dysfunction. Testosterone levels are low. A number of possibilities may explain the phenomenon. Adiponectin is reduced in the obese and may play a role. High leptin levels may interfere with the hypothalamic hypophyseal axis. Greater than normal fat content is associated with higher concentration of the enzyme aromatase which converts androgens to estrogen. The latter may act on the hypothalamic-pituitary axis to reduce circulating levels of gonadotropins. If obese males are given an aromatase inhibitor, their testosterone level increases. The inflammatory cytokines released by the fat cells may affect the endothelial function (release of NO) in erectile tissue and be responsible for the erectile dysfunction.

The Polycystic Ovarian Syndrome (PCOS) best describes the close association between obesity and female infertility. Abdominal obesity with Insulin resistance is found in about 50% of the patients. Hyper androgenism and poor or absent ovulation are the other consistent features. As compared with non obese women with PCOS, the obese patients have a severer form of hyper androgenism. Insulin acts on the ovarian thecal cells to produce excessive amounts of androgens. This may result in the greater degree of androgen secretion in obese PCOS patients. Obesity is also linked to a reduced synthesis of Sex Hormone Binding Globulin (SHBG) by the liver, thus making available more free sex hormones, particularly androgens which in turn are responsible for the hirsutism seen in these patients. The hypothalamic hypophyseal axis may be interfered with in obese PCOS patients to a greater extent than the non obese patients with the same condition. This in turn may be responsible for the more severe forms of ovulatory and menstrual abnormalities seen in obese PCOS patients.

General Introduction to Pathophysiology of Weight Loss

A loss in weight by > 10% of the normal body weight in a person is categorized as weight loss. There are a variety of causes which lead to this problem **(Fig.3)**. In this discussion, pathophysiologic mechanisms of only the more relevant causes of weight loss will be discussed, with emphasis on cachexia

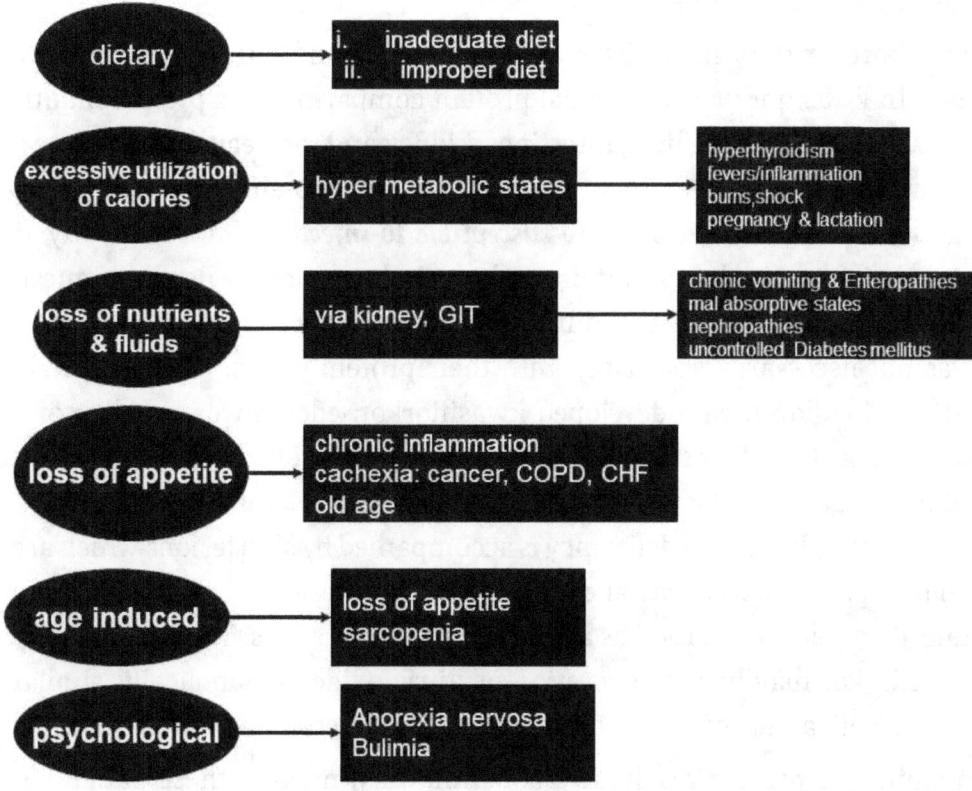

Fig. 3. Causes of weight loss

Protein Energy Malnutrition (PEM) / Protein Calorie Malnutrition (PCM)

The specter of famine and in its wake, starvation, has loomed large over the Indian sub continent for centuries. The 20th century wound its way through the Bengal famine of 1943, the Bihar starvation of December 1966, and the severe drought in Maharashtra during 1970–1973. Since then however, famine as a cause of severe malnutrition on a mass scale has been successfully evaded. Notwithstanding, recurrent natural disasters, the large population of people living under the poverty line, and ignorance will continue to nurture instances of severe malnutrition (under nutrition) and starvation for years to come, particularly in children under 5 years of age where PEM continues to be a major cause of morbidity and mortality in the developing and third world countries. Pathophysiology of starvation and under nutrition therefore has a greater relevance in the Indian context than the pathophysiology of obesity. In the scheme illustrated in Fig 3, this falls under dietary cause of weight loss.

Mal (under) nutrition is classified as Primary when the essential ingredients of the diet such as the proximate principles (proteins, fats and carbohydrates) are deficient along with the vitamins. This type of under nutrition is common in the Indian subcontinent and is mainly responsible for the PEM. In Secondary malnutrition, there is an adequate supply of the essential nutrients but these cannot be assimilated because of a number of reasons such as increased demands, gastrointestinal disorders etc (Fig. 3). As the name suggests, PEM pertains to deficiency of both proteins and caloric intake. If the former predominates, gross manifestations of protein deficiency are seen (Kwashiorkor). At the other end of the spectrum is marasmus caused mainly by a deficiency of calories. More commonly, there is component of both in a typical presentation of PEM.

Proteins are stored in the viscera (Visceral compartment) and in the skeletal muscles (somatic compartment). In kwashiorkor, the visceral protein compartment is predominantly involved. The organs most affected are the liver, intestines, kidney and the heart. More severe the protein loss, more complex the PEM. Edema is a typical feature of protein loss (*Explain why, and also why patients with low serum proteins are susceptible to infection*). More recently, the explicit role of protein deficiency in the causation of edema in kwashiorkor has been questioned. The doubts arose when it was reported that during world war II, nutritional edema seen in German prisoners was not necessarily associated with their protein status. It was later realized that in famine affected children who developed kwashiorkor, edema was rapidly corrected with relatively low protein diets. This suggested that apart from protein deficiency, some other factor was involved in causation of edema. Later studies have suggested that oxidative stress is likely to be involved. Interestingly, protein deficiency is accompanied by skin lesions which are similar to those seen in pellagra. However critical evaluation of these lesions showed that they resembled effects produced by solar irradiation as a result of oxidative stress. The source of the oxidative stress in kwashiorkor may be the generation of nitric oxide. Metabolically, similar situation exists in patients with adult respiratory distress syndrome, shock and multi organ failure.

Non availability of protein results in poor immunity, making these patients susceptible to severe infection. Fatty liver is a common accompaniment of kwashiorkor, and has been attributed to deficiency of synthesis of carrier proteins by the liver. More recently, free radical induced damage has been considered. Fatty liver is not commonly found in patients with predominance of energy deficit (marasmus). The myocardial contractile function is compromised in severe protein deficiency state because of degenerative changes in myofibrils. Other tissue cells (the Body Cell Mass) will also break down eventually in order to provide the protein for gluconeogenesis. Systolic blood pressure and cardiac output are reduced. The cardiac dysfunction also contributes to the development of generalized edema. Protein starvation leads to under proliferation of the bone marrow and thus to anemia. In severe PEM, erythropoietin is reduced. In the initial stages, the haemoglobin concentration may not be obviously low (**Why?**). Stunted growth of the body as also the brain reflects the low protein assimilation over a period of time.

Marasmus at the other end of the spectrum, is mainly a result of deficient caloric intake. In this too, protein depletion occurs because protein breakdown is used to generate amino acids which win turn will be converted in to energy by gluconeogenesis. The skeletal muscles (somatic protein compartment) are broken down to make this provision, and hence the plasma albumin level may diminish only slightly if at all. In addition, fuel for energy generation is provided by rapid fat break down. Thus a marasmus patient typically looks emaciated. Metabolically, PEM secondary to starvation is a hypo metabolic state.

Acute short term starvation may arise as a result of sudden deprivation of food when a person is lost in a hostile environment, during religious fasting and other such situations. In a normal person, about 20% of energy is available as fat, 9–10% as protein and <1% as carbohydrate, mainly as glycogen in the liver. Most of the energy under the circumstances is derived form the liver glycogen and break down of about 75 gm of protein. If the food deprivation exceeds about 24 hours, fat break down begins and FFAs are released in order to provide the energy. In an obese person, sustenance can go on for a few months while energy reserves in a normally built person are likely to be used up in about 2 months. (*What simple investigation would you do to find out whether a person is using fats to sustain energy needs?*).

In contrast with acute starvation/fasting which is a hypometabolic state, some situations demand availability of increased energy (hypermetabolic states). If this additional energy input is not provided, rapid muscle break down and gluconeogenesis are encouraged by high circulating levels of various stress hormones and cytokines such as IL1 and 6 and TNFα. which are released by the oxidative stress secondary to the stressful condition. Typically, hyperthyroidism, burns, acute trauma and septic shock fit in to this category (Fig.3).

Cachexia literally means "bad condition." Hippocrates had called it an illness in which "the shoulders, clavicles and thighs melt away." He had further declared that it was a fatal illness. When chronic disease is complicated with cachexia, there is increased mortality. The obvious finding is severe wasting of body tissues, particularly muscle.

A number of chronic diseased conditions are associated with this syndrome. These include cancer, congestive heart failure (CHF), chronic obstructive pulmonary disease (COPD) and AIDS. The most probable reason for development of cachexia is the release of various inflammatory cytokines prominent amongst which are ILs 1,2,6, TNFα and C reactive proteins. Interferon γ is also raised. The cytokines may act in various ways. One such is the activation of the ubiquitin-proteasome mechanism which is involved in rapid breakdown of muscle protein. The activity of the ubiquitin pathway is enhanced by the increased cortisol level which occurs in stressful situations. The inflammatory cytokines breakdown fats to increase FFA concentration and decrease the activity of the lipoprotein lipase in the liver as well as the fat cells. The FFAs accelerate the secretion of more cytokines. Behavioural changes such as feeling of general ill health, fatigue and anhedonia (inability to appreciate pleasure) have also been attributed to these cytokines. Apart from the cytokines, some other mediators are also recognized as participants in the development of cachexia. Testosterone stimulates myoblasts and is responsible for the proliferation of satellite cells which are the 'stem cells" for striated muscle. It is an inhibitor of inflammatory cytokine release while acting as a promoter of IL 10 which is an anti-inflammatory substance. In elderly males, with a decrease of testosterone levels, these beneficial effects are no longer available, and the process of muscle breakdown is hastened. In the presence of illness, the effect is additive. Insulin like Growth factor I may also be involved. Its concentration is reduced in patients who have a poor nourishment status. It is one of the factors that may be required for muscle protein synthesis.

Even if the basic pathophysiology of cachexia in the inflammatory diseases mentioned above is some what common to all, there are a few subtle characteristics of some of these diseases. TNFα is found in CHF patients having cachexia. One benefit of using Angiotensin II antagonists is that blood levels of various inflammatory cytokines decrease. Surprisingly however, therapeutic use of TNF α antagonist does not seem to improve the cachexia in these patients. Severe weight loss in COPD patients spells poor ability to thrive. The muscle weakness reflected as poor diaphragmatic function and respiratory muscle weakness which contribute to lowered pulmonary ventilation. Inflammatory cytokine antagonists seem to improve weight and exercise tolerance in such patients, something that is not seen in CHF patients. It is difficult to explain this difference.

Cachexia of cancer is the archetypal severe loss in weight. As such, chemotherapy for cancer is notorious for inducing severe anorexia through its actions on the GIT. Added to this, the secretion of cytokines by the tumor cells sets in motion an array of events which produce the cachexia. Peripherally, these substances induce muscle breakdown via the ubiquitin-proteasome pathway, increase the secretion of cortisol, itself a potent proteolytic agent which acts by activating the ubiquitin pathway. Centrally, the inflammatory stress is responsible for increased secretion of the corticotrophin release hormone (CRH) which, apart from stimulating the release of cortisol via the hypothalamic-adrenal connection, is also responsible for directly inhibiting appetite. Added to this is its action in the hypothalamus which reduces the secretion of NPY, a potent orexigenic substance. Tryptophan release by cancer cells may be responsible for the increase in 5 hydroxy tryptamine (Serotonin) and CRH release. Cachexia of cancer can be partly controlled by using cytokine antagonists, and Cannabinoid receptor stimulants.

Cachexia of renal failure is complex and multifactorial. It is usually a combination of reduced energy intake because of severe loss of appetite combined with protein breakdown caused by high levels of circulating cytokines. Zinc deficiency may be involved.

Severe cachexia is seen in patients of AIDS, and is associated with high mortality in these patients. The inflammatory cytokines are the main culprits. Growth hormone and cytokine inhibitors may help in slowing down the process.

Age related loss of weight includes muscle wasting as one of the causes (please refer to the section on Sarcopenia in chapter 9). Here again, inflammatory cytokines are the main players involved. Development of overall muscle weakness may set up a vicious cycle of more inactivity because of the weakness which in turn prevents the elderly from moving around in order to obtain adequate nourishment, especially under a situation where they may be on their own. Apart from sarcopenia, other factors may contribute to age associated disregulation of inflammatory cytokines and its subsequent outcome, weight loss, which may at times progress to cachexia. Various suggestions have been offered to try and explain this. Age related hyposecretion/absence of sex hormones may allow the normally secreted cytokines take an upper hand. Susceptibility to chronic pathology which may be sub clinical, may have a role to play in enhanced secretion of inflammatory cytokines. Age related anorexia could occur because of early satiety in response to increased leptin and CCK levels which act to reduce secretion of NPP and AgRP, both orexigenic in nature (Fig.1). It has been reported that relaxation of the stomach in response to feeding does not occur effectively in the elderly, and brings on early satiety. Depression which is often an accompaniment of senescence may by itself prevent the elderly from obtaining adequate intake of energy and hasten weight loss.

Cachexia, more often in young women, may occur secondary to Anorexia Nervosa. There is deliberate attempt to lose weight because of an overwhelming desire to become thin. The severe dieting is often practiced in combination with heavy exercise. The dual stress rapidly induces weight loss. One of the common findings is amenorrhea with a reduction in estrogen. The latter may be responsible for generation of inflammatory cytokines which then induce various mechanisms which further reduce energy intake. Deaths have been known to occur consequent to myocardial degeneration. Bulimia, again more commonly seen in young women, is binge eating followed by induced vomiting and or purging in some while others resort to follow binge eating by severe starvation. The repeated induced vomiting may be responsible for esophagitis, esophageal strictures. Such a pathology because of the symptoms of pain and epigastric discomfort restrict energy intake and precipitate rapid weight loss. The weight loss though does not match that usually seen in patients with Anorexia nervosa.

In conclusion, malnutrition, whether a consequence ofover nutrition or under nutrition, is responsible for varied pathophysiology. The pathophysiology of either situations is complex. In India, paradoxically, both types of nutritional disorders are prevalent, and need tackling on a war footing.

References

1. Barba C, Cavalli-Sforza T, Cutter J et al. WHO Expert consultation. Lancet. 2004; 363: 157–163

2. Chada DS, Gupta N, Goel K, Pandey RM, Kondal D, Ganjoo RK, Misra A. Impact of obesity on the left ventricular functions and morphology of Asian Indians. Metab.Syndr. Relat. Disord. 2009: 7: 151–158.

3. De Koning L, Merchant AT, Pogue J, Anand SS. Waist circumference and waist to hip ratio as predictors of cardiovascular events: meta-regression analysis of prospective studies. Eurpoean Heart J. 2007; 28: 850–856.

4. Deurenberg P, Deurenberg-Yap M, Gurruci S. Asians are different from Caucasians and from each other in their body mass index/body fat percentage relationship. Obesity Reviews. 2002; 3: 141–146.

5. Deurenberg P, Deurenberg-Yap M, van Staveren W. Body mass index and percent body fat: a meta analysis among different ethnic groups. Int. J. Obes. Relat. Metab. Disord. 1998; 22: 1164–1171.

6. Enas E, Singh V, Munjal YP, Bhandari S, Yadav RD, Manchanda SC. Reducing the burden of coronary artery disease in India: challenges and opportunities. Indian heart J. 2008; 60: 161–175.

7. Flier Js, Maratos-Flier E. Obesity. Chapter 64 in "Harrison's Principles of Internal Medicine." Edited. Kasper DL, BraunwaldE, Fauci AS, Hauser SL, Longo DL, Jameson JL.16th ed. MCGraw Hill, London. 2005; pp 422–429.

8. Gambineri A, Pelusi C, Vicenatti V, Pagotto U, Pasquali R. Obesity and the polycystic ovary syndrome. Intern. J. Obesity 2002; 26: 883–896.

9. Golden MH. The development of concepts of malnutrition. J. Nutr. 2002; 132: 2117S-2122S.

10. Halsted C. Malnutrition and nutritional assessment. Chap. 62 in Harrison's Principles of Internal Medicine." Edited. Kasper DL, BraunwaldE, Fauci AS, Hauser SL, Longo DL, Jameson JL.16th ed. MCGraw Hill, London. 2005; pp 411–415.

11. Hammoud A, Gibson M, Matthew Peterson C, Hamilton B, Carrel D. Obesity and male reproductive potential. J. Andrology. 2006; 27: 619–636.

12. Janssen I, Katzmarzyk P, Ross R. Waist circumference and not body mass index explains obesity related health risk. Am. J. Clin. Nutr. 2004; 79: 379–384.

13. Kane AB, Vinay Kumar. Environmental and nutritional pathology. Chap. 9 in Robbins and Cotran. Pathologic Basis of Disease. Edit. Vinay Kumar, Abbas AK, Fausto N. 7th edn. Elsevier, China; 2005. pp 461–468

14. Kerndt P, Naughton J, Driscoll C, Loxterkamp D. Fasting: History, pathophysiology and complications. West. J. Med. 1982; 137: 379–399.

15. Konturek SJ, Konturek JW, Pawlik T, Brzozowki T. Brain-gut axis and its role in the control of food intake. J Physiol & Pharmacol. 2004; 55: 137–154.

16. Madhavan A, Beena Kumari R, Sanal MG. A pilot study on the usefulness of body mass index and waist hip ratio as a predictive tool for gestational diabetes in Asian Indians. Gynaecol. Endo. 2008; 24: 701–707.

17. McPhee SJ, Hammer GD. Physiology of body weight control in Pathophysiology of Disease: An Introduction to Clinical Medicine. 6th Edn. McGraw Hill, New Delhi; 2010; pp 531–533.

18. Misra A, Vikram NK. Insulin resistance syndrome (metabolic syndrome) and obesity in Asian Indians: Evidence and implications. J Nutrition. 2004; 20: 482–491.

19. Morley J, Thomas D, Wilson M_M. Cachexia: pathophysiology and clinical relevance. Am. J. Clin. Nutr. 2006; 83: 735–743.

20. Nestel P, Lyu R, Low LP, Sheu W H, Nitiyanant W, Saito I, Tan CE. Metabolic syndrome: recent prevalence in East and Soth Asian populations. Asia Pac. J. Cin. Nutr. 2007; 16: 362–367.

21. Patel S, Hu F. Short sleep duration and weight gain: a systematic review. Obersity. 2008; 16: 643–653.

22. Poirier P, Giles TD, Bray GA, Hong Y et al. Obesity and cardiovascular disease: Pathophysiology, evaluation and effect of weight loss. Circulation. 2006; 113: 898–918.

23. Porth C. Alterations in body nutrition. Chapt. 8 in Essentials of Pathophysiology: concepts of altered health states 2nd edn. Lippincot Willaims and Wilkins. London. 2007: pp 165–178.

24. Reaven G. Metabolic syndrome: Pathopysiology and implications for management of cardiovascular disease. Circulation. 2002; 106: 286–288.

25. Redinger R. The pathophysiology of obesity and its clinical manifestation. Gastroenterol Hepatol. 2007; 11: 856–863

26. Shore S, Johnston RA. Obesity and asthma. Pharmacology and Therapeutics. 2006; 110: 83–102.

27. Straznicky NE, Eikelis N, Lambert E, Esler MD. Mediators of sympathetic activation in metabolic syndrome obesity. Current Hypertension Reports. 2008; 10: 440–447.

28. Sugerman HJ. Pathophysiology ofobesity comorbidity: The effects of chronically increased intra abdominal pressure. In Minimally Invasive Bariatric Surgery, Ed. Schauer P, Schirman BD, Brethauser S. Springer Verlag New York. 2007. pp 1–6.

29. Van Cauter E, Knutson K. Sleep and the epidemic of obesity in children and adults. Europ. J. Endocrinilogy. 2008; 159; suppl. 1:S 59-S 66

30. Watson RA, Pride NB. Postural changes in lung volumes and respiratory resistance in subjects with obesity. J. Applied Physiol. 2005; 98: 512–517.

31. World Healthy Organization. Obesity: Preventing and managing the global epidemic. WHO Obesity Technical Reports Series no. 894. WHO Geneva, 2000.

CHAPTER - 9

Pathophysiology of Musculoskeletal Disorders

Plan

1. General introduction
2. Structure and function of the striated muscle
3. Effects of aging on the skeletal muscle
4. Muscle fatigue
5. Hypertrophy, atrophy and injury
6. Fibrillation and fasciculation
7. Rigor mortis
8. Pathophysiology of theneuromuscular junction
    i. physiology
    ii. disturbances in transmission of the impulse across the NM junction
    a. Myasthenia gravis
    b. NM blockade in general anesthesia
    c. Nerve gases
8. Physiology of bone formation and remodeling
9. Pathophysiology of osteoporosis
10. Bone injury: pathophysiology of fractures and physiology of bone repair

General Introduction

The musculo-skeletal system constitutes a large bulk of the body mass, and determines the extent, direction and efficiency of body locomotion. Each of the players involved, the bulk of the striated muscle and the skeleton, has an anatomical and physiological identity of its own. Yet, to accomplish their allotted function they must work in unison. Therefore if either one of the two malfunctions, the other is affected adversely. For example fracture of a limb bone may lead to wasting of the muscle attached to it.

The student may at this point in time review the structure of the striated muscle. The sarcomere is the functional unit of the muscle fibre, and contains the contractile proteins. Myosin constitutes the bulk of these proteins (about 60%). At any instant, a number of contacts between the myosin heads and the actin filaments are taking place. This generates the so called resting tone in the muscles. When a nerve impulse is fired, the number of myosin heads and actin filaments that are brought into play (sliding filament theory) increases dramatically, giving rise to a shortening When a series of nerve impulses in rapid succession encroach on the muscle fibre, each new contraction occurs before the previous contraction is over. This is because additional contractile elements are brought in to play. A stage will come when at a given frequency, each contraction merges into the next one without any relaxation at all. This is the "tetanus" At each contraction, a certain number of motor units are recruited *(What is a motor unit?)*. More the power required, more the number of motor units brought into play.

Functionally, there are two types of muscle fibres-type I and type II. The type I are fatigue resistant fibres with low power output, but are capable of generating sustained force eg in the maintenance of posture. They are rich in mitochondria, and have plenty of enzymes that can carry out oxidative metabolism, and have a dense capillary network. They are also called the slow red fibres. As against these, the type II variety are large, fast acting, with relatively less number of mitochondria. They are capable of carrying out anaerobic metabolism. They fatigue earlier than the type I variety, and have plenty of glycolytic enzymes for quick supply of energy. They constitute the fast white fibres. Different muscle have different distribution of these fibres types. The gastrocnemieus is a muscle that contains both types of fibres, the ocular muscles classically contain fast fibres, while the soleus is the slow fibre type of muscle. *(What type of fibres would most suit i. a sprinter; ii. a soccer player; and iii. a long distance runner?)* The muscle type is determined by the innervation it receives at birth. It has been shown that when a muscle having predominantly slow fibres is transplanted with a nerve which delivers rapid impulses, it develops properties of a fast muscle. The opposite is also true.

More recently it has been shown that skeletal muscles can express both the α and β estrogen receptors (ER) in both the sexes. Men who have undergone exercise training have the propensity to develop more ER, particularly in their capillaries, along with the Vascular Endothelial Growth Factor. In fact a number of activities such as muscle angiogenesis, reduced atheroma formation, peripheral vascular dilatation, increase in muscle cell mass and strength, and some beneficial metabolic readjustments that occur with exercise training have been linked to the effects of estrogen.

There are gender differences in the striated muscles. In males, the power generation and strength are more than in the female. However, in the females, the fatigability is less and recovery after exercise is quicker. It is possible that for a similar percentage output of power, the male muscles contract more strongly resulting in more occlusion of blood flow as compared with the female muscles. This may induce intermittent ischemia with accumulation of metabolites which is of a greater degree than in the females There is also a surmise that females have a greater proportion of the red fibres in their muscles as compared with men.

Muscles are specialized cells and if damaged can not regenerate on their own. Cells known as "satellite cells" which are a type of myoblasts have been recently identified as those which can help in regenerating damaged muscle fibres. These cells are found on the edge of the muscle fibre, between the sarcolemma and the basal lamina. It is possible that the increase in muscle growth induced by steroids is attributable to satellite cell proliferation. Estrogens may also be responsible for activity of these cells. This may take place via the action of Insulin Like Growth Factor I. A combination of these activities may be responsible for the protective influence estrogen may have on preventing exercise induced damage to skeletal muscles.

Effects of Aging on the Skeletal Muscle

Age related loss in muscle is called "Sarcopenia." After the age of about 50 years, this is likely to happen at a rate of 1–2% per year. With a steady increase in life expectancy world wide, this problem is likely to occupy a prominent niche in the management of the elderly. In the US of A, about US$ 20 billion were spent on tackling this issue. The mechanism which causes sarcopenia is as yet ill defined. Some of the contributory factors include neuronal damage, hormonal changes with senescence, chronic inflammatory states, and poor nutrition combined with poor physical activity.

Growth Hormone, Insulin and Insulin Like growth factor I and testosterone are anabolic in function. With increasing age, the functional effects of these hormones is on the decline. At the same time, there is an increase in cortisol level and various cytokines which are catabolic in nature. The synthesis of proteins may be reduced because of activity of free O_2 radicals which damage muscle protein forming DNA. There is a simultaneous decline in the levels of anti-oxidant enzymes. Apoptosis is the frontrunner amongst the possible causative mechanisms. There is mounting evidence that individual nuclei in the multinuclear skeletal muscle cell may be undergo a phenomenon like apoptosis. The possible interaction of the various factors is depicted in **Fig. 1**. (the algorithm is constructed based on Proctor et al 1998, Roubenoff 1999, Marcell 2003). The clinical manifestations are easily explained. There is a reduction in muscle strength with time, generally starting around the 5th and the 6th decades. Typically, the walking speed reduces, and more effort is required to carry out daily chores. With advancing age, inability to maintain balance may supervene, resulting in falls and injuries. The immobilization further exaggerates muscle disuse. Sarcopenia is to be distinguished from muscle atrophy which is an adaptive mechanism to nerve damage, ischemia, disuse etc. The latter is likely to occur at any age. Some of the affected fibres degenerate while some may remain normal while some may even hypertrophy in response to the physiotherapy offered. Most importantly, in certain situations, atrophy may be reversible while sarcopenia, at best may only be slowed down with physical training and other supplementary methods. To date there is no evidence that the phenomenon is reversible. It has been observed that if the starting muscle mass is high, effects of sarcopenia are less evident.

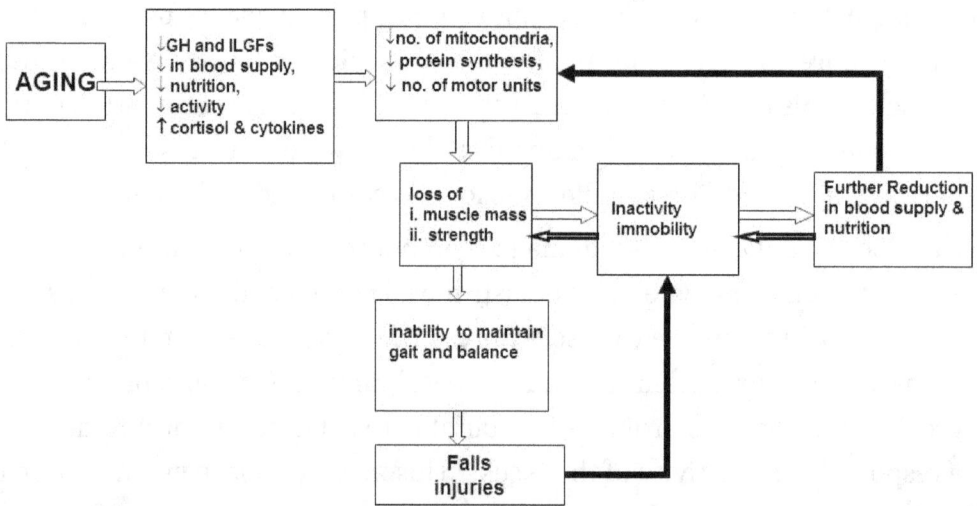

Fig. 1: Sarcopenia and its effects- proposed algorithm for a vicious cycle

Fatigue of Skeletal Muscles

The ability of muscles to contract effectively declines after heavy exercise. This is muscle fatigue. Traditionally, fatigue has been attributed to i. accumulation of metabolites (lactic acid); ii. depletion of energy sources (ATP); and iii. exhaustion of acetylcholine at the nerve terminals. A mechanism of "central fatigue" has also been suggested, and could be attributed to the excessive afferent input from the working muscles into the brain. Increase in body temperature which occurs with severe exercise may be relevant to the occurrence of fatigue.

Universally, the increase in lactic acid which occurs during strenuous exercise has been blamed for the onset of fatigue. This dissociates into H^+ and lactate. It is the rapid increase in the intracellular H ions that has been considered to be the main culprit. The acidosis may hinder generation of energy molecules (ATP), reduce the availability of Ca^{++} for excitation/contraction coupling by slowing/reducing its release from the sarcoplasmic reticulum. The reduction in the pH of the extra-cellular fluid may excite muscle afferents which signal the sensation of discomfort. More recently it has been postulated that there is a rapid breakdown of creatine phosphate during high intensity exercise, increasing the levels of inorganic phosphate. This may be responsible for the fatigue by acting via intracellular mechanisms. It has also been suggested that high concentrations of lactic acid may partially block the neuromuscular junction and cause fatigue.

Depletion of acetylcholine (Ach) at the neuromuscular junction though a plausible reason for onset of fatigue does not usually occur under physiological conditions until and unless the rate of exercise is so intense that the extremely high frequency of nerve impulses exhausts the acetylcholine at the nerve terminals.

Glycogen depletion because of high intensity exercise (>75% of maximum exercise capacity) or prolonged exercise of lower intensity causes fatigue. Carbohydrate loading before undertaking an athletic event has been shown to delay onset of fatigue and improve muscular performance.

An interesting presentation of fatigue is the Post Polio Syndrome. A person who has recovered from Polio Myelitis, develops muscle fatigue many years later. Recovery after polio has been associated with sprouting of axons from some of the motor neurons which may have survived the disease. These axons innervate muscle fibres to form a new cluster of motor units. After a few years, these motor units may denervate, causing a loss in the number of functional units, giving rise to fatigue. The exact mechanism of this pathophysiology is unclear. Various factors such as vascular lesions, immune mechanisms precipitated by the original disease, or even reactivation of the virus have been considered. Fortunately Polio Myelitis has since been eradicated from our country.

Muscle Hypertrophy and Atrophy

Excessive work load leads to an increase in size of skeletal muscle cells, not their number. The protein synthesis in muscles increases during hypertrophy. As understood mostly from studies on heart muscle, this is brought about by release of various growth factors. ***(What do you think should be the cellular changes in a hypertrophied muscle?).***

When underused, or not used at all, muscles undergo atrophy. The commonest cause is immobilization of a limb because of injury. Such atrophy is easily reversible. Similarly ischemia of a muscle or application of continuous pressure may result in atrophy, the latter by producing ischemia. However if the muscle is deprived of its nerve supply, the atrophy is permanent. In such a situation, it may be delayed or its extent may be reduced by physiotherapy and electrical stimulation of muscle. Some hormones may induce muscular atrophy. Thyroid hormone when secreted in excess, breaks down muscle to obtain protein for conversion into energy. Lack of Insulin results in muscle wasting for the same basic reason. Glucocorticoid hormones break down protein and cause atrophy. It is for this reason that muscle weakness and atrophy are clinically evident in Hyperthyroidism, Type I Diabetes Mellitus and Cushing's Disease. At the cellular level, the ubiquitin-proteasome pathway may be involved in Hyperthyroidism and Cushing's Disease. Muscle wasting that occurs in patients with heart failure and cancer is probably brought about by the secretion of TNFα.

Muscle Injury

Muscle injury (Strain) which may be mild (Grade I), moderate (Grade II) or severe (Grade III) occurs most commonly during exercise because of sudden excessive stretch. Histological appearance is seen as torn sarcolemma with damaged myofibrils. The injury results in the appearance of high levels of circulating marker enzymes such as Creatine Kinase (CPKMB is the typical isoenzyme), lactate dehydrogenase, and myoglobin. This is followed by an inflammatory response involving various substances namely 5HT, histamine, bradykinin and prostaglandins. These are responsible for the swelling and pain of injury. **(How?).** After 3–4 days of the acute inflammatory stage, repair mechanisms are brought into play. There is increased fibroblast activity and capillary growth. The fibroblasts lay down collagen which is suitably remodeled by controlled physiotherapy as healing occurs. The sequence may take up to 3–4 weeks before function is restored. Sometimes, the microtrauma may not be severe enough to cause acute strain,. The pathologic effects surface after about 24–48 hours as delayed onset of muscular soreness (DOMS) and may persist for about 3–4 days. At the cellular level, the frank disruption of muscle cells is absent, but the micro trauma is enough to initiate local inflammatory reaction which explains the relative delay in the onset of symptoms in this condition.

Muscle Fibrillation and Fasciculation

Injury to the motor nerve supplying a muscle will result in paresis/paralysis of that muscle and later atrophy. At the same time, the muscle becomes overtly sensitive to whatever Ach that may be available locally. The result is fine contractions of individual muscle fibres which cannot be seen and are called fibrillation. Physiologically these are excitatory post synaptic potentials, enough to cause local disturbances which may be picked up on an EMG but not enough to summate to generate a full fledged action potential. The probable reason is an up regulation of the Ach receptors on the atrophying muscle. These are typically seen in lower motor neuron lesions. Fasciculations on the other hand are visible jerky contractions which are a result of abnormal discharge of the spinal motor neurons.

Rigor Mortis

ATP is required not only for the excitation contraction coupling in a muscle, but also for separating the myosin heads from the actin filaments when relaxation occurs. When the muscle is devoid of ATP, this process cannot occur and the muscle goes in to a state of contracture. In the same context, when death occurs, ATP is no longer available to undo the prevalent actin-myosin head combination which then surfaces as developing contracture of various groups of muscles. This is rigor mortis. In Forensic Medicine and Medical Jurisprudence, the phenomenon is helpful in determining the time of death. The post mortem contracture disappears after a few hours when degeneration of muscle proteins sets in.

Pathophysiology of the Neuromuscular Junction

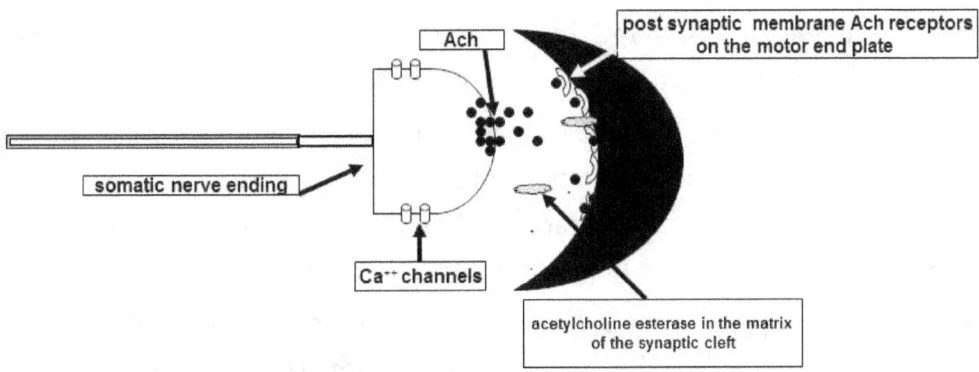

Fig. 2. Levels at which functioning of the NM junction may be interfered with

The student may recall the normal anatomy and physiology of the neuromuscular junction on the striated muscle. The functioning of the NM junction may be interfered with at various levels **(Fig.2)**. If the nerve is sectioned, the impulse will not arrive at the junction at all. The permeability of the presynaptic membrane to Calcium increases with the arrival of the nerve impulse, and Ca^{++} enters through its channels at the tip of the nerve terminal to facilitate presynaptic release of Ach. Antibodies to these channel proteins may interfere with Ca^{++} uptake. Because of this, the release of the neurotransmitter ligand at the myoneural junction is affected. The condition is called the Lambert-Eaton Myasthenia Syndrome (LEMS), and presents in a fashion similar to that of the classical NM junction disorder-Myasthenia Gravis (MG). Contrary to the latter, LEMS patients show improved response with repetitive nerve stimulation as against the decremental response classically seen in patients of MG *(How do you explain this?)*. Also, LEMS is associated with a variety of autonomic dysfunctions.

Acetylcholine receptors which are located on the motor end-plate are nicotinic in nature. In the striated muscle this receptor consists of four subunits: α, β, γ and δ. Of these, $\alpha\gamma$ and the $\alpha\,\delta$

are involved in the binding of receptor agonists. In the same context, they also bind antagonist ligands. If sufficient amount of the neurotransmitter (Acetylcholine) is released at the somatic nerve terminal, it generates an end-plate potential when it combines with the receptors on the synaptic membrane. This in turn fires off a muscle action potential. If the quantum of the Ach is less, only excitatory post synaptic potentials (EPSP) will be generated, and notan action potential. Carrying the argument further, the molecules of Ach must occupy the receptors on the motor end plates. Logically, if enough numbers of these receptors are not available, transmission of the nerve impulse will be interfered with *(Explain)*. Typically, this happens in **Myasthenia Gravis** where anti-bodies which are formed against the receptor proteins destroy the receptors, thus reducing the availability of functional receptors. Apart from this, the antibody after it combines with a receptor, also promotes the combination of this complex with neighbouring receptors. This is followed by endocytosis of the complex with a further reduction the number of available receptors. The exact mechanism involved in the pathogenesis of the disease is unclear. The thymus is probably involved. Some cells in this gland which have Ach receptors may invite the initial autoimmune reaction. Clinically, the syndrome may vary in severity from mild (grade 1) to life threatening (grade 4) involving respiratory muscle paralysis). The myasthenia immunoglobulin is an IgG which can cross the placenta. As a result, neonates born to myasthenia mothers may suffer from a transitory Myasthenia syndrome.

The physiological principle used in the diagnosis and management of Myasthenia Gravisis governed by the fact that if cholinesterase is inhibited, the Ach will repeatedly engage the available receptors and strengthen the contraction. The diagnosis may be established electro-physiologically by repeated stimulation of a motor nerve. *(Explain)*.

Cholinestrase inhibitors of some types may have dangerous non-therapeutic effects, and have formed the bulk of various insecticides such as Malathion. More alarmingly, nerve gases such as Tabun, Sarin and Soman belong to this category of drugs, and in the real world of international terrorism, it is important to be aware of this aspect of cholinestrase inhibition. The main pathophysiological effects of such Cholinesterase inhibitors reflect hyperactivity of the parasympathetic branch of the autonomic nervous system. However the neuromuscular junctions of the skeletal muscles are also adversely affected. Normally, with each nerve impulse, the released ligand occupies the Ach receptors to produce the depolarization while the free acetylcholine(Ach) in the synaptic cleft is rapidly hydrolyzed by the cholinesterase which is a membrane bound enzyme in the postsynaptic membrane. The bound ligand is also metabolized. But when the esterase inhibitor neutralizes the enzyme, the free Ach is available to diffuse in the area for rebinding with the other available receptors. When the next impulse arrives at the junction and releases another batch of Ach aliquots, the generation of endplate potentials becomes asynchronous, rendering them inadequate to summate into an action potential. Thus the toxic effects of cholinesterase inhibitors also extends to the neuromuscular junction producing paresis/paralysis.

Iatrogenically, striated muscle may need to be paralyzed during surgery. This is best done by blocking the neuromuscular junction by using competitive inhibitors for Ach. Alternatively, if the depolarization of the end plate is maintained for a long period, the Ach released will cease to induce the rapid change in the excitatory post synaptic potential. The depolarizing agent changes the resting membrane potential of about -80 mv to -55 mv which is the potential required for triggering the AP. If this level of potential is maintained continuously, then the new batch of the ligand (Ach) which may occupy the receptors, fails to induce the initial change (from -80 mv to -55 mv) which is required to trigger off the action potential because the membrane has already achieved that state. As a consequence, the transmission of the impulse ceases, bringing about a paralysis. Drugs which cause this are classified as "Persistent depolarizers," and Succinylcholine is the example. These drugs are able to achieve their effect as they are relatively resistant to breakdown by cholinesterase. One of the more serious pathophysiological concerns of long term infusion of persistent depolarizers is the loss of intracellular K^+, leading to a life threatening hyperkalemia.

Fig. 3 Formation of Bone

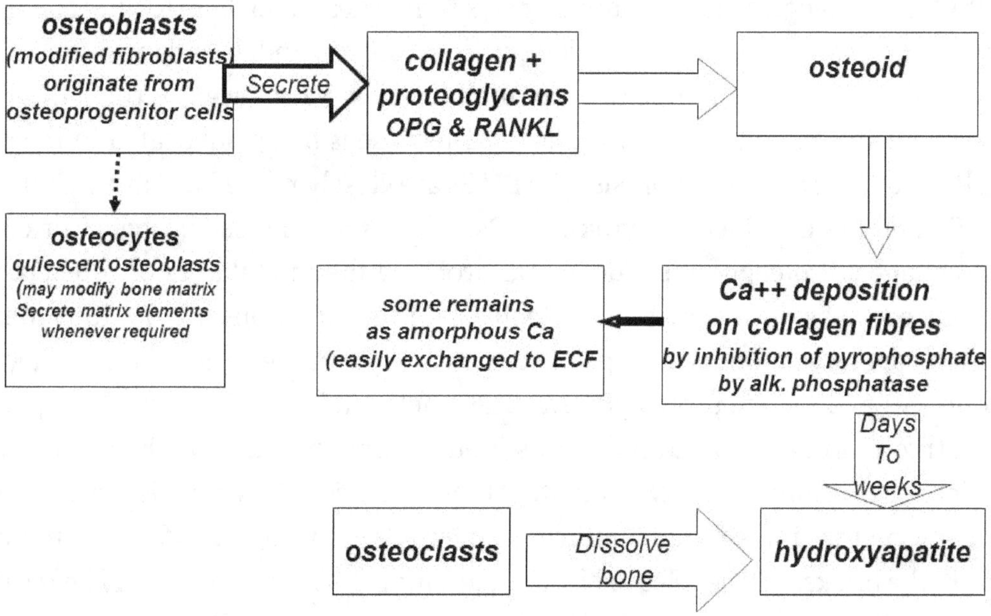

Physiology of Bone Formation (Modeling) and Remodeling

Table I. Bone matrix constituents

1. Collagen fibres
2. Proteoglycans
3. Glycoproteins
   Osteocalcin; Sialoprotein; Bone albumin; A Glycoprotein; Laminin; Osteonectin
4. Bone Morphogenic Proteins BMP 2, 6 & 9
5. Calcium & Phosphorus as Hydroxyapatite [Ca 10 (PO4)6 (OH)2]

Anatomically, there are two types of bone tissue. Cortical bone which is the covering layer, particularly in the long bones, and cancellous bone which is trabeculated and spongy. Trabecular bone is metabolically more active and covers a much larger bone surface area as compared with the cortical bone. Because of this, effects of bone loss are more evident in this variety of bony structure. Osteocytes are located randomly within the trabecule. The vertebra are a classical example of the latter though it is also found at the ends of long bones. As against this, the cortical bone has a basic unit called the osteon. This is cylindrical in shape and surrounds a blood vessel. The osteocytes are enmeshed around this unit and are connected via tiny canaliculi. The thickness of the bone is represented by the cortical covering.

The bone matrix is constituted by collagen (about 35%) and inorganic material (65%) which contains the Calcium and Phosphate. Bone is being continuously laid down and reabsorbed. In the early phases of life, more bone is laid down than reabsorbed, and is called "**Modeling.**" A typical example is the bow legs of young cowboys and cavalrymen of yester years. Bone density reaches its maximum in early adulthood, and thereafter from the mid thirties, it starts to decline. The process of **Remodeling** (see below) then takes over in maintaining as far as possible, bone density as close to the desired level. The organic matrix, collagen, gives the bone its elasticity and tensile strength when it is pushed and pulled while the mineral elements give it the hardness. Collagen is made up of collagen fibres, Proteoglycans, Glycoproteins and Bone Morphogenic Proteins **(Table I)**. Proteoglycans are complex polysaccharides which help to bind collagen fibres, and may control transport of charged particles such as calcium and their deposition. The various proteins are involved in deposition of calcium in the matrix. More recently, the BMPs have been identified as powerful bone formation inducers. Their use is being advocated in the treatment of unfilled bone defects and osteoporosis. The BMPs are closely related to Transforming Growth Factor$_\beta$ (TGF$_\beta$) and are classified as cytokines. Osteoblasts are the osteoprogenitor cells which secrete the collagen, Proteoglycans **(Fig. 3)** and probably the modulators of osteoclast activity (see below). Osteocytes may be quiescent osteoblasts. Their functions are not fully identified as yet. Osteoclasts are large monocytes type of cells derived from the bone marrow. They function like macrophages and secrete proteolytic enzymes, lactic and citric acid which help to dissolve the bone matrix and release calcium from its binding. Simply put, bone is like a cement wall with the bricks as the mineral entity bound together by the cement which is the organic matrix. It stands to reason that if the organic matrix (cement) of the wall) is eroded, the minerals held together by it (the bricks) will be released from their binding. This anomaly may be recalled later.

Bone Remodeling

The process of bone formation matching bone reabsorption is known as remodeling, and is a continuous process which is put into place after the mid thirties.It is carried out by "remodeling units" which consist of a layer of osteoclasts surrounding which is a layer of osteoblasts. In the centre of the unit is a capillary. To begin with, the osteoclasts absorb the bone in the area and migrate to a neighbouring zone. This is followed by the osteoblasts laying down new bone to fill in the void created by the osteoclast activity. Remodeling occurs in a cycle of various stages namely activation, reabsorption, reversal, formation and resting. There are various advantages of this process. The bone strength and shape gets adjusted to bone stress. The old matrix is continuously replaced with the new, and hence the bone strength remains constant. It helps in the repair of the micro architecture of the bone whenever it is damaged. Bone remodeling is a source of releasing calcium in to the blood in order to maintain normal blood calcium levels. One of the most important stressors which direct bone remodeling is gravity and the constant pull and push of muscles attached to the bones. Long term space flight is associated with bone demineralization. It is also a well documented fact that prolonged immobilization too induces demineralization. Bone remodeling is essential for smoothening out the rough surfaces created with new bone formation in a fractured bone.

The possible mechanism of bone laying and reabsorption is outlined in **Fig.4.** PTH receptors are available only on the osteoblasts and not on the osteoclasts. The former secrete two ligands- the osteoprotegrin (OPG) and the Receptor Activator of Nuclear factor kB Ligand (RANKL). OPG is responsible for bone laying and inhibits osteoclast activity while the latter, RANKL promotes osteoclast activity. Under normal circumstances, both the factors balance out each others' actions, but when PTH is secreted in excess, the osteoclastic activity dominates and bone reabsorption exceeds bone formation. There are numerous other factors which affect the process. These include estrogen and androgens, Vitamin D, Insulin like growth factors, interleukins, prostaglandins, and TNF

Fig. 4. Possible mechanism of bone formation vs bone reabsorption

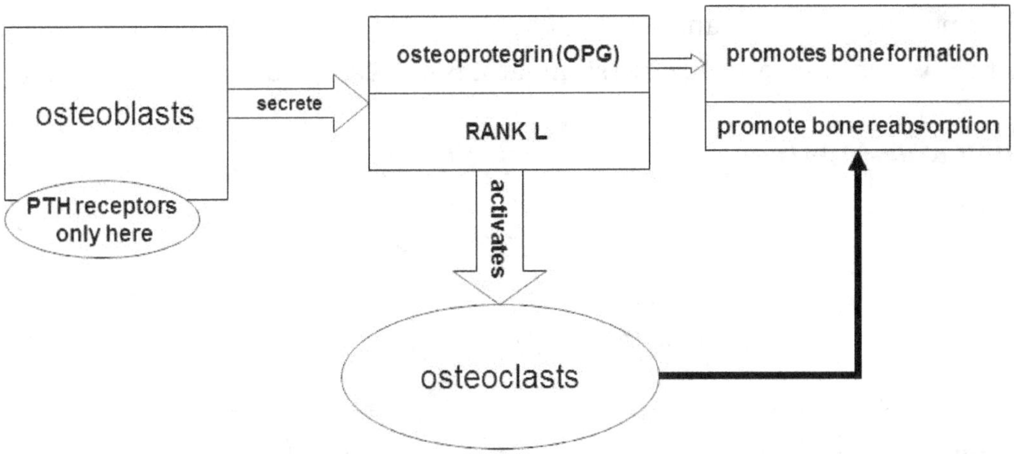

Pathophysiology of Osteoporosis

Maximum bone mass is present during young adulthood-to about the mid thirties. It is determined by the Vit D status of an individual, hormones such as Insulin, Growth hormone, Insulin Like Growth factors and the number of estrogen receptors. Other factors that contribute are the degree of physical activity, dietary intake of calcium, and over all nutrition status and heredity. Osteoporosis is said to occur when there is a decrease in bone mass. As per WHO guidelines, if the measured bone density is lower by > 2.5 the standard deviation of young healthy adults, osteoporosis is present. WHO has also categorized grades of bone density as normal when it is $833mg/m^2$, and osteoporosis when it is $<648 mg/cm^2$. A decreased bone mass (Osteopenia)is designated as that which lies between the two.

The basic pathophysiology of excessive bone loss may be because of excessive bone reabsorption or inadequate new bone formation.

Aging is one of main causes of primary osteoporosis categorized under inadequate new bone formation. With aging there is decrease in activity of osteoblasts, a decrease in the formation of osteoblasts, decrease in the activity of various growth factors which are required for new bone formation, and an overall reduction in the physical activity. The decline in bone density at about 0.7% per annum occurs with an increase in age beyond the mid thirties in both males and females.

Loss of bone density is accelerated in females during the first decade of menopause and may occur at about 2% per year. There are a number of possibilities which explain the acceleration of the bone density loss when estrogen levels decline. Lack of estrogen promotes an increase in the levels ofcytokines such as IL1, 6 and TNF. These in turn stimulate the release of RANKL and decrease that of OPG by the osteoblasts and Osteocytes, thereby promoting the RANKL sponsored activity of osteoclasts which eat into the bone. Because of this the new bone formation actions of osteoblasts is overwhelmed.

Pull and push on bone because of muscle contractions is a potent mechanical stimulus to bone laying activity. This is diminished with physical inactivity. The importance of gravity as a stimulus to bone formation is evident by the fact that bone demineralization is a serious side effect suffered by astronauts after along duration space flight. At least partial prevention is obtained by putting the astronauts through an in flight exercise regimen. However weight training exercise which is a more effective stimulator is not feasible during space flight. *(Why?)*.

Bone Injury and Repair

Fig. 5. Development of osteoporosis

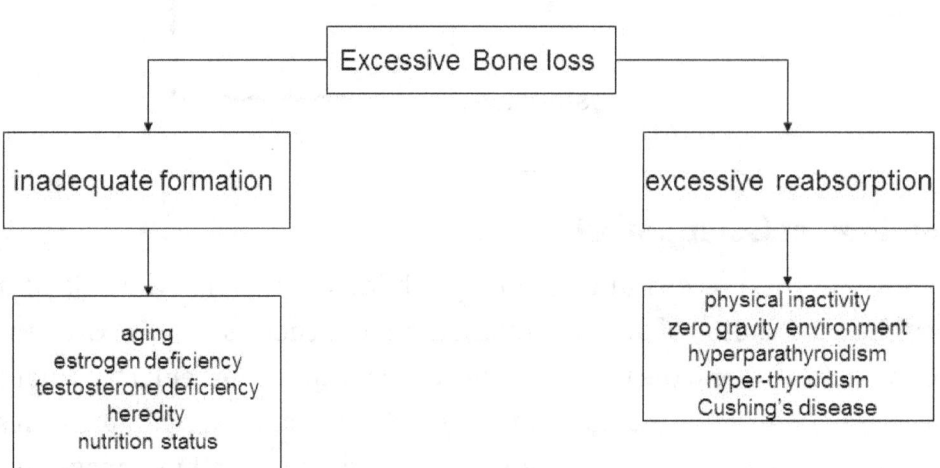

Fractured bone goes through a sequence of events during its repair. The process begins with the formation of a **hematoma**. The fibrin meshwork is implanted with various hemopoietic growth factors. This is followed by deposition of fibroblasts, and osteoblasts which begin to lay down the new collagen. The process is aided by various proteins such as the BMPs, osteogenin. Insulin and Insulin like growth factors are also required. The end result of this activity is the formation of the **procallus.** The latter is then strengthened to form the **callus** with the deposition of calcium and phosphate. **Remodeling** then takes over. This may be likened to the use of the smoothener (plainer) a carpenter uses after two wooden pieces are hammered together. The hematoma occurs within hours, the procallus formation takes a few days, the callus formation occurs over a period of a few weeks, while the final process of remodeling is slow, and takes months to years. Occurrence of infection, paucity of blood supply and deficiencies of growth and thyroid hormones and other essential nutrients delay the healing process.

References

1. American Medical Assoc. Module 3. Osteoporosis management. Pathophysiology of osteoporosis. 2004

2. Barbosa de Lira CA, Vasncini RL, Cabral FR et al. Post polio syndrome: renaissance of poliomyelitis? Crit. Review Phys. Rehabil. Med. 1995; 7: 147–183.

3. Best and Taylor's Physiological Basis of Medical Practice, 13th edn.; Ed. West JB. William & Wilkins, London; 1990.

4. Bijlani RL. Understanding Medical Physiology: A Text Book for Medical Students. 2nd edn. Jaypee New Delhi.1997.

5. Brown SP, Miller WC, Eason JM. Exercise Physiology: Basis of Human Movement in Health and Disease. Lippincott, Williams and Wilkins, Baltimore. 2006.

6. Davidson's Principles and Practice of Medicine. Edt. Haslett C, Chilvers ER, Boon NA & Colledge NR. 19th Edn. Churchill Livingstone., London. 2002

7. Ganong WF. Review of Medical Physiology. 22nd edition. Lange Medical Books/McGraw-Hill, London. 2005.

8. Goodman and Gilman's Pharmacological Basis of Therapeutics. Edited. Brunton LL, Lazo JS, Parker KL. 11th edn. McGraw Hill London; 2006.

9. Gray's Anatomy: The Anatomical Basis of Medical Practice; 40th Edn. Ed. Standring S; Elsevier, Churchill Livingstone, London; 2008.

10. Guyton AC, Hall JE. Textbook of Medical Physiology.11th edition. Elsevier, Philadelphia;2006.

11. Harrison's Principles of Internal Medicine. Edited. Kasper DL, Braunwald E, Fauci AS, Hauser SL, Longo DL, Jameson JL.16th ed. MCGraw Hill, London. 2005

12. Marcell TJ. Sarcopenia: causes, consequences and prevention.J. Gerontology. 2003; 58A: 911–916.

13. Marzetti E, Leeuwenburgh C Skeletal muscle apoptosis, sarcopenia and frailty at old age Experimental Gerontology. 2006; 41: 1234–1238.

14. McCance KL, Huether SE. Pathophysiology: The Biologic Basis for Disease in Adults and Children. 5th edition. Mosby; St. Louis USA; 2006.

15. Merck Medicus Modules: Osteoporosis-Pathophysiology. www.merc.medicus.com/disease modules/osteoporosis/pathophysiology /2009. Pp 1-16.1-16.

16. Proctor DN, Balagopal P, Nair KS. Age related sarcopenia in humans is associated with reduced synthetic rates of specific muscle proteins. J. Nutrition. 1998; 128: 351–355s

17. Robbins & Cotran. Pathologic Basis of Disease. Edited. Kumar V, Abbas AK, Fausto N. 7th edn. Elsevier Saunders, China. 2005.

18. Roubenoff R. The pathophysiology of wasting in the elderly. J. Nutr. 1999; 129: 256s-259s

19. Short KR, Bigelow ML, Kahl J, Singh R et al. Decline in skeletal muscle mitochondrial function with aging in humans. Proceedings of the National Acad. of Sciences (USA) 2005; 102: 5618–5623.

20. Westerblad H, Allen DG, Lannergren J. Muscle fatigue: lactic acid or inorganic phosphate the major cause? New Physiol. Sci. 2002; 17: 17–21.

21. Wilk A. Estrogen receptors in skeletal muscles: Expression and activation. Thesis, Depart. of Laboratory medicine, Karolinska Institute, Stockholm, 2008.

CHAPTER - 10

Circadian Rhythms and Pathophysiology of Stress

Plan

1. General Introduction
2. Normal circadian rhythms
   i.General; ii. the HPA axis and circadian rhythm. iii. manifestations of circadian rhythm disturbances;iv. sleep and circadian rhythm; v. age and circadian rhythm disturbances;vi. principles of managementof circadian disturbances.
3. Stress:
   i definition; ii. the HPA axis in stress; iii. sympathoadrenal system and stress; iv. endocrine responses to stress;
   v. acute vs chronic stress; vi. systemic responses to stress.
   vii. pathophysiology of some important stressors;
   viii.strategies of adaptation; ix. some manifestations of stress

General Introduction

The complexities of modern day to day living, fierce competition in the professional field and other such activities impose considerable stress and strain on individuals. Fast air travel across continents, unscheduled working hours after, shift work, and the usually forgotten fact that the advent of electricity can change night in today, are often responsible for disturbing the normal day/night cycle which is a part and parcel of daily life of a human being.. All this may stretch the normal physiological and psychological resources of a person to their limits, leading to a breakdown of the physiological system. It is the aim of this chapter to look in to the endocrinological aspects of such phenomena and highlight some of the more recent issues pertaining to the pathophysiology they generate.

Normal Circadian Rhythms

In humans as in most other species, the daily cycle of exposure to light and darkness affects body physiology and behaviour patterns. Some of the often studied parameters which reflect these daily changes are body temperature, varying levels of some hypothalamic and adrenal cortical hormones, heart rate and blood pressure fluctuations, and performance. These changes occur over a period of about 24 hours (or a day), are regular and predictable and are described under the term "Circadian" changes. The origin of the term comes from Latin where circ means about and diem (day). The science dates back to the 1700s when a Frenchman, de Mairan first described changes in leaf opening and closing patterns over a period of 24 hours. Modern studies in chronobiology were initiated on the fruit fly by Pittendrigh. Jurgen Aschoff though has been considered the founding Father of studies on human circadian changes.

Circadian rhythms exhibit certain characteristics. The rhythm persists even when humans are made to live in the absence of the day/night cycle ("Free running cycle"). Therefore an internal, biological clock, independent of the day and night cycle, controls these changes. Under these circumstances, the clock may runa few hours on either side of the classical 24 hour cycle. This biological clock helps the body to prepare for the changes that it is subjected to during the day and night cycles, and at the same time, tries to ensure that there is synchrony between the various physiological changes that occur. (Pittendrigh has aptly commented that the rhythms are such that they aptly match the "Day within" with the "day outside" to enable the animal to cope. It is now believed that individual organs have their own circadian rhythm, and all these rhythms need to be synchronized to ensure smooth physiological function. It has also been suggested that other unknown geomagnetic influences may be responsible. A derangement in this internal clock may then result in disturbed well being of the individual. Typically, a minor change in the clock may be brought about by an alteration in exposure to light and darkness which occurs in day to day living. Thus the light/darkness and any other circumstances which may alter the rhythm act as external cues or "Zeitgebers." If there is a large difference between the "the day within and without," disturbance of the rhythm occurs and manifests itself clinically. The symptoms slowly recede as the two come together and finally synchronize. This is exactly what happens when a person feels tired and disturbed when s/he travels across continents, but gets back to one old self after a couple of days.

The rhythm operates within certain parameters. The **Amplitude** of the rhythm depicts the difference between the peak and trough of a particular paradigm during the 24 hour cycle. This is particularly relevant to Endocrine physiology where in measurement of the peak and trough of the blood level of a hormone help to depict the physiological response of the gland to the 24 hour cycle. The peak or the trough may be related to a particular event in the day-night cycle. For example, ACTH peak is usually related to the early morning hours. This relationship is the **Phase**. The time interval between two peaks of hormone level in blood is the **Period**. A change in the pattern of these attributes may be brought about by a change in the day night exposure periods which may result in **Resetting** of the cycle as in changes in time zones. An example is a day worker changed to night shift feels awful for a day or two, but by 4–5th day is as good as s/he was during the day shift. This also occurs in endocrine disease such as excessive secretion of growth hormone, ACTH and cortisol.

The seat of origin of circadian rhythms has been identified as the suprachiasmatic nucleus (SCN) in the hypothalamus. Light signals which fall on the retina, are conducted to the SCN via the IInd cranial nerve in the retino-hypothalamic pathway that emerges from the optic tract. Glutamate is a likely neurotransmitter candidate. Possibly a second connection is made via geniculo-hypothalamic tract emerging from the same retinal cells that give rise to the retino-hypothalamic tract.γ aminobutyric acid (GABA) and Neuropeptide Y are the neurotransmitters involved. This pathway also goes to the SCN. The day to night change centres around the hormone Melatonin which is secreted by the Pineal gland. Presence of light makes the pineal stop secreting Melatonin while absence of this stimulus increases the secretion of the hormone, making the individual feel sleepy. Usually, melatonin secretion occurs between about 9 pm to 8 am the next morning. The SCN simultaneously synchronizes various other functions such as body temperature, hormone secretion, to name the more important ones, with the sleepiness/wakefulness cycles brought about by melatonin secretion. Experimentally induced damage to the SCN is known to disrupt various rhythms. It has been seen that in new born rats, the rhythm becomes evident after about a week of birth suggesting that the SCN needs to develop afferent and efferent neural connections after being exposed to external cues before it can establish the various rhythms. As to whether this happens in human neonates is not clear. No other brain structure is known to take over its function. The SCN communicates with the Pineal gland via neural pathways. The mechanism which is used to communicate with other brain structures and body parts is not known. This signal could be hormonal or neural. The internal clock in the SCN is probably set by expression of various proteins from certain genes. The presence of these so called "Clock" genes has been established in mice. The expression by these genes is light dependent. Some of the genes, namely CRY2, MTNR 1B and PER3 are located in the retina. Their expression has been related to the day/night light cycle.

Hypothalamic Pituitary Axis (HPA) and Circadian Periodicity

Hormones are secreted throughout the day in varying concentrations which make them reach either their zenith for the day, or their nadir. Such activities are hormone specific. For example, adrenocorticotrophin (ACTH) concentration is lowest during the night, and peaks sometime in the early morning hours while the growth hormone (GH) reaches its peak concentration during deep sleep at night. More recently, a Cortisol Awakening Response (CAR) has been described. Here, the cortisol concentration may attain a value which is about 30% higher than the usual peak level about 30 minutes after awakening. In acutely stressful situations, the CAR may be upgraded by as much as 50% though it maintains the rhythm. A disturbance in this pattern of hormone secretion may also be reflected in some hormonal disorders. Investigation of the circadian and pulsatile patterns of hormone secretion may thus be helpful while dealing with hormonal disease.

The secretion of ACTH has been extensively studied in this regard, and the findings have contributed immensely to the understanding of circadian and pulsatile secretion phenomena of hormones. These variations are important in ensuring the criticality of hormonal function, and are likely to influenced by a number of factors which may be neural, hormonal and environmental. The hormone, like many others, is secreted in a pulsatile manner throughout the 24 hour cycle, and as many as 15–18 pulses may occur over this period. The frequency is known to increase in stressful situations. The hormone concentrations released during these pulses are variable and are an important indicator of hormonal activity. A change in the circadian secretory pattern of hormones may occur in diseased states, circadian rhythm alterations and stress, and thus the estimation of these hormones forms an integral part of investigation of the pathophysiology involved.

The circadian pattern of Hypothalamic Pituitary axis (HPA) hormone secretion is under a number of influences which may be internal or external. The effect of light and darkness on suprachiasmatic nuclei and the influence of food availability (the so called Zeitgebers) via the ventromedial hypothalamic nuclei have been implicated in the generation of circadian rhythms. Corticotrophin Release Hormone (CRH), the Anti Diuretic Hormone (ADH), Leptin and IL6 have been implicated in circadian alterations of HPA activity. Leptin replacement in patients diagnosed with Leptin deficiency has been known to restore HPA rhythm disturbances. Alterations of the activity of the HPA may be responsible for the puslatility.

Plasma levels of a number of endocrine secretions have been used as markers to study circadian rhythms. These include melatonin, growth hormone, prolactin, ACTH and cortisol. Body temperature has been used as a non-endocrine marker. Aging, and behavioural changes because of mental disorders disrupt circadian secretory pattern of CRH, ACTH and melatonin.

Manifestations of Circadian Rhythm Disturbances

The usual symptoms which occur with a disruption of the normal rhythm include malaise, fatigue, daytime sleepiness, difficulty getting to sleep, and when sleep comes, difficulty in continuing with it; impaired performance, reduced day time alertness, and increased irritability. Gastrointestinal symptoms such as loss of appetite, disturbances in digestion and bowel movements are common. When circadian rhythms were disrupted experimentally in rats, a change in their body temperature patterns was observed. Obesity, Insulin resistance and raised leptin levels in blood were recorded. Similar metabolic patterns are observed in human metabolic syndrome. The influence of the circadian rhythms on the brain of these mice was reflected as some remodelling in brain structures which could account for some of the behavioural functions. Human cognitive functions are adversely affected by the day/night cycle change, and the observations made on mice may suggest that similar pathophysiology could be possible in human brains. Symptom intensity reduces as the rhythms within and the new external cues start coming closer together with the passage of time, and finally disappear when they synchronize fully. The bottom line as far as the effects of circadian change is concerned is the decrement in performance.

Situations which are known to disrupt circadian pattern are given in **Table I**.

Table I: Situations which alter normal circadian rhythm

1. Change of work shifts-day to night
2. Jet lag
3. Change in sleep pattern due to various causes such as aging, chronic pain and illness, behavioral disorders
4. Space flight

The introduction of electric lighting has probably been responsible for wide disruptions in sleep wakefulness patterns of human beings though this observation may not be easily recognized. Medical and military personnel, and other shift workers are often exposed to effects of circadian rhythm disturbances. Disturbances in functional capabilities produced in the earlier phase of the change may induce operational hazards. For example, it has been observed that medical doctors may make errors in interpretation of clinical and laboratory data when exposed to sudden circadian changes because change in their duty pattern, say from day to night. Reduced functional capacities in an air traffic controller exposed to a disruption of circadian rhythm may become the underlying cause of an air craft accident. In the same context, long distance international flights have always been an issue raising concerns of flight safety as aircrew are subjected to large circadian variations very rapidly. This may result in the undermining of their efficiency. This problem was nonexistent in the days of intercontinental travel by ships.

One of the more interesting aspects of circadian rhythm change that captures the fancy of most people is the "Jet lag." The earth rotates on its longitudinal axis of 360^0 in 24 hours. Hence, every one hour, there is a change of 15^0 in its longitudinal axis. The longitudinal coordinates of India lie between about 65^0(West) to about 95^0 (East) which is a span of about 30^0. Typically, then, the eastern most part of India should be 2 hours ahead of the western most one. However, countries have standardized their time and thus both these areas will exhibit the same Indian Standard Time (IST) though physically, the sun will shine 2 hours earlier on the eastern side. A difference in the latitude (north/south movement) seems to make no difference to the circadian rhythms.

Modern air travel in fast jet liners cross the time zones rapidly. It has been observed that for a disturbance of the rhythm to occur, a rapid change of about 3 to 4 hours in the time zone is necessary. *(Why is it that the circadian disturbances were not seen in the pre-air travel era?).* When this happens, there is a mismatch between the body's circadian rhythms with the external environment prevalent in the new setting. The symptom complex that ensues is popularly called the jet lag. The symptoms are basically the ones that have been described earlier. The body may take many days to synchronize with cues offered by the new environment. Greater the time difference between the old and the new locations, more severe the symptoms and longer it takes to regularize the disturbed rhythm to its new setting. Travel from the east to the west requires body physiology to adjust to cues in less time than what it is usually used to and this produces greater disturbance. Circadian disturbances do not occur if one travels from north to south or vice-versa as time zones are altered. Passengers and aircrew are affected alike. The relevance of jet lag in aircrew is obvious. The elderly are more severely affected by jet lag.

Space flights have been known to disrupt circadian periodicity. This happens because of a number of factors such as increased work load with reduced sleep time, rapid changes in exposure to light and darkness on different sides of the spacecraft (the solar sides being always bright) in the vacuum of space, unscheduled work demands. Attempts are being made to find ways to avoid such disruptions as far as possible during space flights so that the functional efficiency of cosmonauts is not compromised.

Sleep and Circadian Rhythm

One of the most prominent effects of de-synchronization of the circadian rhythm is that on a person's sleep. Though mentioned earlier in the symptoms complex, it is discussed here as a separate issue because of its relevance to day to day functioning. The more common reason for circadian induced sleep disturbance is the change in environment because of change in longitude to a new geographical environment, or change of work shifts. The more complex one however is the so called primary disorder where the circadian clock itself-viz the SCN- is disturbed. Some of these patients develop a delayed onset of sleep and after that sleep late (the Delayed Sleep Phase syndrome) while in others, the sleep schedule is advanced as compared with the accepted sleep timings-the Advanced sleep phase syndrome. Mutations in some clock genes probably the PER 2 gene, expressed in the SCN have been implicated in this condition.

Sleep disturbances occur in a number of clinical conditions, and because of that, the normal circadian pattern of physiology gets disturbed. The hall mark of bipolar depressive illnesses is insomnia. The cause and effect relationship of these illnesses and circadian rhythm disturbance is as yet not clearly worked out. It has been postulated that there is a phase advance in depression. Some of the hormonal disorders which may be concomitant with depression for example, the reduction in prolactin secretion or the TSH response to TRH stimulation, may be as a consequence of the circadian shift, and not because of the primary pathology in the brain. Alcoholism is a major impediment to good sleep hygiene and with it brings about disturbing circadian changes. Other central conditions which are associated with sleep disturbances are epilepsy and Parkinsonism. Recently it has been shown that Melatonin secreting cells were found depleted in some patients with Parkinsonism. Alzheimer's disease is another condition in which the rhythm is disturbed.

Age and Circadian Disturbances

The circadian clock seems to be advanced in the aged. The onset of sleep is early, and so is the awakening. Sleep therefore is disturbed and unsatisfactory. These features are akin to patients who have the Advanced Sleep phase syndrome. The normal circadian rhythm amplitudes are also reduced in the elderly, and are may account for the altered circadian and sleep pattern. Body temperature amplitude of healthy elderly people during the day/night cycle is lower than that of healthy young adults, and may be associated with sleep disturbances in the elderly. The association between activity and body temperature during the day is disturbed in the aged. The cortisol level at its nadir during the 24 hour cycle is much higher in the elderly as compared with the young. Also the baseline cortisol level is higher in the aged. As to whether this reflects circadian rhythm liability or an overall lowered general health status which acts as chronic stressor is not clear. The secretion of the cytokine IL6 is higher in the elderly during the night time as compared with young adults in whom this cytokine follows a more established circadian pattern. Higher IL6 blood levels have been associated with reduced sleep. IL6 secretion is associated with many factors which are more prevalent in the elderly. These are lower levels of sex hormones, higher fat percentage, and greater possibility of chronic inflammatory conditions. On the basis of this, it may be hypothesized that the derangement in the circadian secretion pattern of IL6 may be linked to the sleep disturbances in the elderly. On its own accord, IL6 is one of the modulators of the HPA axis, and this in turn may explain the high cortisol level in the elderly. Hence it could be argued that the IL6 involvement is an indirect one, while it is the cortisol that is the bane of the circadian changes.

Principles of Management of Circadian Disturbances

A number of strategies have been tried to overcome the adverse effects of circadian rhythm disturbances. These include adjustment of sleep time schedules at the new destination by using Chronotherapy. With this technique, the time at which a person goes to sleep may be advanced or delayed by an hour each day until it matches the desired time at the new place. For example, going west would need delaying of the time of going to sleep, while the opposite is true for adjusting to an eastward travel. Phototherapy is another method which involves the exposure of an individual to light intensity of 2500 flux during the morning hours if the phase has to be advanced, or during the later part of the afternoon if it has to be delayed. If the visit to the new location is short (less than 3 days) it is best to keep the sleep pattern as close to that which was being followed at the home base. Drugs have been used to modulate effects of circadian shifts. These Chronobiotics are namely the Diazepam group of drugs which act on the GABA/benzodiazepam receptors in the SCN, and melatonin. The latter is by far the more promising. Its mode of action is not clear. It may help either by hastening a phase shift, or generate asleep producing effect.

Neuro-endocrinology of Stress

Stress has been variously defined. The more classical definition was given by Hans Selye as "The non-specific response of the body to any demands made upon it." Others have called it a "Perceived threat to well being" of an individual, or "A mutual action of forces that takes place across any section of the body." A more recent definition calls it "A general reaction of the mammalian central nervous system (CNS) which plays a vital role in the way an organism monitors internal conditions, as well as conditions in the world around it, in order to attempt to survive." Even the thought that a threatening situation may be encountered is enough to generate stressful reaction. In principle, stress threatens homeostasis.

Various stressors may impinge on to the body as neuro sensory signals (which may be nociceptive, auditory, visual etc), infections, trauma, both physical and psychological.**(Table I).**Application of restraint /immobilization has been used often as an experimental model to study the effects of stress. Stress induced input is

Table II:Generally recognized stressors

A. Psychological which cause fear, anxiety

B. Unpalatable sensory/physical stimuli such as pain; immobilization; isolation; heat and cold; insomnia

C. Cardiovascular stressors- hypovolemia ofhemorrhage, dehydration;orthostatic stress, exercise

D. Hypoxia and hypoglycemia

coordinated by a number of central groups of neurons **(Fig 1)** which in turn direct the traffic to the hypothalamic-pituitary-adrenal (HPA) axis. The hypothalamic Corticotrophin Release Hormone (CRH) and Arginine Vasopressin (AVP) neurons play a major role in the activation of the stress response system. These neurons are reciprocally connected to opiod secreting neurons as well as the catecholaminergic (CA) neurons of the locus ceruleus. The opiod peptides provide the analgesia which must ensue under stressful conditions, while at the same time, exercising restraint over the HPA axis in order to optimize the stress response. The neurotransmitters involved are serotonin and acetylcholine which excite the CRH and AVP neurons while γ-aminobutyric acid/ benzodiazepine and glucocorticosteroids function as their inhibitors. It is the latter that is mainly responsible for controlling the activity of the HPA viacentral control mechanisms. The various central controllers (Fig 1) have CRH receptors (CRH R1 receptors). The CRH neurons in turn express α_1 adrenergic receptors for the norepinephrine (NE) released by the catecholaminergic neurons of the locus ceruleus (LC) as their neurotransmitter, thus completing a rapid response feedback loop which mutually enhances the activity of both sets of neurons. At the same time, collateral connections inhibit this process via other CRH and α_2 adrenergic receptors on the LC and hypothalamic CRH neurons. Apart from the R1 subtypes, CRH R2 receptors which are located in the limbic system and in subcortical regions as also in some peripheral tissues such as the myocardium, the gut and skeletal muscle, are thought to play a part in the overall stress response.

There are two sources of Arginine Vasopressin (AVP) which is released during the stress response. The magnocellular neurons of the hypothalamus secrete the hormone directly via the posterior pituitary in response to hypovolemia caused by various conditions, while the parvocelluar neurons secrete the AVP which influences the HPA axis by acting in synergy with the CRH. This action is particularly relevant to chronic stress situations. The AVP helps CRH to secrete ACTH by the pituitary to enhance the overall corticosteroid response. Like the CRH, AVP connects with the hypothalamic pro-opio-melanocortin neurons to promote secretion of hypothalamic opioids such as β endorphin which in turn help to allay pain during stress, and also to regulate the intensity of the overall stress induced effects.

The HPA axis in the Stress Response

This axis is an important mediator of the stress response. The CRH binds to CRH R1 subtype of receptors on the pituitary ACTH secreting cells. This normally happens in a circadian fashion, and the activity is further influenced by the day-night cycle, with peak levels of CRH→ACTH→Cortisol being attained in the early morning hours. During exposure to acute stress in particular, the main change in this cycle is the increase in its amplitude, while the circadian rhythm is maintained. It is the glucocorticoid which exerts its actions to help physiology cope with the stressful situation by immunosuppression, blunted reproductive and thyroid activity (the latter in order to minimize energy utilization). At the same time, cortisol reciprocally inhibits the CRH and associated activity in order to optimize the stress response.

AVP supports CRH but has only a little influence in the direct stimulation of ACTH cells in the anterior pituitary. Other neurotransmitters which are involved in response are Angiotensin

II, and various cytokines (TNFα, ILs 1 & 6). These act on the HPA to enhance its activity. The type of mediator that is released is governed by the type of stress encountered.

The Brain Renin Angiotensin system (RAS) and Stress

The RAS has been linked to both short term and long term adaptation to stress. Studies in rats have shown that neurons which express AngioII AT1 receptors help in modulating the sympathoadrenal system response to acute as well as chronic stress. AT1 receptor antagonists were seen to significantly attenuate the hormonal and sympathoadrenal responses to stress induced by isolation. The findings have a bearing on the use of such medications in control of stress induced clinical manifestations in patients.

The Sympathetic Response and the Sympathoadrenal System during Stress

This system constitutes the main neuronal pathway which operates during stress, the HPA axis described above is mainly a humoral one. One of the central modulators of the stress response are the catecholaminergic (CA) neurons **(Fig. 1)**.These neurons are widely spread throughout the CNS and the ascending pathways. They make reciprocal contacts with the HPA as also with other brain areas such as the amygdala and the brain stem. The latter impinge on the intero-medio- lateral (IML) neurons found in the thoracic segments of the spinal cord to innervate the sympathetic ganglia and the adrenal medulla. It is this part that constitutes the sympathoadrenal pathway. The catecholamines thus released are responsible for the various sympathetic effects on organ systems **(Fig. 2)**. The main stimulus for the release of catecholamines by the adrenal medulla is the presynaptic release of acetylcholine. A. medulla also has receptors for ACTH, histamine and Angiotensin II which help in releasing catecholamines. ACTH helps in the secretion mainly of epinephrine.

Fig. 1. Possible sequence of activation of the HPA axis and the stress system (based on Tsigos et al 2004)

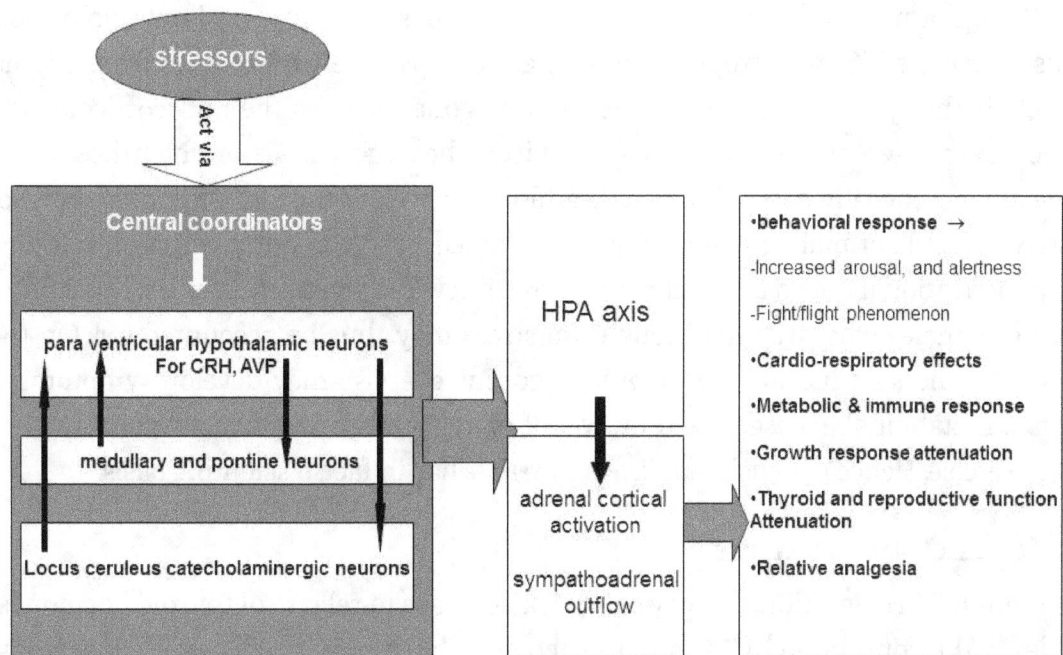

Fig.2. Descending catecholaminergic sympatho-adrenal pathways in stress derived from Fig 8 of **Kvetnansky** et al 2009 . CA stands for catechoalminergic

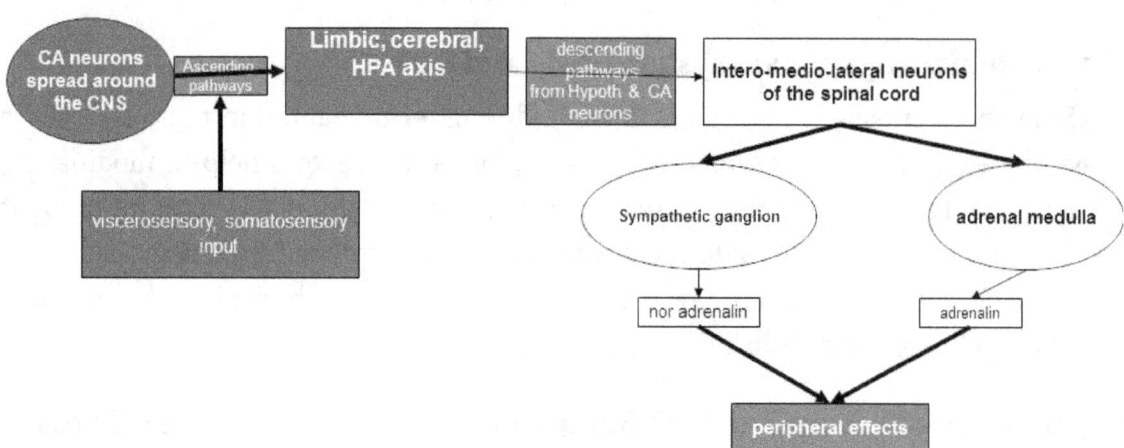

Meso-prefrontal-Cortical Dopaminergic System in Relation to Stress Endocrinology

Studies in rat models have shown that the neurons of this system are activated by stressful environments. They make connections with the HPA system. More recently, serotonin has been implicated in the neuroendocrinology of stress. Studies in rats have shown that serotonin (5 HT) neurons in the raphe nuclei are connected to the hypothalamus and the amygdala. 5 HT receptor agonists in these animals increase the secretion of various stress hormones such as oxytocin, ACTH, glucocorticoids, renin and prolactin. It has also been observed that patients suffering from anxiety disorders respond favourably to serotonin reuptake inhibitors thus supporting the involvement of 5HT in the pathophysiology of stress.

Systemic Effects of Stress

Reproductive system and stress

The HPA axis, a major player in the processing of physiology and pathophysiology-of stress, inhibits the release of Gonadotropin hormone Release hormone (GnRh). Apart from inhibiting the GnRh cells in the hypothalamus as also the pituitary gonadotrophs, the glucocorticoids which are released in response to stress, reduce the sensitivity of the target tissues on which the sex hormones act. The amenorrhoea in some elite female athletes, as well as the low leutinizing hormone and testosterone levels of male athletes may be explained by the endocrine manifestations of the HPA axis activation induced by the chronic stress of severe, long term exercise. Infertility in both males and females who are under long term stress may thus be accounted for. On the other hand, those who stop indulging in regular rigorous exercise may develop symptoms such as insomnia, irritability because of withdrawal of opiod secretion which is a constant feature of regular exercise. Hence the cliché, *addicted to exercise* has in fact, a scientific basis.

Thyroid function and stress

Stress inhibits TSH secretion, thereby causing a decrease in release of thyroid hormones. At the tissue level, the conversion of T4 in to the active T3 hormone is reduced. Philosophically, the aim of this response is to conserve energy. High levels of cortisol induced by CRH, somatostatin and cytokines are likely to be responsible for this action.

Growth is adversely affected by chronic stress. This has hormonal connotations. High CRH level during stress may be responsible for inducing secretion of somatostatin by the hypothalamus. As a result, there is a reduction of growth hormone (GH) secretion by the anterior pituitary, and consequently, the Insulin Like Growth Factor I (IGF I). Children exposed to chronic stress may develop growth retardation as explained. This has been recently called Psycho-social Dwarfism. Removal of the stressful environment restores the normal rate of growth in such children.

Metabolic effects of stress

Hyperactivity of the HPA axis in chronic stress situations produced experimentally in monkey models has led to the development of a metabolic syndrome type of disorder with Insulin resistance, obesity, hypertension, osteoporosis and growth hormone deficiency. The underlying pathophysiology is the stress induced hypercortisolemic state. Stressful situations in diabetics increases the severity of the condition (*Explain?*)

Immune mechanism and stress endocrinology

Inflammation activates cytokines Tumour Necrosis factor-α, and Interleukins 1 and 6 (IL 1&6). All of these are recognised activators of the HPA axis via release of CRH and AVP. The most prolific of them all is IL 6 responsible for hypercortisolemic state which in turn suppresses the inflammatory activity. Poor control of infections is thereby explained. In the same context, decreased ability to fight neoplasms during chronic exposure to stress is also explained. Ordinarily, a balance needs to be maintained in the degree of activation of the HPA axis during chronic stress. Its hyperactivation can be the cause of immune function disorder. Conversely, hypoactivity may result in ascendance of autoimmune disorders or inflammation.

Gastrointestinal system during stress

The main effects seen are a decrease in gastric motility and an increase in colonic motility. This may account for the symptoms of nausea, vomiting in some patients, particularly when exposed to acute stress, and diarrhoea in others. In chronic stress, persistent bowel disturbances and vague alimentary tract pain are the usual features. While the raised CRH level influences the vagus nerve, the locus ceruleus and the central adrenergic neurons act via the sacral parasympathetic innervation to increase activity of the colon. This accounts for the recurrent diarrhoea and abdominal pain. One of the GIT presentations of acute severe stress such as burns, surgery, severe trauma, septicemia is the occurrence of acute gastritis and stomach/duodenal ulcers. The ulcers may be single or multiple, often located in the fundus or the proximal duodenum. When gastritis occurs, there is a reduction in acid secretion, and hence excess acid secretion is not the causative factor of the ulcers. High levels of glucocorticoids released during stress suppress the formation of prostaglandins which are required for the generation of the mucus gel barrier by the gastric epithelium. When this occurs, the mucosal barrier breaks down and the existing acid diffuses into the epithelium to damage the mucosa. At the same time the sympathetically induced vasoconstriction produces hypoxia of the tissue, enhancing the mucosal damage resulting in a decrease in the other protective mechanism such as the secretion of bicarbonate which normally helps to buffer the H^+ ions secreted in the stomach.

Insomnia and stress

Insomnia may occur in any one who is exposed to stress, whether acute or chronic. It is also possible that insomnia may be the cause of stress. Typically, such patients may take a long time to fall asleep (increased sleep latency),have frequent awakenings which are longer lasting than usual, and feel fatigued the following morning (A feeling of physical and mental tiredness). When acute stress is the cause, removal of the stressor usually restores a person's normal sleep pattern. Chronic stress is a greater problem. Even a minor degree of stress if applied chronically, may cause insomnia. Chronic psychological disturbances are more associated with producing sleep disturbances as compared with medical illnesses. The HPA axis is involved in the pathophysiology of insomnia. Experimentally, sleep disturbance has been known to be precipitated by infusions of CRH. Various studies have looked in to cortisol levels in patients with insomnia. The results have been conflicting, though the consensus of opinion seems to be that there is an overall increase in cortisol levels, indicating HPA over activity. Cortisol and ACTH values were highest in the evening and early night. A disturbance in circadian patterns of the cytokines IL6 and TNFα have also been noted.

Pre-optic hypothalamic nuclei involved in the sleep process are connected with various parts of the limbic cortex. Hence emotional disturbances may on their own be responsible for insomnia of stress. EEGs of stressed patients usually have a higher degree of fast activity signifying a higher level of arousal. In the long run, because of the hypercortisolism, the prevalence of Insulin resistance and obesity is higher in chronic insomniacs. Confounding this finding is the fact that insomniacs have a higher metabolic rate, and heart rate. *(Why is this finding confounding?)*. Stress induced insomnia has been directly linked with the appearance of anxiety and depression in patients and vice a versa.

The response to acute and chronic stress

The various components of the neuroendocrine system respond differentially to acute and chronic stress.

In acute stress, the HPA axis is driven mainly and rapidly by the CRH mechanism while AVP which is released by the parvocellular neurons is helpful. The hall mark of the response is the increase in amplitude of the CRH, AVP pulsations which however continue to maintain the circadian rhythm of their secretion. Other mediators such as cytokines increase the activity of the HPA. The brain RAS and the catecholaminergic neurons stimulate the sympathoadrenal pathway. The intensity of the latter also determines the duration over which the acute response lasts, the reciprocal connections from the HPA with the central coordinating neurons coordinating the attenuation. The endocrine response is accompanied by a behavioural one- the fight or flight response because of activation of the limbic and cerebral cortical activity. In chronic stress the main source of activation of the HPA is the AVP. The circadian pattern of secretion of CRH and AVP is disturbed and the expression of genes controlling the various secretions is enhanced probably because of permanent changes in them brought about by the chronicity of the stressor. Classically the changeover of the secretory pattern from circadian to non circadian is made out from the absence of the Cortisol Awakening Response (CAR) by measuring the salivary cortisol. The amplitude of the CAR is increased in acute stress, while its periodicity is maintained.

Endocrine engineered manifestations of stress

Exposure to stress invites the classical "General adaptation Syndrome" described by Selye. There are the three classical stages which he described: Alarm, resistance and exhaustion.

The endocrine response to acute stress is manifested by the activation of the sympathoadrenal system- the alarm stage or the classical fight or flight reaction of Cannon. This involves the cardio-respiratory system, pupillary dilatation, dry mouth, increased coagulability and a state of hyper arousal. The effectors are i. epinephrine (E) and norepinephrine (NE) from the adrenal medulla, and postganglionic sympathetic release of NE. The student should review the cardiovascular, respiratory, metabolic, effects of this activity from any standard text book Physiology. Normally, epinephrine forms the bulk of the endocrine secretion. However the ratios may be somewhat altered by the type of stress the individual is exposed to. For example, hypoxia, hypotension and psychological stress increase the release of NE in relation to E while hypoglycaemia is responsible mainly for the sudden surge of E. This explains the clinical findings of a full bounding pulse and raised blood pressure in a person who has sudden hypoglycaemia. A number of neuropeptides are released from the adrenal medulla. Some of these are: Endorphins and Encephalins, Adrenomedullin, Neuropeptide Y, Vasoactive Intestinal Polypeptide (VIP). They may exert their effects in various ways such as influencing catecholamine release by the A medulla, regulation of adrenal blood flow, modify vasoconstrictor or dilator responses. Some of them, for example the opiod peptides may help by producing analgesia as well as regulating the HPA axis in order to optimize the stress reaction while Neuropeptide Y may be helpful in situations such as septicaemia. The response of the adrenal medullary hormones released during stress is potentiated by the glucocorticoids which are also released at the same time. The latter are known to increase the vasoconstrictor effects of catecholamines by promoting the expression of α adrenergic receptors in the blood vessels, and perhaps influence the function of β adrenergic receptors.

Table III: Factors determining response to a stressful situation

1. Genetic/hereditary
2. Physiological/psychological reserves
3. Nutritionaland general health status
4. Age

Together with E, they make available more metabolic substrate by promoting glycogenolysis and gluconeogenesis. Ideally, the response invoked by a stressor should be adequate enough to contain that stressor in a way that the various suppressor activities that are aroused should be of benefit to the subject during that period of time, and not be responsible for producing any adverse effects. The degree of stress given/ experienced may be plotted against the physiological response attained, the relationship is best described as a S shaped curve. An upward-leftward shift denotes an excessive reaction while a downward-rightward shift reflects a response which is under par. A number of factors may determine as to where the response in a given individual would fall (Table III).

Stress and performance

Performance is affected by stress **(Fig. 3)**. It may be noted that graded, controlled application of stress improves performance (probably because of an increase in the arousal state) while excessive stress will make it deteriorate. This curve can also shift to the right during hypo arousal, while in hyper arousal, it moves to the left. This relationship becomes particularly relevant in many work place situations. In military aviation, a number of factors on a day to day basis may impinge upon the aviator which may shift the curve to either side of the expected norm. Multiple stressors acting in unison may rapidly bring an individual's performance to the descending limb of the curve, and be a major contributory factor in an aircraft accident.

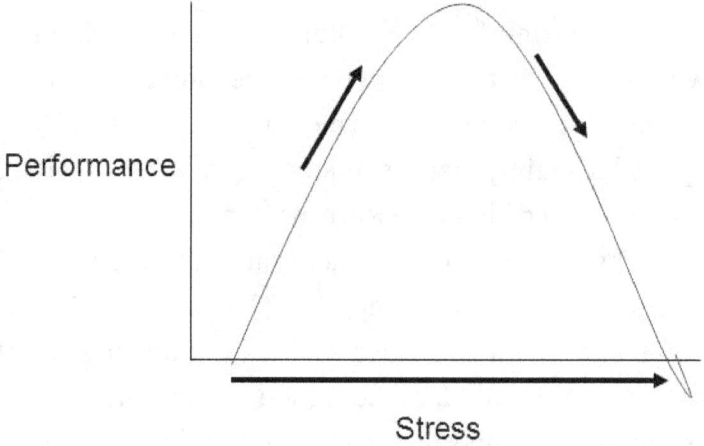

Fig.3 Stress vs performance

Pathophysiology of some important stressors

Stress of immobilization has been considered to be one that induces both severe physical as well as psychological stress. It is responsible for activation of the HPA axis as well as the catecholaminergic neurons in the brain and the periphery. In the brain, apart from the CA neurons, neurons in the hippocampus, hypothalamus and the cortex are brought in to play. The effector response goes via the sympatho-adrenal-medullary pathway and the HPA axis. Clinically, there is invoking of anxiety and fear, with involvement of the cardio-respiratory and other physiological effects. Immobilization may occur under various circumstances such as forced captivity, application of whole body plaster casts because of injury, being bedridden after a sudden spinal injury and paralysis. In the context of military aviation, long duration flights in single seat fighter aeroplanes where the pilot is subjected to near immobilization as s/he is strapped in to the ejection seat, are likely to generate the anxiety and discomfort which may result in producing severe stress in the aircrew and thus compromise flight safety.

Pain is an accepted stressor. The painful stimuli finally activate the efferent output to the HPA axis and the sympathoadrenal pathway via the paraventricular neurons in the hypothalamus. The limbic system and the somatosensory cortex form an integral part of the circuit involved. Other than physiological manifestations of stress induced hypercortisolism, chronic pain is often responsible for increasing anxiety levels, causing cognitive impairment, depression and mental confusion.

Ageing seems to increase basal activity of catecholaminergic neurons in the brain, and there by the activity of the sympathetic outflow. Nevertheless, the adrenal medullary output of epinephrine is definitely blunted in the elderly. The reason for this is not as yet deciphered. In the heart, there is an increase in the NE spill over attributed to a reduced neuronal uptake of the neurotransmitter after sympathetic activity. This may be related to an increased propensity to myocardial irritability in the elderly when subjected to acute stress. The overall ability of the elderly to cope with stress is definitely reduced, probably because of a depletion in body reserves-whether physiological or otherwise.

Strategies of adaptation to stress

By far the simplest strategy would be to remove the individual from the source of the stress or vice a versa. But this may or may not be feasible at all times. One of the most important mechanisms in use by the body is the physiological and psychological reserve that a person has. This is typically demonstrated by the ability of a physically fit person to cope with an exercise load because s/he has a higher cardiac reserve than the average individual. Also the oxygen carrying capacity is higher because of a higher RBC count etc. A psychological "hardiness' has been associated with a better ability to cope with stress. Such individuals may accept a stress as a challenge rather than as a threat, and therefore respond better to a given situation. An overall satisfactory nutrition status helps in meeting with demands at times of stress. From Fig. 3 it may be deduced that controlled stress given as a part of training will help in keeping performance as close to the optimal as possible. Repeated exposure to the same stress will eventually help an individual to overcome it to a large extent.

The Neuroendocrine responses are genetically influenced to a large extent. Genes may be responsible for ensuring that an appropriate physiological response is produced when an individual is exposed to a stressful environment.

Some manifestations of stress

A number of situations may give stress to an individual. The usual response is the classical one comprising of cardiovascular, respiratory and metabolic changes. Of the latter of particular importance is the hyperglycaemic response. These are often accompanied by psychological features such as anxiety and insomnia. Removal of the stress by management of the condition that has brought it on will rapidly resolve the problem. Chronic stress situations are of a greater concern. One of the major conditions is the Post Traumatic Stress Disorder. The advent of Positron Emission Tomography (PET) and Functional Magnetic Resonance Imaging scanning has highlighted the involvement of the amygdale and he hippocampus in this disorder.

References

1. Benca R. The impact of stress on insomnia. MedScape Neurology@2007 medscape.

2. Carrasco G, Van de Kar L. Neuroendocrine pharmacology of stress. Europ. J. Pharmacol. 2003; 463: 235–272

3. Cattaneo A. Sleep disturbances in aging: circadian clock disruption or altered perception of environmental light? www.geragogia.net/editoriali/sleepdisturbances.html/9/9/2011

4. Chrousos G, Vgontzas A, Kritikou I.HPA axis and sleep. Chap. 31 in Adrenal Physiology and Diseases. Edt. Chrousos G. Endotext.com 2004

5. Fahie-Wilson MN, John R, Ellis AR. Macroprolactin; high molecular mass forms of circulating prolactin. Ann. Clin. Biochem. 2005; 42: 175–192.

6. Karatsoreos I, Bhagat S,Bloss E, Morrison J, McEwen B. Disruption of circadian clocks has ramifications for metabolism, brain, and behavior. www.pnas.org/cgi/doi/ 10.1073/pnas.1018375108

7. Kvetnansky R, Sabban E, Palkovits M. Catecholaminergic Systems in Stress: Structural and molecular genetic approaches. Physiological Reviews. 2009; 89: 535–606

8. Larry Jameson L. Principles of Endocrinology. Chap. 317 inHarrison's Principles of Internal Medicine." Edited. Kasper DL, Braunwald E, Fauci AS, Hauser SL, Longo DL, Jameson JL.16th ed. MCGraw Hill, London. Pp 2067–2075.2005.

9. Latta F, van Couter E. Sleep and biological clocks. Chapter 13 in Hand Book of Psychology. Ed. Gallgher M, Nelson R, Weiner J. John Wiley & Sons, London. 2003.Pp. 355–377.

10. Magri F, Cravello L, Barili L, Sarra S, Cinchetti W, Salmoiraghi F, Micale G, Ferrari Stress and dementia: the role of the hypothalamic pituitary-adrenal axis. Aging Clin. Exp. Res. 2006; 18: 267–270

11. Magri F, Locatelli M, Balza G, Molla G, Cuzzoni G, Fioravanti M, Solerte SB, Ferrari E. Endocrine circadian rhythms as markers of physiological and pathological brainaging. Chronobiology Int. 1997; 14: 385–396.

12. Maitra A, Abbas AK. The Endocrine system. Chap. 24 in Robbins & Cotran Pathologic Basis of Disease. Edt. Kumar V, Abbas AK, Fausto N. 7th Edn. Elsevier Saunders, Penn. Pp.1155–1225; 2005.

13. Mallis M, DeRoshia CCircadian rhythms, sleep, and performance in space. AviatSpaceEnviron Med. 2005; 76; Suppl. 6: B94 - B107.

14. McEwen BS. Neuroendocrine interactions. Psychopharmacology- The Fourth Generation of Progress. Edt. Bloom F, Kupfer D. Raven Press, New York. 1995; pp 715–718

15. Mohan Kumar V. Sleep and sleep disorders. Indian J Chest Dis. Allied Sci. 2008; 50: 129–135

16. Molina PE, Ashman R. Endocrine Physiology. 3rd Edn. Mc Graw Hill, London. 2010

17. Paz-Filho G, Wong Ma-Li, Licinio J. Circadian rhythms of HPA axis and stress. Chapt. 27. In Adrenal Physiology and Diseases. Edt. Chrousos G. Endotext.com 2004

18. Pani L, Porcella A, Gessa G. The role of stress in the pathophysiology of the dopaminergic system. Mol. Psychiatry. 2000; 5: 14–21.

19. Porth C. Stress and Adaptation. Essentials of Pathophysiology: Concepts of Altered Health States.3rd Edn. Lippincott, Williams and Wilkins. 2011; pp 209–223.

20. Saavidra J. Brain Angiotensin II: new developments, unanswered questions and therapeutic opportunities. Cell Mol. Neurobiology. 2005; 25: 485–512

21. Seals DR, Esler MD. Human ageing and the sympathoadrenal system. J Physiology. 2000; 528: 407–417

22. Souhami RL, Moxham J.Text Book of Medicine; 2nd Edn. Churchill Livingstone, London; pp 657–665; 1995

23. Stratakis C, Chorusos G. Neuroendocrinology and pathophysiology of the stress system. Annals of the New York Acad. Sciences.2006; 771: 1–17. Wiley Online Library of 17 Dec 2006

24. Tsigos C, Kyrou I, Chrousos G. Stress, Endocrine Physiology and Pathophysiology. Chapt. 8 in Adrenal Physiology and Diseases. Edt. Chrousos G. Endotext.com 2004:

25. Turek FW, van Reeth O. Circadian rhythms. Chapter 58 in The Handbook of Physiology. Section 4. Environmental Physiology. Vol. II. Edited Fregly MJ, Blatteis CM. American Physiological Society. Oxford.1996; pp 1329–1359.

26. Young J, Landsberg L. Synthesis, storage and secretion of adrenal medullary hormones: physiology and pathophysiology. Chapter 1 in Coping with the Environment: Neural and Endocrine Mechanisms. Hand Book of Physiology. Section 7. Volume IV. Edt. McEwen B, Maurice Goodman H. Oxford Univ. Press. 2001. pp 3–19.

27. Van Someren E. More than a marker: Interaction between the circadian regulation of temperature and sleep, age related changes, and treatment possibilities. Chronobiology Int. 2000; 17: 313–354.

28. Vitaterna M, Takahashi JS, Turek FW. Overview of circadian rhythms. National Institute of Alcohol Abuse and Alcoholism. //pubs.niaa.nih.gov/pub./arh25–2/85–93.htm dated 31/5. 2011

29. Waterhouse J, Reilly T, Atkinson G. Jet-lag – a review. Lancet.1997; 350: 1611–1617.

30. Wirz-Justice A. Biological rhythms in mood disorders. in Psychopharmacology- The Third Generation of Progress. Edt. Bloom F, Kupfer D. Raven Press, New York. 1995; pp 999–1017.

CHAPTER - 11

Pathophysiology of Common Blood Disorders

General Considerations

Plan

1. General considerations
   i. Constitution and functions of blood; ii.Blood rheology
2. Pathophysiology involving red blood cells
3. White blood cells in the inflammatory response
4. Response of WBCs to altered body environment: Leukocytosis, leukopenia
5. Pathophysiology of Disseminated Intravascular Coagulopathy

Blood which constitutes about 8% of our body weight, by itself is a peculiar fluid for it contains not only extracellular fluid (plasma 55%) but also intracellular fluid within the cells (45%) that float in it. At this point of time, the reader may review the normal constituents of blood **(Fig.1)**. It is a fluid that can be displaced rapidly from one part of the body to another with change of posture and gravitational force. Plasma probably originates as a secretion of intra-embryonic coelom which enters the veins because the lining of the coelom in the region is particularly thin. The yolk sac as the origin of the red blood cells (RBC) is well documented, but the origin of the hemopoietic cells from which the myeloid and lymphoid stem cells arise is still debatable. It has been suggested that these may arise from the mesoderm of the aorta/gonad mesonephros. In the postnatal life, the bone marrow is the source of all the cells concerned.

A quick referral to the details of properties and functions of blood outlined in **table I** here may be made whenever required from any standard book. Most of this chapter is devoted to the more common elements in blood viz the Red Blood Cells, and their pathophysiology. However, an important aspect of physical properties of blood is its viscosity which has considerable influence on blood flow. This aspect is generally not addressed in standard text books of Physiology and Pathophysiology. An attempt has been made to highlight it here. Only some of the salient features of white blood cells and their role in infection have been covered

Table I (a). Properties of blood

1. At 37°C Specific gravity of whole blood around 1.051, and plasma at about 1.021, and relative viscosity about 1.8 times that of water.

2. pH 7.35-7.45

3. Constitutes about 20% of ECF

4. Volume about 5 l (70 ml/Kg body wt of which RBCs constitute about 30ml/Kg.

   May be sampled by venipuncture, finger prick/heel prick, and arterial puncture

Table I (b). Functions of blood

1. Transportation of respiratory gases, nutrients, hormones, waste products etc

2. Protective

   i. Phagocytic and Immune functions

   ii. Coagulation

3. Body temperature control

4. pH regulation

5. Blood group antigens

Fig. 1. Constitution of blood. Almost all the cell volume is made up by RBCs. The buffy coat in a centrifuged tube between the plasma and the RBCs contains 1-2% of the WBCs and platelets.

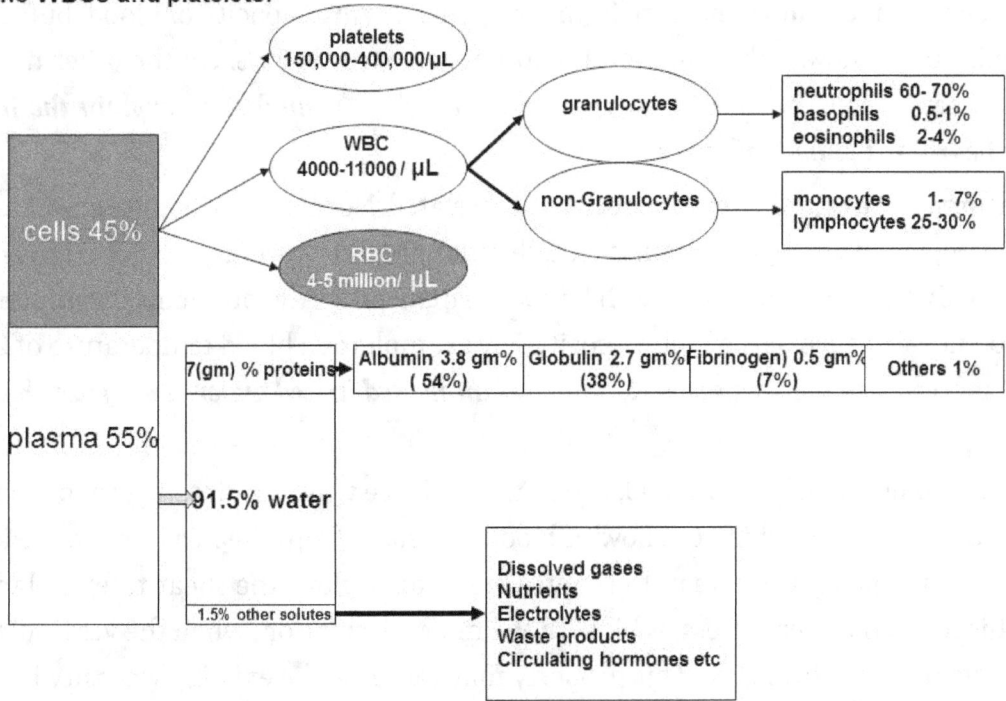

Rheology of Blood: The Relevance of Blood Viscosity to Physiology of Blood Flow

When a fluid flows through a tube, its various layers rub against one another to generate a frictional force which imparts viscosity to the fluid. Viscosity (measured as centipoises-cP in the cgs system, or as pascal-seconds (Pa.s) in the SI system is responsible for contributing to the resistance to free flow of fluid/liquid through a tube. In some fluids eg water, the viscosity, and hence the resistance to flow, is independent of velocity of fluid flow as long as the flow is not turbulent. Such fluids are called Newtonian fluids. Blood on the other hand is termed a non-Newtonian fluid, thereby implying that changes in its velocity of flow will be associated with changes in its viscosity, and hence the resistance to flow. This affects blood flow through tissues of humans in health and disease.

Plasma is mostly water but contains organic and inorganic substances such as proteins and electrolytes which interact with each other and the water molecules. This increases its viscosity relative to that of water by 1.8, but this figure does not vary over a wide range of hematocrit, thus making plasma a near Newtonian fluid like water. Whole blood on the other hand because of its cells in addition to the plasma, is far "thicker." The relationship of viscosity to the hematocrit is curvilinear, with a hematocrit of 20% recording a viscosity of about 2 times that of water, while that of 60% measures about 8 times that of water, with the normal hematocrit value of 40–45% resulting in a viscosity which is about 4 times that of water. Therefore, **hematocrit of blood** is an important factor which affects viscosity. Higher the hematocrit therefore, greater is the resistance offered to flow. Hence the polycythemia induced by hypoxia in physiological situations such as high altitude exposure or in Chronic Obstructive Pulmonary disease, or to that matter the abnormally high RBC count inbone marrow disease such as Polycythemia vera, may induce a compensatory increase in the oxygen carrying capacity of blood, but is likely to compromise blood flow if the hematocrit approaches/exceeds 60%. On the other hand, lower the hematocrit, faster the blood flow as seen in anemia. *(How do you explain the functional murmur heard in a patient of severe anemia).*

Blood **temperature** and viscosity are inversely related. Normally the body temperature range being limited, blood viscosity remains virtually unaltered. However in accidental hypothermia when the body temperature falls to well below normal, one of the problems encountered is the slow blood flow because of the high viscosity. For example, at a blood temperature of 20° C, the viscosity increases to 10centipoises. *(If a hand is immersed in iced water, it may turn blue. Please explain)*

When blood flows through a vessel, it generates a force which is directly proportional to the viscosity and the rate at which the flow of blood increases from the near zero value along the vessel wall to its maximum rate in the centre of the axial flow (the shear rate). In large blood vessels this force, the shear stress, is high. In the microcirculation, when the vessel diameter is small, the velocity of flow is low. This property may increase the existing viscosity because the slow flow permits rouleaux formation which further reduces the flow of blood. This natural tendency of RBCs to adhere to one another may increase under pathological conditions such as sickle cell disease, malaria and Diabetes Mellitus and after myocardial infarction. However, the extent to which this phenomenon actually takes place and changes blood flow adversely in the situations listed above is as yet undecided. White blood cells (WBCs) are far more rigid as compared with the RBC and hence do not deform easily during their passage through the capillaries. This has been attributed more to the high viscosity of fluid inside the WBCs.

Of particular significance is the increase in rigidity and viscosity of activated neutrophils. These may set up a vicious cycle of increased release of free oxygen radicals which in turn may further increase the viscosity and rigidity of the already deformed neutrophils, worsening the local situation. Under normal physiological conditions, the **Fahraeus-Lindqvist** phenomenon comes in to play. By this, the hematocrit at the small arterioles and capillaries is automatically reduced so that even at the low flow rates, the increase in viscosity because of the reduced hematocrit, is not allowed to occur. This helps in the maintenance of blood flow through the smallest diameter blood vessels. In fact in states of hypo-volemic shock, one of the factors that may play a large role in producing hypoxic cell damage is the fact that peripheral circulation becomes highly compromised because this phenomenon fails. Intravenous fluid administered in time help to overcome it and re-establish peripheral circulation.

The basic pathophysiology of an increase in blood viscosity is the decrease in blood flow. This may occur in various conditions.

Exposure to high altitude terrain induces an increase in hematocrit. If this increase exceeds 60%, the cardiac output decreases because of an increase in viscosity which may reach a value which is 8–10 times that of water, and about 2ce that of the viscosity of normal blood. This is one of the major factors that constitutes the syndrome described as Seroche Monge Disease (Failure to acclimatize to high altitude or Chronic Mountain sickness).

Table II:

A. Morphologic characteristics of RBC

Volume	Cell size	Hb content (colour)
Normal	normocytic	normochromic
Increased	macrocytic	normochromic
Decreased	microcytic	hypochromic

B. Indices

1. Mean corpuscular volume: MCV

    = PCVper 100ml/RBC count (78-94 μm^3)

2. Mean Corpuscular Hb (MCH); (averageamount of Hb per RBC)

    = Hb gm%/RBC count (28-32 pgm)

3. Mean Corpuscular Hb Concentration (MCHC)

Hb concentration in a single RBC 32-38%(average33%)

normo-chromic cells: MCHC within range; hypo-chromic:<32%

(No such thing as HYPERCHROMIC anemia)

Pathophysiology of RBC

Normal Erythropoiesis

All formed elements of the blood in the postnatal life originate from the pluripotent hemopoietic stem cells which have a multiple potential to develop into either the myeloid or the lymphoid series of cells. Some of the precursor stem cells of the bone marrow do find their way in to the blood stream, but do not proliferate. The latter happens only in the bone marrow and therefore the specialized environment of the bone marrow is mandatory for this process. The RBCs are a descendant of the committed stem cells of the myeloid series which are acted upon by IL 11 and thrombopoietin which help in the formation of the Colony Forming Unit Erythrocyte (CFU-E). The initial differentiation seems to be an inherent property of the precursor stem cell itself, and cytokines involved probablyact only as catalysts. Only at the stage of CFU –Eformation, the influence of erythropoietin (EPO) as a regulatory factor becomes established. The first recognizable precursor of the RBC is the proerythroblast (**Fig. 2**). (*Reference to Fig. 2: Why is it that cells in stages 4 and 5 during development are unable to synthesize DNA nor undergo mitosis?*). The RBCs released in to circulation are specially terminally differentiated, specialized cells with a fixed life span. It is obvious therefore that arrangements exist to replenish them and their precursor cells in a regular fashion. EPO is required availability results in a decrease in RBC production (*Patients of chronic renal failure suffer from fairly severe anemia-Explain*). Other factors necessary for production of RBCs are i. iron ii. Folic Acid, Vitamins B12, C, Copper, testosterone. The physical characteristics of the RBC may be revisited from a standard text book. It is important to realize that the cell membrane needs to be highly flexible if the RBC has to wind its way through capillaries which may be smaller than their diameter of about 7μ. Other physical attributes of these cells knon as blood indices are given in **Table II.** Erythropoiesis is carried out in the fetus in the yolk sac and in later months in the liver and spleen. After birth, until the age of about 20 years, it occurs in the long bones. Thereafter, this function is carried out by the flat bones. In adults, the liver and the spleen may take on the responsibility of erythropoiesis under severely stressful situations (Extra medullary erythropoiesis).

Fig. 2. Steps involved in the maturation process of RBC

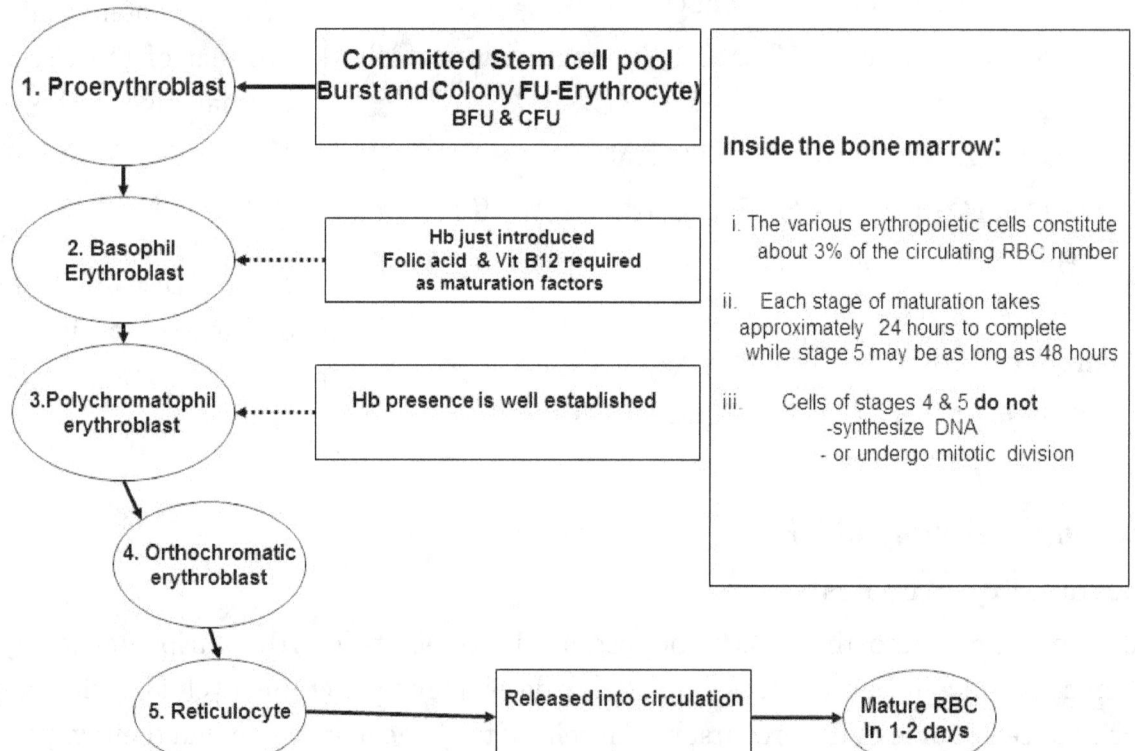

Anemia is by far the commonest blood disorder worldwide. It has been defined variously as a
i. reduced level of circulating haemoglobin (Hb) or RBCs; ii. reduced quantity and quality of Hb
and number of circulating RBCs; or iii. as a deficiency of Hb as result of reduced number of RBCs
or a reduced synthesis of Hb. Various ways in which anemia may be classified is given in **Fig.
3.** A patient with anemia needs to be investigated without delay in order to confirm the type of
erythropoietic disturbance that is involved. This should be done before instituting a treatment
because the initial status is likely to be altered after treatment is started. This in turn may
mislead the direction of ongoing investigation being done to identify the erythropoietic defect.
A clinical classification may be in order to circumvent the problem. The approach is based on
derivation of either **the Corrected reticulocyte % index** and under certain circumstances, the
Reticulocyte production index.

Relevance of Reticulocytes in Anemia

The reticulocyte is the last stage in the development of the RBC. It contains remnants of a
pyknotic nucleus which is incapable of DNA synthesis. Once the reticulocyte enters the
blood stream, the condensed nucleus is thrown out and the cell changes in to a mature RBC.
Hemoglobin synthesis in the cell is still ongoing, and gradually tapers off within 2–3 days.
Normally, about 1% of the RBCs are reticulocytes. They have to be specially stained using a supra
vital dye if they are to be visualized on a peripheral smear. The presence of reticulocytes is an
indicator of effective RBC production by the bone marrow in response to erythropoietin (EPO)
and confirms that the bone marrow is functioning normally with adequate supply of iron.
Variations in the reticulocyte count therefore may be used to derive the bone marrow response
to anemia. It is possible that the reticulocyte count may have increased without an increase
in absolute RBC count, and may thus be a fallacious presentation of bone marrow response to
anemia. In order to correct for this possible fallacy, the % reticulocyte count is calculated for
the prevalent hematocrit (corrected Hct) of the patient) to provide the **Corrected reticulocyte
index** (Reticulocyte % in the patient's RBC) x (patient's Hct)/ 45% (which may be taken as the
normal Hct, given the age and sex). This gives a true picture of the bone marrow response to
anemia, and may be used to classify the type of anemia clinically. Higher the corrected index,
more active the bone marrow.

With a normal Hct of 45%, the developing erythrocytic cells spend about 4 days in the bone
marrow, and 1 day in the peripheral circulation before they become mature RBCs. But when a
demand is made on the bone marrow to speed up the manufacture of RBCs, the ratio tends to
reverse. For example with anemia severe enough to produce a Hct of about 25%, the RBC passes
through the bone marrow development stages to that of a reticulocyte in about 2–2.5 days, but
remains a reticulocyte in the blood for about 1.5–2days.So even if the total maturation time of
the RBC remains unaltered (about 4–5days) there is a paradigm shift in the number of days it
lives in circulation as a reticulocyte before it matures into a RBC. This is called a **shift factor**,
and is normally taken as 2 (days) because the reticulocyte does not take more than 2 days to
mature into a RBC. It may be used to calculate an index called the **Reticulocyte production**

index (Absolute reticulocyte %) x (corrected Hct /shift factor). If this index is <2, it indicates a hypo-functioning bone marrow, while if > 3, it indicates a hyperactive bone marrow responding well to EPO. However for this parameter to be derived, it is necessary that the peripheral smear must show the presence of premature polychromatophil cells. If not, the corrected reticulocyte index may be used. Based on this, anemia may be classified as hypo or hyper proliferative (Fig. 3).

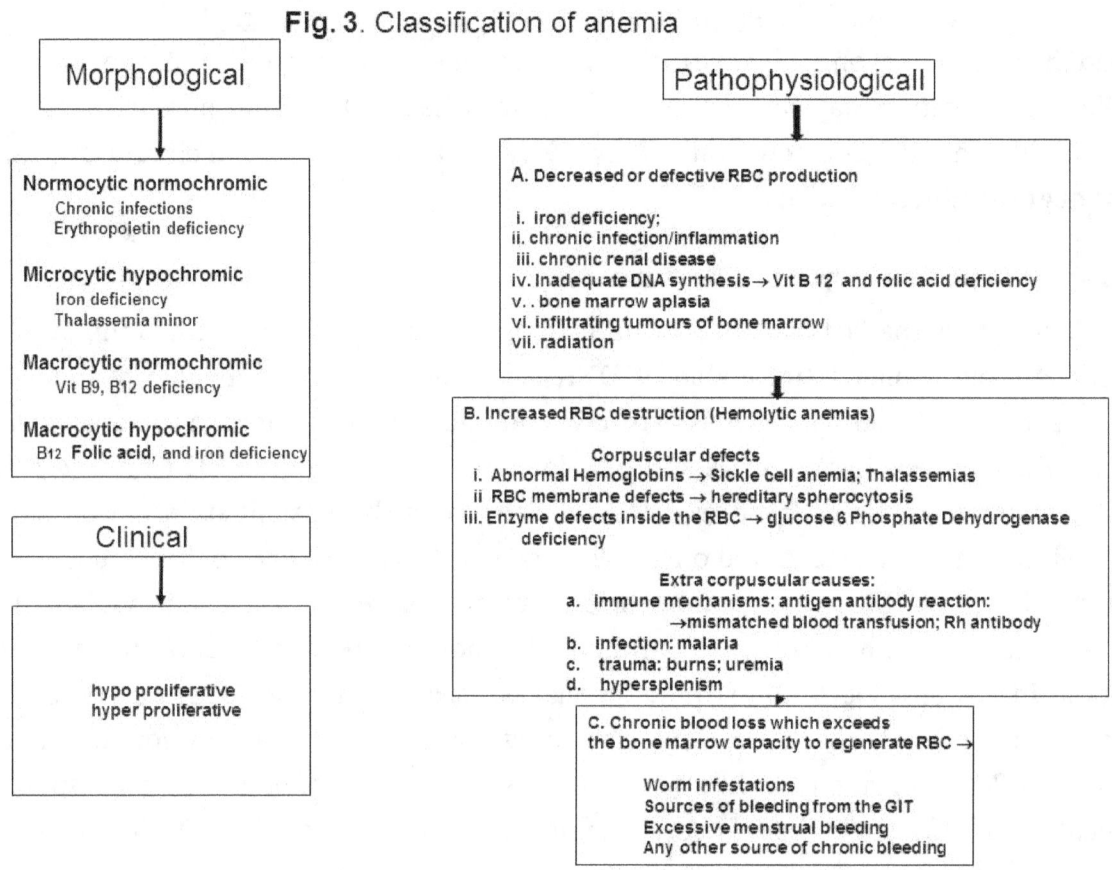

Fig. 3. Classification of anemia

Pathophysiology of Anemia and Its Correlation to the Symptom Complex

Whatever the cause of anemia, the basic alteration in physiology which is responsible for the symptom complex is tissue hypoxia. (***What is the difference between hypoxia of anemic origin and hypoxic hypoxia?***). Anemia is said to have set in when the Hb in males falls to <13 gm% and in females to <12 gm%. The pathophysiological effects and the clinical symptoms and signs which follow as a result are determined by i. the severity ii. the rapidity of onset and iii. the age.

Compensatory mechanisms camouflage symptoms until the Hb reduces by 2–3gm%. The three main mechanisms of compensation are i. increase in cardiac output. ii. redistribution of blood flow. iii. shift of the Oxy-hemoglobin dissociation curve to the right. More blood is diverted to the heart, brain and musculature occurs from the skin, GIT and renal vascular beds. The redistribution from the skin is one of the first compensatory mechanisms. This accounts for the pale conjunctiva and the skin. There is an increase in cardiac output because of arteriolar dilatation secondary to tissue hypoxia. There is some compensatory increase in plasma volume to compensate for the reduced RBC mass. Therefore the total blood volume remains unchanged.

Table III:

Common symptoms of anemia graded as per severity

A. **Mild (Hb between 11-13gm%)**

   i. early fatigue and shortness of breath during moderately severe exercise

B. **Moderate (Hb <~ 11 to ~ 8 gm%)**

   i. Easy fatigue

   ii. Breathlessness on mild to moderate exercise

   iii. GIT symptoms: anorexia; epigastric pain (often when worm infestation present)

   iv. Irritability; Lack of concentration

C. **Severe(Hb <7- 8 gm%)**

   i. Claudication pain in muscles because of ischemia

   ii. Dizziness; Weakness and feeling of ill health (malaise)

   iii. Inability to do exercise because of rapid onset fatigue and severe breathlessness

   iv. Breathlessness at rest when high output heart failure occurs

   v. Dysphagia (more common in women)

Symptoms/signs associated with some specific types of anemia

D. **Hemolytic anemias**

   i. jaundice; ii. Hematuria iii. ischemic pain (sickle cell anemia); iv. hepato-splenomegaly

E. **Anemia because of hypo functioning of bone marrow**

   i Petechial hemorrhages due to associated thrombocytopenia; ii Infections due to associated leucopenia

At the same time, the oxy-hemoglobin dissociation curve shifts to the right because of increase in 2,3 Biphosphoglycerate (2,3 BPG), facilitating off loading of O_2 at higher O_2 tension (shift of the P50 to the right). During this phase, the deficit in tissue oxygenation may be clinically reflected as breathlessness and relatively early fatigue during severe exertion which was being easily tolerated earlier. The end result of these adjustments is that the presence of anemia may become obvious only under stressful circumstances such as exercise.

The Hyper-dynamic Circulation in Anemia

In the early stages of anemia, exercise tolerance is reduced. This is evident from the onset of fatigue and breathlessness at a heart rate which is below the predicted heart rate for the given level of exercise. The typical presentation in moderately severe to severe anemia is a resting tachycardia with increase in systolic pressure and a lowering of the diastolic pressure. Anemia reduces blood viscosity, increasing the blood flow velocity. Clinically, a systolic cardiac murmur is the result. Peripheral vasodilatation occurs because of tissue hypoxia and the accumulation of lactic acid. The increase in pre load increases the stroke volume and therefore the systolic blood pressure, while the peripheral vasodilatation brings about a decrease in the diastolic pressure. The tachycardia is a reflex response to deactivation of carotid baroreceptors because of widening of the pulse pressure. With increasing severity of anemia, the sympathetic drive induces Na^+ and fluid retention by the kidneys via the Renin Angiotensin II mechanism. There is thus a further increase in cardiac output. The sustained increase in this parameter with high levels of Angiotensin II are responsible for remodelling and hypertrophy of the left ventricle and the arterioles, which promote the development of failure. A patient with pre-existing myocardial ischemia is subject to greater morbidity and mortality if anemia co-exists.

So far, anemia as a cause of heart failure has been described. The converse is also true. Patients with heart failure have been known to develop anemia. A number of possible mechanisms have been described to explain this aspect. Heart failure is often accompanied by a cachexia syndrome which is attributed to the secretion of inflammatory cytokines of which TNFα is by far the most talked about. The secretions are known to reduce appetite and thus reduce intake of nutritional elements required for normal manufacture and maturation of RBCs. They may also inhibit bone marrow function. Low cardiac output in CHF may be responsible for poor renal perfusion and renal malfunction, giving rise to EPO deficiency and anemia. Angiotensin converting Enzyme (ACE) inhibitors have been used extensively in the management of heart failure. There is mounting evidence that use of these agents may be associated with the development of anemia in CHF patients. It has been shown that ACE normally hydrolyses a compound N-acetyl-seryl-aspartyl-lysyl-proline (Ac-SDKP) which is a natural inhibitor of erythropoiesis. When ACE becomes deficient as result of treatment with ACE inhibitors, hydrolysis of this compound is compromised, and its inhibitory activity of erythropoiesis dominates, contributing to the production of anemia. Higher circulating levels of Ac-SDKP have been found in patients who have an otherwise unidentifiable cause of anemia. The condition develops insidiously over a period of time.

Respiratory Effects of Anemia

The disproportionate perception of respiratory discomfort during exercise is one of the earliest symptoms of mild to moderate anemia. Simplistically put, the system is unable to cope with the increased O_2 demand of exercise in a situation already compromised with tissue hypoxia. It is however difficult to explain the pathophysiology which is responsible. The arterial blood PO_2 in such patients is within the normal range, and the PCO_2 may actually reduce with maximal exercise. Hence stimulation of peripheral chemoreceptors is not the cause of the exertional dyspnea. It is possible that tissue hypoxia which is already present is exaggerated by the increased demands on the working muscles and is signalled by muscle afferents to produce the sensation of respiratory discomfort. The muscle ischemia may release chemical mediators such lactic acid which in turn may stimulate the peripheral chemoreceptors to increase the ventilation out of proportion to the required effort and thus produce the symptom. Lactic acid is capable of utilizing glucose transporters to cross the blood brain barrier. If this happens, it may be surmised that central chemoreceptor stimulation by the H^+ ions carried in by the lactic acid may also play a role in stimulating ventilation. Patients with severe anemia may have increased ventilation at rest. The pathophysiology involved here is more difficult to explain. Normally, on a long term basis, excessive production of lactic acid is easily metabolized by the liver. But in very severe anemia the liver's capacity for this action may be overwhelmed, and lactic acidosis may prevail over a period of time. Only a few cases of such a combination have been reported. Under the circumstances, ventilation may get stimulated via excitation of peripheral and central chemoreceptors and account for the increased rate and depth of ventilation (Kussmaul breathing) even at rest.

Anemia of sickle cell disease (SCD) brings its own problems as far the lungs are concerned. These may present as either acute or chronic lung disease rather than issues concerning control of ventilation. Acute manifestations may present as asthma, acute chest syndrome or thromoboembolism. Sudden onset of sickling may cause aggregation of the RBCs in the lung vasculature, with release of various inflammatory cytokines. These may be responsible for a bronchospasm. The aggregation of RBCs may induce an infarction with damage to the lung tissue with fluid accumulation which in turn is responsible for a diffusion deficit. There is a tendency towards a state of hyper coagulation initiated by the RBC clumping and enhancement of thromoboembolism setting up a vicious cycle which extends the area of lung parenchymal involvement. The overall effect is the clinical presentation of the acute chest syndrome consisting of severe chest pain, bronchospasm, tachypnea and hypoxemia. In the chronic form, the airway often becomes hyper reactive, and the chronic inflammation set up in the lungs continues the airway obstructive disease pattern. The patches of recurrent lung parenchymal damage probably heal by fibrosis. The outcome is a combined obstructive, restrictive lung disease pattern.

The Central Nervous System (CNS) in Anemia

Vague symptoms pertaining to the CNS are often reported by patients with anemia. These are listlessness, irritability, inability to concentrate, headache, and dizziness, and are particularly noticeable in school-going children. The pathophysiology involved is obscure. The headache may be partly explained by the hypoxic vasodilatation of the brain vasculature. Some other symptoms are thought to be caused by hypoxia of the brain tissue in general, but there is no definite explanation as to the pathophysiology involved. In very severe anemia, brain edema may be responsible for the more severe symptoms and signs of CNS involvement. CNS (spinal cord) disorders in Vit B_{12} deficiency anemia are however better documented. There is myelin degeneration in the posterior column and the lateral columns though the spinothalamic tract sensations are less commonly involved. A mixed clinical picture of both sensory and motor involvement is seen (that is why "Combined degeneration of the spinal cord.") and it is at this stage that the anemia becomes "Pernicious." The mechanism as to how the demyelination occurs is still debated. S-adenosyl methionine (SAM) is required for the formation of phosphatidyl choline necessary for myelin synthesis. SAM comes from methylation of homocysteine, a reaction in which Vit B$_{12}$ acts as a coenzyme. When the vitamin is deficient, myelin deposition is affected adversely. In severe deficiency, brain de-myelination may also occur and present as irritability, dementia or psychosis. More recently, it has been postulated that absence/deficiency of Vit B$_{12}$ leads to formation of neurotoxic cytokines such as IL6 which may be responsible for the demyelination.

Gastrointestinal Symptoms in Anemia

Many of these are vague. For example, anorexia, epigastric pain and discomfort are often reported by patients with iron deficiency anemia (IDA) and may be caused by worm infestation and not the anemia per se. Similarly, symptoms of chronic gastritis occur with Vit B$_{12}$ deficiency, but again, it is the basic stomach pathology that is responsible for the symptoms rather than the anemia. For unknown reasons, IDA in children is associated with "Pica" which is predilection to eating mud. Increased fondness for consuming ice (Pagophagia) is also reported. Severe IDA, particularly in females, is known to cause dysphagia because of development of a lower pharyngeal membrane or web (The Plummer-Vinson syndrome). In severe anemia, other epithelial tissue involvement may present as a glossitis, again in patients with IDA.

Symptoms/signs Associated with Some Specific Types of Anemia

A modest degree of **Jaundice** is often a classical clinical sign of rapid hemolysis of RBCs. It is of the pre-hepatic variety and the indirect bilirubin level is usually less than 4 mg%. If the jaundice is more severe, hepatic or renal disease may be associated. *(Why does jaundice occur at all in haemolytic anemia, even when the liver may be quite normal, and why is it mostly because of an increase in the indirect bilirubin?)*

Hemoglobinuria may follow a bout of rapid extensive INTRAVASCULAR hemolysis.

Splenomegaly may be the presenting sign in certain types of haemolytic anemias (hereditary spherocytosis). Precipitation of gall stones may occur in haemolytic anemias, particularly those in which chronic repeated bouts of hemolysis happen. *(Explain?)*

Pathophysiology of Hemolysis in Hemolytic Anemia

The three common issues that engulf hemolytic anemias in general are: 1. premature destruction of RBCs resulting in their reduced life span. 2. increase in erythropoiesis by the bone marrow as a compensatory mechanism triggered by an increase in erythropoietin secretion by the kidneys. 3. increase in blood levels of hemoglobin breakdown products. Hemolysis may be INTRAVASCULAR or EXTRAVASCULAR. In the former variety, the RBC destruction is caused by trauma to the RBC as in i. cardiac valvular defects which generate abnormal eddies and currents in the blood flowing through the damaged valves; ii. continuous physical trauma to the blood in long distance running, walking (March hemoglobinuria); iii. parasitic infections (Falciparum malarae); iv. release of toxins during sepsis and v. burns. In patients with severe burns, the RBCs may undergo fragmentation and also change shape to spherocytes, which in turn will hasten their breakdown. Other forms of RBC destruction may be precipitated by immune mechanisms such as mismatched blood transfusion, presence of other circulating antibodies, direct trauma to the red cells, or hyperactivity of the spleen. Typical features of intravascular hemolysis are hemoglobinuria and jaundice. Normally the free Hemoglobin (Hb) released on breakdown of RBCs is promptly mopped by a protein which is an alpha 2 globulin. The combined product is easily dealt with by macrophages. This prevents the escape of Hb in to the urine. However for unknown reasons, the α2 globulin becomes deficient when intravascular hemolysis occurs. Most of the free Hb which then cannot be treated by macrophages, gets oxidized to the brown colored methemoglobin. Much of it is reabsorbed in the proximal convoluted tubule, but some of it escapes in to the urine along with the free Hb, giving it a brown color.

Extra vascular hemolysis occurs whenever the RBCs are altered in shape. This makes their membranes rigid which makes it difficult for them to negotiate the narrow sinusoids in the spleen by altering their shape as the normal RBCs do.. This damages the cell membrane of the RBC and hence they undergo hemolysis. Such RBCs often contain abnormal hemoglobins as in Sickle cell disease and are classified as hemoglobinopathies. As the RBC destruction takes place mainly in the spleen, splenomegaly is the result. Suffice to say that splenectomy is often a mode of treatment in hemolytic anemias which have cell membrane defects as in hereditary spherocytosis, sickle cell anemia. Many of these patients develop overactive bone marrow which may lead skull and skeletal deformities.

Hemoglobinopathies and Hemolytic Anemias in India

Hemoglobinopathies are conditions in which the hemoglobin produced is abnormal. Sickle cell disease (HbS), Thalassemia A and B, are the well recognized conditions. Other Hb variants such as Hb E and Hb D (Punjab) have also been reported.

Sickle cell anemia is found quite commonly in the tribal belts of South India, Assam, Bihar, Orissa and Maharashtra, and as many as 20 million people are affected by it. The average frequency of this disease in India has been reported to be4.3% with a maximum of 9.3% in Orissa. The sickle cell trait as a single gene probably originated as a protective measure against Falciparum malaria. It is believed that these cells responded to malaria infestation by sickling. The sickled cells were promptly destroyed along with their parasites by the reticulo-endothelial system where the environment was hostile to the malarial parasite. The genetic defect was probably imported to India in African soldiers who worked as body guards for Indian princes. It has also been reported that the sickle cell gene haplotypes of affected Indians and habitants of the Persian Gulf are identical, but no satisfactory explanation has emerged for this observation. It has been observed that the overall severity of the disease is less in some countries including India. It is possible that other genes present in these populations are responsible for this phenomenon.

The HbS which characterizes this condition is a beta chain defect. The glutamic acid in the β chain is replaced by valine. In homozygous sickle cell disease, this abnormal Hb may constitute between 60–80%of the Hb present while in heterozygous disease or sickle cell trait, the HbS constitutes only about 20–40% while the remaining is the normal HbA. The severity of the disease is determined by the percentage of HbS. Sickling occurs when the affected person undergoes a stressful situation. Hypoxia is by far the most important of these stresses. If the degree of sickling that occurs is high, the damaged cells clog the peripheral blood vessels which results in the painful crisis often seen in patients. Deoxygenation results in the cytoplasm changing in to a rigid semisolid gel which makes the cell wall change its shape. In the early phase of the disease, when relieved of the stress, the cells revert to normal shape. But after repeated exposures, some cells irreversibly retain the S shape, and rigidity, and may be seen in a peripheral smear. Other stresses which are known to induce sickling are cold, dehydration, acidosis, severe exercise, and infections. One would expect that Sickle cell disease being an inherited condition, may manifest easily at birth. However this does not happen because HbF which is present at birth protects against the disease for several months because it is able to reduce the degree of polymerization of HbS whenever it occurs.

By far the most frequent hemoglobin abnormality seen in India is the β Thalassemia trait. Its frequency ranges from 3–17% in different parts of the country, being the highest in the Punjab. In 2007, there were an estimated 30 million Thalassemia trait carriers in India, with about 66000 patients with obvious disease. It has been estimated that as many as 10,000 are added to this list every year.

The normal Hb molecule consist of 2 α chains and 2 β chains (The adult Hb A (Hb $\alpha_2\beta_2$) Thalassemia occurs when there is mismatch between the number of either the α and β chains in the Hb molecule. If the α chain is deficient, the disease is called α Thalassemia, and vice-a-versa. In the former, there is a deletion of the α genes which are four in number either partially or completely. The latter state is fatal. In this situation, the β chains remain unattached, and cause cellular damage. In β Thalassemia, it is the unattached α chains are responsible for the damage to the cell. Because the globin production is less, the Hb per cell is less, and this makes the cell hypochromic. The decreased amount of Hb gets accommodated in a smaller sized cell. Hence these cells are microcytic. Therefore, Thalassemia is responsible for releasing microcytic hypochromic RBC in to circulation. If a microytic hypochromic anemia typically caused sued by iron deficiency does not respond to treatment with iron compounds, Thalassemia may be suspected. The deficiency of the concerned protein chains may vary from complete absence to varying degrees of reduction. In α Thalassemia, the unattached β chains are relatively soluble to help to produce Hb variants which are viable to a certain extent. On the other hand, in β Thalassemia, the accumulation of insoluble α chains in the cell results in extensive hemolysis. Henceβ Thalassemia is a combination of hemolysis as well as defective RBC formation. This makes it clinically a more aggressive disease. In carriers, the unattached α chains are taken up by δ chains. The combination is known as HB A$_2$. The percentage of HbA$_2$ does not exceed about 2.5% of the total Hb in the normal population. In people with β Thalassemia this variety of Hb is in excess of 3% and may be considered almost diagnostic of the condition. Some of the unattached α chains in this condition alsoget attached to γ chains which these RBC may produce. The combination results in the generation of some HbF (Fetal Hb). Hence Hb F levels in β Thalessemia are often high. Their presence accentuates the tissue hypoxia (***Why does HbF not release its oxygen easily?***) and this in combination with the anemia is responsible for increased secretion of erythropoietin. The bone marrow thus becomes hyperactive and deformed- a typical feature of this disease and also other severe haemolytic anemias. The deformed RBCs are easily destroyed in the spleen, accounting for the massive splenomegaly which is often present in such patients. Deranged erythropoieisis results in reduced formation of a liver peptide called hepcidin because of the down regulation of the concerned gene. Hepcidin is responsible for limiting iron absorption in the small intestine, probably via duodenal hepcidin receptors. In the absence of this peptide, iron absorption increases rapidly. This coupled with the iron released by extensive RBC breakdown, is responsible for the hemochromatosis. (The pathophysiology holds true for other severe haemolytic anemias as well). One of the benefits of repeated blood transfusions in such patients therefore is the restoration of normal hepcidin level and regulation of iron absorption. Other details of the various Thalassemia may be found elsewhere in standard text books.

Hb E and Hb D. These hemoglobinopathies are also found in India.

The Hb E disease is seen commonly in the eastern regions-Bengal and Assam, with a frequency of about 11%. In this condition, the glutamine is replaced by lysine in the β chain of the Hb molecule at position 26. The trait is asymptomatic, but when in the homozygous condition, a mild anemia which may be accompanied by splenomegaly is the usual clinical manifestation.

Hb D is found mostly in Punjab while the overall frequency in India is low. The glutamic acid in position 126 of the β chain is replaced with glycine. The electrophoresis spread of this Hb is the same as that of HbS. But this variety of Hb does not sickle on exposure to stress. It may be distinguished from HbS because of its greater solubility. Other than this, the D variety is similar to Hb A. The anemia which occurs if the disease is a homozygous state, and is mild and asymptomatic. Some times the disease is found as a heterozygous condition with HbS.

Of the other RBC abnormalities which result in RBC breakdown, **Hereditary spherocytosis** is well recognized in India. The disease is a familial inherited disorder which may remain asymptomatic or present with anemia of various degree (seen in about 80% of the patients), hepatomegaly (about 50% of the patients) splenomegaly (about 93%), jaundice (83%) and gall stones. The primary abnormalityin this condition is a decrease in red cell membrane surface area which may result from deficiency of proteins such as spectrin, ankyrin, protein 4.2 and band 3 proteins. The deficiency of the latter destabilizes the lipid by-layer of the cell membrane, and changes the normal shape of the cell in to a spherocyte. Such RBCs are easily trapped and destroyed in the spleen, resulting in splenomegaly. The spherocytes can be seen on a peripheral smear. The RBC break up early during a fragility test in decreasing saline concentrations.

Glucose 6 phosphate dehydrogenase deficiency results in hemolysis in patients with this inherited sex linked recessive disorder only if they are exposed to certain drugs such as aspirin, acetaminophen, certain antibiotics, sulpha drugs, nitrofurantoin, and anti malarials like primaquin, quinine and mepacrine. Severe bacterial and viral infections may also precipitate a hemolysis of RBC with this enzyme deficiency. About 5% of the Indian population may be affected. Females are asymptomatic carriers as the G6PD gene is located on the X chromosome. Hemolysis when it occurs is usually self limited and may not require any drastic treatment measures. It is however advisable that this enzyme deficiency is eliminated before starting treatment for malaria.

White Blood Cells (WBC) and Their Role in Infection and Inflammation

The reader may refer to any standard text book to review the general information of the various types of WBC and their function.

The neutrophils (so called because their granules do not stain either with basic or acidic dyes) or Poly morpho nuclear leucocytes (PMNs) because their nuclei have 3 to 5 lobes), circulate in blood for a short while (about 6 hours) after which they migrate in to the tissues. Once inside the tissues, they either die a natural death in a few days, or die in the process of performing their phagocytic function.

The journey of the various WBCs to the site of injury in the tissues is a complex one and is carried out in three phases. The WBCs are found along the margins of the blood vessels **(Margination)** where they "tumble" along as they keep making loose adhesions with the endothelial surface by means of proteins called Integrins which are adhesion molecules, present on the WBC cell surface. Near to or at the site of damage, the endothelial margins use Selectins to tether the WBCs so that they are concentrated in the injured area of the tissue. Once stuck to the blood vessels in the injured area, the WBCs extend their pseudopodia to crawl through the inter-endothelial spaces to enter the tissue space. In this manner, the WBCs find their way to the damaged area in order to take part in the inflammatory process. This process of **"Emigration"** is used by all the granulocytes, the monocytes and the lymphocytes and occurs mainly in the venules. In the first few hours of the inflammation/infection, the PMNs are the main cells which undergo the emigration while in the later stages (2–3 days), the monocytes are the majority cells. It is however the macrophage (a monocyte converted to this form when it leaves circulation) which forms the first defensive barrier to bacterial invasion.

The process of adhesion between the endothelium and the various leucocytes is dependent upon the adhesion molecules as mentioned above. The Selectins expressed on the endothelial surfaces are of various types; E Selectins (for those exclusively on the endothelium), P Selectins which are also found on platelets, and L Selectins expressed mainly on lymphocytes. Chemically, these are proteins and extend across the membrane. Integrins on the other hand are expressed as glycoproteins on the surfaces of various WBCs. They also extend across the membrane, and are of two types-α and β. Apart from the Selectins and Integrins, the extra cellular matrix and the WBC exhibit adhesion molecules which help to keep the WBCs in place when they enter the extra cellular space. These are Mucin-like glycoproteins. Of the Integrins, the β type are associated with helping the WBCs to adhere to the extra cellular matrix.

The process of adherence of the WBCs is influenced by a number of chemical mediators which are released locally. The more important ones are some of the cytokines, leukotrienes, and products of the Complement system. One of the more important cytokines which helps in the chemo-attraction of WBCs (the PMNs in particular) and their attachment to the adherence molecules is Interleukin 8 (IL 8) earlier called the Neutrophil Chemotactic factor. Other factors identified are monocytes chemo-attractant protein-1, platelet factor 4. In addition, IL8, IL1 and tumour necrosis factor have been found to increase the expression of Selectins on the endothelial surface and also facilitate WBC emigration. The WBCs need to be activated if they are to perform their task when confronted with infection/inflammation (**Leukocyte activation**). The main aim of this process is to increase the cytosolic calcium and activate various proteolytic enzymes. At the same time, there is an increase in the expression of the adhesion molecules and an increase in the secretion of inflammatory cytokines. When activated, the WBCs generate various receptors with specific functions on their cell surfaces. The Toll Like receptors are meant for tackling microorganisms, the α helical trans-membrane receptors deal with microorganisms as also chemical mediators that are released in response to tissue damage, while opsonin receptors help in the process of phagocytosis.

Once the WBC (mainly the neutrophils and the macrophages) are activated, they are ready to phagocytosis the offending microbes/material. This is done in stages. Initially, the defending cells must ensure that the offending agents are attached to their surfaces for further treatment. The Mannose and the Scavenger receptors do this duty. The former is able to attach itself to the mannose and fructose moieties contained in the bacteria, and also identify these bacteria as non-self. These receptors are found mostly on the macrophages. The scavenger receptors on the macrophages attach to various proteins on the bacteria in order to facilitate their destruction. The whole process is enhanced by opsonization of the unwanted materials by attachment to opsonin receptors. These receptors respond to particles prepared by coating with various substances such as Ig G anti-bodies, and complement proteins. Once this is over, the phagocytic membrane throws out pseudopodia to **engulf** the foreign material. The complex then fuses with the granules containing the lysosomal enzymes which then proceed to digest the material. The lysosomal enzymes act by production of free oxygen radicals (Reactive oxygen species) to degrade the ingested material. This method of bacterial destruction is therefore oxygen dependent. The activity can also occur independent of oxygen by an increase in the permeability of bacterial membranes caused by proteins from the granules of the WBCs. Defensins, major basic protein (of the eosinophils), lactoferrin and elastase fall under this category of bactericidal proteins. At the end of this action, the acid hydrolases which function best at a pH of about 4.5, help to finally degrade the invasive material. During the process of phagocytosis, free oxygen radicals, lysosomal enzymes, prostaglandins often leak out in to the extracellular spaces and cause both acute and chronic inflammation. Thus the savior in fact may become the "destroyer." The WBCs which have participated in the phagocytic process undergo a natural apoptosis and in turn get digested by the macrophages. It is therefore understandable that the inflammatory process that is initiated has to be controlled and stopped in order to prevent unwarranted damage to tissues. This happens in several ways. The released chemical substances as such have a short life span and get rapidly degraded after their release. The WBC change their function of releasing inflammatory mediators to anti-inflammatory substances such as lipoxins, and Transforming Growth Factor (TGF). The mechanism as to how the roles are reversed is as yet ill understood.

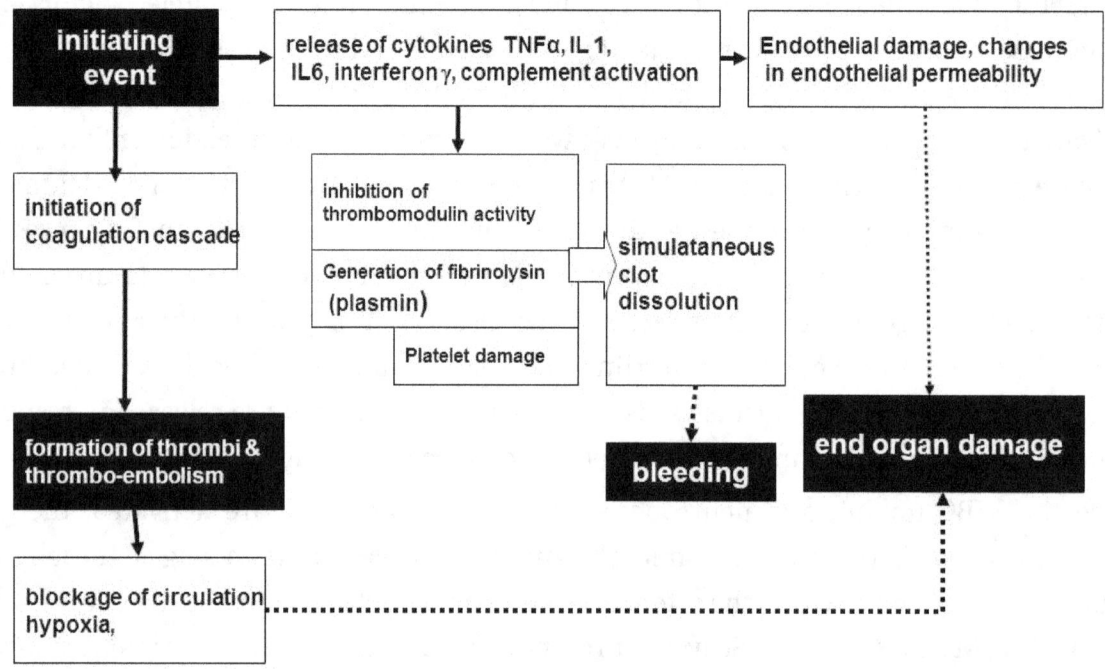

Fig. 4 Possible triggering mechanisms in DIC

Table IV: Possible causes of deficient WBC function

1. Bone marrow suppression
2. Inability of WBCs to adhere
3. Limitedtransfer of lysosomal enzymes from WBCs/macrophages to the phagocytic vacuoles
4. Inability to generate ROS

There is always a possibility that the myriad functions which the WBCs perform to defend the body against infection can go awry because of either genetic or acquired deficits. This then may lead to severe infections. The various possibilities which may result in such a situation are given in **Table IV.**

White Blood Cell Response to Altered Body Physiology Environment

In general, the WBCs may increase in number beyond the physiological normal limits (Leukocytosis) or decrease in number (Leukopenia).

The WBCs respond to infection by an increase in their number beyond the upper normal limit of about 11,000 /µlitre (Leukocytosis). This phenomenon could also be a prelude to cancerous conditions involving these cells. Trade mark leukocytosis occurs because of an increase in neutrophils (Neutrophilia) in response to pyogenic infection. The bone marrow releases rapidly large numbers of leukocytes on stimulation by various cytokines such as Interleukin 1 (IL1) and Tumor Necrosis Factor (TNF). The eosinophils increase in number when the body is accosted by allergy, and worm infestations. This is in response to stimulation by IL5. Chronic infections, collagen disorders result in an increase in monocytes usually accompanied by a lymphocytosis. In physiological states such as exercise, leukocytosis occurs because the increase in circulating adrenalin reduces the margination of the WBCs, allowing more cells to go into circulation. On the other hand, corticosteroids reduce the passage of WBCs in to extra cellular spaces, thus allowing the circulating cells to increase and show an apparent leukocytosis.

Leukopenia (Total leukocyte count below 4000/ μlitre) is again mostly a Neutropenia or a granulopenia. Reduced number of lymphocytes is relatively uncommon. This may occur on two counts: A suppression of bone marrow activity involving the WBC precursor stem cells or a rapid removal or sequestration of the WBC from circulation. On occasions, this may occur if the rate of utilization of the cells at tissue level because of severe infections is faster than their replenishment in to the circulation. Splenomegaly may be the reason of entrapping a large number of WBCs, enough to produce a leukopenia.

Disturbances in Hemostasis

The reader may recall that the normal process of hemostasis involves i. vasoconstriction. ii. platelet activation iii. coagulation and iv. clot lysis to ensure that the flow of blood is re-established after a suitable interval. A disturbance in any of the first three or their combination may be responsible for derangement in hemostasis.

Local vasoconstriction to limit blood loss after injury has been associated with release of endothelin by the injured endotelium as well as of serotonin by the platelets at the site of injury. More important is the increase in blood vessel fragility. This may occur in infections, Vitamin C deficiency (deficiency in collagen deposition needed for support of the endothelial lining). This also happens in Cushing's Disease, possibly in the elderly, and as a part of adverse drug reactions. The micro-hemorrhages which may occur in skin and mucus membranes are usually innocuous, but may also occur occasionally in muscles and joints. Rarely this may present as haematuria and small gastro-intestinal bleeds. **(What do you think will be the bleeding and clotting time in such patients (eg. those with Vit C deficiency and why?)** The mechanisms involved differ under varying conditions. For instance, infections may damage blood vessels because of the ensuing vasculitis, or Disseminated Intravascular Coagulopathy (described later) while the drug induced damage may be secondary to antibodies against the concerned drug. Vitamin C deficiency prevents the laying down of collagen which supports the vessel wall. This in turn increases blood vessel fragility. In Hereditary Telangiectasia, the blood vessels become thin, tortuous and bleed readily. The hemorrhages appear in mucus membranes and may present as frequent nose bleeds.

Platelets are an integral part of the process of hemostasis. Platelet dysfunction attributed to a variety of causes may present asa bleeding disorder. The dysfunction could be congenital or acquired. Congenitally, the platelets may not be able to i. adhere, ii. secrete the substances which help their activation, or iii. fail to aggregate because of the absence of a glycoprotein. The latter has been called thromboasthenia. The acquired platelet deficiencies known to cause bleeding disorders, occur commonly because of consumption of anti-inflammatory medicines (aspirin being one such), and uremia. The mechanism of the former is the inhibition of cyclo-oxygenase which results in the reduced availability of thromboxaneA$_2$ (TxA$_2$) and prostaglandins which are required for platelet activation. However the mechanism by which uremia produces platelet dysfunction is undetermined. (***Why is aspirin contraindicated in dengue fever?***)

Deficiency of any of the coagulation factors may produce a bleeding disorder, the only exception being Factor XII. Most of these disorders are single factor deficiencies such as Hemophilia (factor VIII deficiency), and Christmas Disease (Factor IX deficiency). There is a combination of factors which is essential for adequate hemostasis. Factor VIII (synthesized by the liver) and von Willebrand's factor (synthesized by the endothelium and the megakaryocytes combine to form a complex which prolongs the half-life of Factor VIII to about eight hours while without the von Willebrand's protein it is just about two hours. The deficiency of the latter (von Willebrand's disease) thus affects not only platelet function but also coagulation. In this disease, the platelet count is normal and yet the bleeding time is prolonged.

After this generalization of hemorrhagic diathesis, an attempt is made to briefly explain in the ensuing paragraphs an aspect of bleeding disorders which does not usually find its way into Physiology textbooks.

Disseminated Intravascular Coagulopathy (DIC)

Table V:

Some clinical conditions in which DIC is likely to occur

1. Obstetric emergencies
2. Severe burns
3. Hemolysis
4. Severe viral and bacterial infection
5. Extensive crush injuries

This is a complex syndrome which involves the generation of intravascular fibrin and the activation and consumption of procoagulants and enhanced usage of clot inhibitor proteins. As a result, there is a contradictory situation of intravascular coagulation accompanied by hemorrhage. The confounding feature is a micro or macro embolism with end organ damage as the final outcome.

The condition may accompany a number of well recognized clinical entities listed in **Table V,** the most prominent amongst them being obstetric accidents (Amniotic fluid embolism). The general course of events, whatever the precipitating cause is almost similar to most situations (**Fig. 4**). However the events that lead up to the possible "Final common pathway" varies from situation to situation. For example, the initiating event in Fig 4 when there is intravascular hemolysis may be the release of RBC membrane phospholipoprotein or the adenosine diphosphate from the broken down RBCs. The triggering mechanism in severe bacterial infections for example meningococcal infections, is probably the endotoxin from the bacteria. These factors promote activity of procoagulants which is then followed by the rest of the cascade which may result in release of various cytokines and damage to the capillary endothelium. Necrotic tissue from burns and from dead retained fetuses may be responsible for initiating the procoagulant cascade and endothelial damage. The thrombin which is formed in order to convert fibrinogen in to fibrin and hence the clot, also simultaneously combines with thrombomodulin to generate plasmin (Fibrinolysin) via a series of actions involving tissue plasminogen activator (tPA). Plasmin once generated, because of its overall proteolytic activity, not only breaks down fibrin (The fibrin degradation products are useful as bio markers in DIC), but also inactivates procoagulant factors V, VII, IX and XI, thus reducing availability of coagulant material. Thus a vicious cycle of clot formation and dissolution sets in. The thrombosis produces tissue damage while the inactivation of the many procoagulants induces hemorrhagic tendencies. Thus, DIC is a combination of both coagulation (thrombosis) as well as hemorrhages. Complicating the issue is the tissue damage caused by the hypoxia, and the various damaging cytokines in circulation. Products released during fibrin degradation (FDPs) are responsible for platelet dysfunction, and this may be clinically manifested as petechial hemorrhages.

Clinical Presentation of Pathophysiology of DIC

The condition occurs subsequent to the known clinical conditions /syndromes which are known to be the "Risk factors" for development of DIC. Hence whenever such conditions are present **(table IV)** an effort must be made to look for the symptoms and signs which indicate the presence of DIC. Appearance of petechial hemorrhages, unexplained oozing from surgical incisions, venipuncture and arterial puncture sites, and appearance of subcutaneous hematomas should draw the attention of a clinician to the onset of DIC. In severe cases, internal bleeding may induce hypotension. Micro embolism in multiple organ systems such as the heart, brain, kidney may induce the various manifestations associated with the affected system.

References

1. Adamson J, Longo D. Anemia and polycythemia.Chapt.52 in Harrison's Principles of Internal Medicine." Edited. Kasper DL, BraunwaldE, Fauci AS, Hauser SL, Longo DL, Jameson JL.16th ed. MCGraw Hill, London. 2005. pp 329–336

2. Ambekar S, Phadke M, Mokashi G, Bankar M, Khedkar V, Venkat V, Basutkar D. Pattern of Hemoglobinopathies in Western Maharashtra. Indian Ped. 2001; 38: 530–534.

3. Becker J, Brenner B, Kumar A, Shaaban H Disseminated Intravascular Coagulation in Emergency Medicine. Med Scape Ref. Drugs Dis and Procedures. 26 Oct2011

4. Best and Taylor's Physiological Basis of Medical Practice, 13th edn.; Ed. Best JB. William & Wilkins, London; 1990.

5. Bick R. Disseminated intravascular coagulation; Current concepts of etiology, pathophysiology, diagnosis and treatment. Hematology Oncology Clin of N.Am. 2003; 17: 149–176

6. Bijlani RL. Understanding Medical Physiology: A Text Book for Medical Students.2nd edn. 1997.

7. Chernoff A. The hemoglobin D syndromes. Blood. 1958; 13: 116–127

8. Davidson's Principles and Practice of Medicine. Edt. Haslett C, Chilvers ER, Boon NA & Colledge NR. 19th Edn. Churchill Livingstone., London. 2002

9. Desai D, Dhanani H. Sickle Cell Disease: History and Origin. Internet J Hematology. 2010; ISSN: 1540—2649; pp 1–4

10. Eckmann D, Bowers S, Stecker M, Cheung A. Hematocrit, volume expander, temperature, and shear rate effects on blood viscosity. Anaeth. Analg. 2000; 91: 539–545

11. Erslev A, Gabuzda T. Pathophysiology of Blood. 3rd edn. WB Saunders, London. 1985

12. Essex D, Jin D, Bradley T. Lactic acidosis secondary to severe anemia in a patient with paroxysmal nocturnal hemoglobinuria. Am. J. Hemat. 1997; 55: 110–111.

13. Feghali C, Wright T. Cytokines in acute and chronic inflammation. Frontiers in Bioscience2. 1997; d 12–26.

14. Gallaher P, Jarolim P. Red cell membrane disorders in Hematology: Basic Principles and Practice.Ed. Hoffman R, Benz E Jr,Shatti SJ et al. 4th Edn. WB Saunders. Phil. 2005

15. Ganong's Review of Medical Physiology. 23rd edition. Ed. Barrett K, Barman S, Boitano S, Brooks HL. Lange Medical Books/McGraw-Hill, London. 2010

16. Guyton AC and Hall JE. Textbook of Medical Physiology 12th edition. Elsevier, Philadelphia; 2011.

17. Gray's Anatomy. Early Embryonic Circulation. Chap. 13 in The Anatomical Basis of Clinical Practice. 40th edn. Edt. Standring S. Churchill Livingstone, London. 2008

18. Jain AK. Manual of Practical Physiology. Avichal Publishers, N. Delhi. 2000

19. Kar R, Rao S, Srinivas U, Mishra P, Pati H. Clinico-hematological profile of hereditary spherocytosis: experience from a tertiary care centre in North India.. Hematology. 2009; 14: 164–167.

20. Klabunde R. Viscosity of blood in Cardiovascular Physiology Concepts. 2nd Edn. Lippincott Williams and Wilkins. 2011. www.cv.physiology.com/Hemodynamics/H011. htm 12/2/2011

21. Kutter D. Hereditary spherocytosis is more frequent than expected: What to tell the patient? Bull. Soc. Sci. Med. 2005; 1: 7–22.

22. Maheshwari M, Arora S, Kabra M, Menon P. Carrier screening and diagnosis of β Thalssemia.Indian Pediatrics. 1999; 36: 1119–1125.

23. Mandot S, Khurana VL, Sonesh JK. Sickle cell anemia in Garasia tribals of Rajasthan. Indian Pediatrics. 2009; 46: 239–240.

24. Minter K, Gladwin M. Pulmonary complications of sickle cell anemia. Am. J. Crit. Care Med. 2001; 164: 2016–2019

25. Nadkarni A, Phanasgaonkar S, Colah R, Mohanty D, Ghosh K. Prevalence and molecular characterization of alpha-thalassemia syndromes among Indians. Gen Test. 2008; 12: 177–180

26. Nash G. Blood rheology and ischemia. Eye. 1991; 5: 151–158.

27. Pathophysiology: The Biologic Basis for Disease in Children and Adults. Ed. McCance K and Huether S. 5th Edn. Mosby, Missouri; 2006

28. Rao S, Kar R, Gupta S, Chopra A, Saxena R. Spectrum of Hemoglobinopathies diagnosed by cation exchange HPLC & modulating effects of nutritional deficiency anemias from North India. Indian J Medical Research. 2010; 132; 513–519

29. Robbins and Kotran's Pathologic Basis of Disease. 7th Edn. Edt. Kumar V, Abbas AK, Fausto N. Chap.13.Elsevier Saunders, Penn. 2005.

30. Shah A. Hemoglobinopathies and other congenital hemolytic anemias. Indian J Medical Med. Scineces. 2004; 58: 490–493

31. Sharma S, Sood S, Colah R, (Late),Bhatia H. Frequency of β-thalassemia trait and other hemoglobinopathies in northern and western India. Indian J. Human Gen 2010; 16: 16–25.

32. Text Book of Medicine. Ed. Souhami RL and Moxham J. 2nd Edn. Churchill Livingstone London; 1994

33. Thacker N. Prevention of Thalassemia in India. Indian Pediatrics. 2007; 44: 647–648

34. Trudnowski R,Rico R. Specific gravity of blood and plasma at 4 and 37° C. Clin. Chem. 1974; 20: 615–616

35. Van der Meer P, Lipsic E, Daan Westenbrink B et al. Levels of hematopoieisis inhibitor N-acetyl-seryl-aspartyl-lysyl-proline partially explain the occurrence of anemia in heart failure. Circulation. 2005; 112: 1743–1747

36. William's Hematology. Kaushansky K, Seligohn V, Lichtman M et al.. Hemostasis and Thrombosis. Chap 113; 8th Edn. McGraw Hill New York. 2010 pp. 1721–1734.

37. World Health Organization. Report: Sickle cell Anemia. Fifty Ninth World Health Assembly 2006; Report no. A59/9.

38. Yaish H, Coppes M. Pediatric Thalassemia. Medscape Reference 30/4/2010

CHAPTER - 12

Some Aspects of the Pathophysiology of the Central Nervous System

Plan

1. General Introduction
2. Natural protection of the brain:
   i. Anatomical: external and internal, and effects of head injury
   ii. Physiological: CSF; Blood brain barrier.
3. Intra cranial pressure, and pathophysiological effects of its changes.
4. Pathophysiology of stroke
5. Pathophysiology of the post-concussion syndrome
6. Pathophysiology of spinal cord injury
7. Disturbances in the Autonomic Nervous System: hyper and hypo reflexia
8. Pathophysiology of common behavioural abnormalities
   i. Mood disorders-Depression and Mania ; Anxiety
   ii. Thought disorders- Schizophrenia

General Introduction

The chapter deals in a limited manner with some aspects of the pathophysiology of the human nervous system. The human nervous system detects changes in the environment, analyses them, and reacts to them by ordering activation of various control mechanisms by way of the motor and the autonomic nervous systems. The perpetrator of the nervous system activity is the neuron. The neurons constitute only about 10% of the hundreds of billion cells in the nervous system while the supporting cells, the neuroglia make up the remaining 90%. The neuroglia support neuron function and can modify neuronal activity but do not generate electrical impulses. They help in the repair and maintenance of the nervous system, support the neurons metabolically, and assist in the process of development of the CNS. Anatomically, they do not have axons. Of these, the microglia sub serve immunological function as the brain is devoid of the classical immunity giving cells. The remaining types are the macroglia (astrocytes, oligodendrocytes, Schwann cells).The glial cells can now be individually identified by the specialized proteins they produce. More recently, astrocytes have been found to be critical to neuronal function. They can generate receptors and transporters which interact with neurotransmitters, and may also affect vascular smooth muscle activity and blood flow in the brain. Glial cell tumours are by far the commonest tumours in the brain.

The neurons generate electric impulses which travel to other neurons/target organs along the axons. The latter make contact with the target organ of the neuron through junctions called synapses. *(Q. Why is there no retrograde transmission across a synapse?)*. Neurons can be rapidly activated by passage of ions through ligand gated and voltage gated channels. The ligands that operate the channels are neurotransmitters namely acetylcholine, glutamate (both excitatory) and gama aminobutyric acid (GABA) which is inhibitory. Glycine is excitatory as well as inhibitory. It stands to reason that alterations in neurotransmitter secretions can produce dysfunction in neuronal discharge- for example epilepsy.

The brain and the spinal cord are naturally protected both externally and internally.(The student should recall the various layers of tissue that cover the brain from the outside to the inside). Injury to the brain may occur with or without a disruption of the protective layers (though sometimes it pays to have a thick skull!).The latter may happen because of transmission of a shock wave. A subdural hematoma is a likely outcome of injury to the skull. The brain which "floats" in the cerebro-spinal fluid is relatively mobile while its venous connections are tethered to the skull. Transmission of the shock wave accelerates the brain which then tugs at the veins. When this stress is severe enough, the cortical veins are likely to tear at their moorings as they traverse the subdural space, resulting in haemorrhage which gets released in to the subdural space, resulting in a subdural hematoma. The veins may tear with even with minor trauma to the skull. The adverse effects of the hematoma are caused by the pressure it exerts on the brain tissue. Usually, because of the relatively low pressure in the veins, accumulation of blood occurs over a period of time. Hence the onset of symptoms may be delayed. The clinical presentation is determined by the area involved, the size of the hematoma, and the rapidity with which it accumulates. In the elderly, the size (volume) of the brain is relatively reduced because of atrophy. Hence there is more space for the brain to accelerate/decelerate with sudden impact, increasing the chances of injury. The epidural hematoma on the other hand occurs because of a tear of the middle meningeal artery after a blow to the temporal region. Onset of symptoms after an epidural injury occur faster because the haemorrhage is arterial, and hence under high pressure. Needless to say that increasing the protection of the head by wearing a helmet will go far in preventing serious head injury in case of an accident while riding vehicles such as motorcycles.

The hematoma needs to be drained by making a hole in the cranial vault (burr hole or trephine).

Post concussion syndrome is an entity with which the undergraduate medical student is usually not familiar. A symptom complex which includes irritability, insomnia, lack of concentration, depression, anxiety, easy fatigability, lack of libido, passing of copious,. dilute urine may follow concussion. The pathophysiology involved is probably diffused axonal injury which leads to calcium influx, phospholipase release and lipid peroxidation, a sequel to free oxygen radical generation in the injured areas. Circulatory disturbances and brain edema may cause hypoxia, release of excitatory amino acids and inflammatory substances with similar result. The symptoms often reflect the damage to specific brain areas. It is paradoxical that patients with severe head injury may on occasions get away with only a mild post concussion injury syndrome while those with mild injury may have a severe disorder. Factors such as advanced age, co-morbidities, alcohol consumption may be responsible.

One of the major concerns involving the pathophysiology of the post head injury syndrome is hypopituitarism. This may happen because of a disruption of the hypothalamico-hypophyseal axis causing gonadotrophin, growth hormone, and corticotrophin secretion deficits. A serious manifestation is a secondary hypoadrenalism which though relatively rare, occurs in the acute phase of the injury. It is also possible that the somatotrophs and the gonadotrophs being located postero laterally, are injured directly by a shock wave traversing the brain. Corticotrophs and thyrotrophsare less directly affected by the passing shock wave as they are located antero-medially. Anti Diuretic Hormone (ADH) secretion disorders may be seen in a high percentage of patients with head injury who may present with Diabetes Insipidus or as Syndrome of Inappropriate ADH secretion (SIADH). Complete recovery from this complex syndrome may take between 1–3 years while the damage may be permanent in some patients.

Internal Protection of the Brain

Inside the cranial vault, the brain is protected by both, anatomical and physiological means. The inner layer of the dura from both sides joins centrally to form a thick membranous plate which is anchored to the base of the brain at the ethmoid bone in front along the longitudinal fissure. Posteriorly it merges with the tentorium cerebelli, another membranous form of the dura which separates the cerebral hemispheres from the cerebellar lobes below. In this manner, the brain is firmly tethered to the vault in the longitudinal axis, and helps to limit its sudden acceleration/deceleration in the antero-posterior direction. At the same time, the tentori and the falx are arranged into various compartments in which the brain tissue is distributed. Severe trauma to the head may cause a laceration of these membranes.

The intracranial pressure (ICP) in a normal person is generated by the combined volumes of the brain tissue, blood and the cerebro spinal fluid (CSF) enclosed in a hard, closed cavity- the cranium. Delicate inter dependent adjustments are made continuously within these factors to maintain the ICP, an increase in one factor being compensated by a decrease in another. This phenomenon is called the Monroe-Kelly doctrine. The brain tissue and the CSF are deemed incompressible. So any changes in the intra cranial tension will have to be compensated for by a change in the blood flow in to the cranial vault. This doctrine helps to siphon arterial blood in to the brain to overcome the pulling down of venous blood out of the cranial cavity during positive accelerations (+ve Gz) experienced by fighter pilots during manoeuvres. It is possible that in pilots who have had a burr hole in the skull to relieve a subdual hematoma, the Monroe-Kelly doctrine may not act effectively during exposure to +ve Gz. In the event, such pilots may not be able to withstand +ve Gz effectively. Any addition to the intra cranial volume by tumour, intra cranial haemorrhage or excessive secretion of the CSF will raise the ICP to abnormal levels and generate clinical symptoms and signs.

Table I:Causes of raised ICP

1. Brain oedema/inflammation
2. Space occupying lesions
3. Excessive secretion of CSF/ blocked reabsorption
4. Benign idiopathic

About 150 ml of the CSF is present in the subarachnoid space at any point of time, while about 600 ml is formed each day. (The student should recall here the sites of formation of CSF, its circulation and reabsorption). Apart from acting as a shock absorber, the CSF helps to "float" the brain in the cranial cavity. This may effectively reduce the functional weight of the brain (Archimedes principle). During the high positive acceleration forces (+Gz) met with during fighter aircraft flying, the body weight increases in direct proportion to the amount of Gz applied. For example, at +6Gz, the brain which normally weighs about 1.2 kg, is likely to weigh about 7.2 kg. Under these circumstances the CSF probably reduces the effective weight of the brain and prevents it from herniating through the foramen magnum. The integrity of the vault is one of the factors that helps to maintain the intra cranial pressure/volume relationship at the normal level.

In the lying posture, the hydrostatic pressure of the CSF (a reflection of the ICP) measured by lumbar puncture in between L3 and L4 vertebrae, is between 80–180 mm of H_2O pressure(average 110mm). ***(What will happen to this pressure in the sitting posture?).*** The rate of secretion and absorption of CSF are equal at around the average pressure. Thereafter, the rate of absorption increases almost linearly with increase in the pressure, while the absorption ceases when the CSF pressure drops to below about 70 mm of H_2O pressure. This is thus a safety factor in the dynamics of CSF circulation. An abnormal increase in CSF pressure (which results in hydrocephalus) can occur either because of too much of secretion of the fluid or blockage of its absorption. The end result is the rise in intra cranial tension with its pathophysiological effects. **Table I** lists the various situations in which ICP may increase to abnormal levels.

The initial increase in ICP is compensated for by compliance of the falx cerebri and application of the Munroe Kelly doctrine (compensatory decrease in blood flow and the CSF. The compensatory mechanisms fail when the ICP rises to $\geq$ 250 mmH$_2$O. The two main effects of abnormally raised ICP are a fall in the brain blood flow leading to ischemia followed by hypoxia, and cerebral herniation. These in turn are responsible for the classical clinical features **(Fig. 1)**.

The probable cause of the head ache associated with raised ICP is the stretch and distortion of the dura and the arteries. Headache is typically worse in the morning and is exaggerated by bending forward, coughing and sneezing. All these factors tend to increase the blood volume in the cranial cavity and promote further stretch of the dura and blood vessels. While such patients are asleep, their arterial blood PCO$_2$ increases because of respiratory depression. This in turn may induceCheyene Stokes breathing. There may also be a decreased reabsorption of

CSF which may increase the distortion of the dura. ***(What will happen if a Valsalva manouevre is done in a patient with raised ICP and why?)***

Fig. 1. Evolution of raised ICP

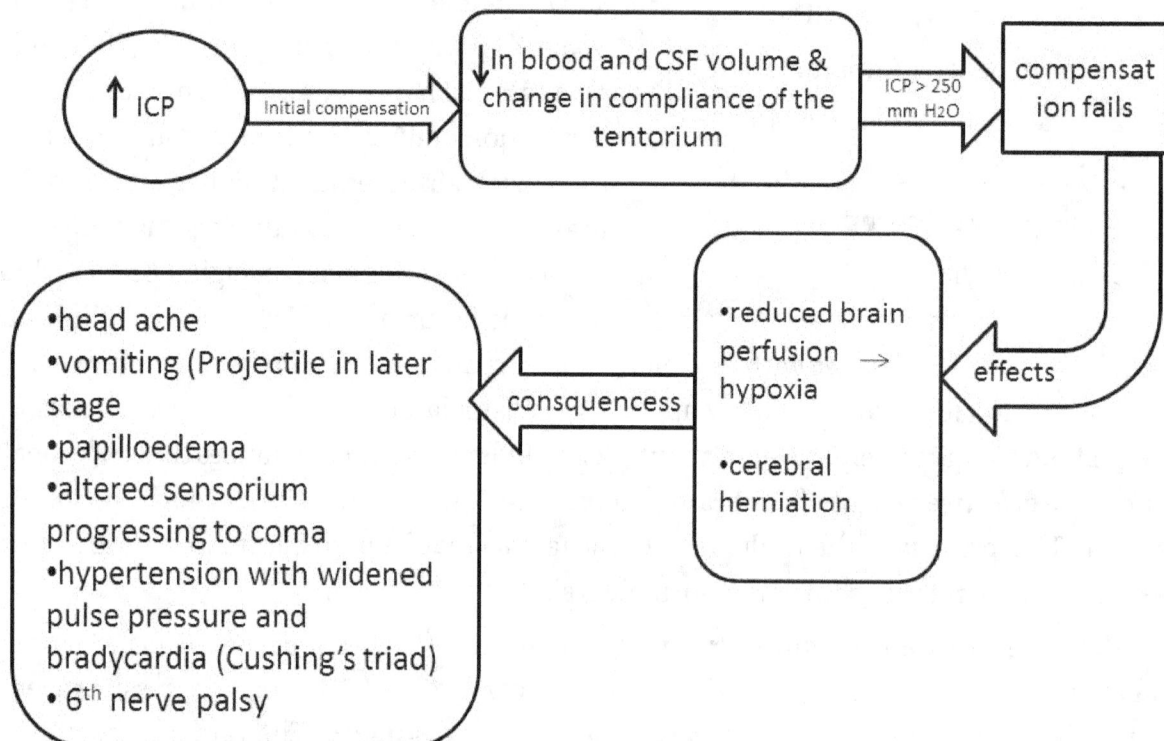

The vomiting which occurs with raised ICP is without nausea. The exact mechanism of its occurrence is not clear but experimental evidence from animals suggests that the chemoreceptor trigger zone in the area postrema may be involved as its ablation reduces it. Vomiting becomes projectile in advanced stages. Papilloedema is the third typical feature of raised ICP. An extension of the arachnoidmater normally wraps around the optic nerve head. The raised ICP transmitted through the CSF under the arachnoid may then compress the optic nerve head and cause papilloedema. Its long term persistence may lead to optic atrophy, probably secondary to interruption of axonal transport in the nerve. Ocular palsies and changes in sensorium which may range from irritability, mild alteration in consciousness to frank coma may be seen. These effects are more often than not the result of brain tissue shifts caused by pressure gradients between the brain compartments produced by the pathology rather than absolute increase ICP. In severe cases, brain contents may herniate through the foramen magnum to produce crippling CNS manifestations.

In the initial phases of raised ICP, auto regulation helps to regularize cerebral blood flow. But this mechanism is disrupted by increasing ICP which hampers blood flow to the brain. The resulting ischemia causes hypoxia which in term sets up a powerful reflex sympathetic drive to try and re-establish the cerebral circulation (Cushing's reflex). As a result there is a systolic hypertension, widened pulse pressure and a reflex bradycardia secondary to stimulation of the arterial baro receptors (Cushing's triad). This is a manifestation of severe increase in intra cranial pressure and must be watched out for especially in patients of head injury.

The Blood Brain Barrier

This may be deemed as the "Physiological" means of internal protection of the brain and is illustrated in **Fig.2.** It is also known as the Brain Microvascular Endothelial Cells. Its most important constituents are the tight junctions in the brain capillary endothelium, the astrocytes that lie in close proximity and the pericytes which envelop about 20% of the capillary circumference. A decrease in the number of pericytes is associated with an increase in capillary permeability. The extra cellular matrix helps to hold together the Brain Microvascular Cell complex with the help of proteins which have receptors on the tight junctional cells. It has been noted that damage to the matrix component of the basement membrane of the endothelial cells is associated with increased permeability of the BBB. A similar barrier exists at the ependymal epithelium as the Brain-CSF barrier. The overall objective of the system is probably the maintenance of the functionality of the tight junctions in a dynamic manner. Calcium is associated with cell signaling in the tight junctional cells.

Fig. 2 Blood brain barrier (BBB)

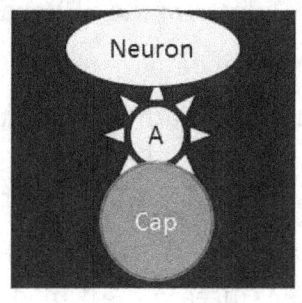

This is made up by

  i. tight junctions between capillary endothelial cells
 ii. astrocytes (A) between brain tissue and capillaries

BBB is deficient in a set of organs called circumventricular organs

Breakdown of BBB may be caused by:
   1. excessive level of circulating bilirubin
   2. infections
   3. heat stroke

The most important function of the BBB is to prevent proteins from leaking out of the capillaries while at the same time it also promotes the brain to blood egress of certain substances. The barrier permits the diffusion of the respiratory gases, alcohol, volatile anaesthetics, peptides and regulatory proteins. It controls the entry of nutrients and glucose as well as neurotransmitter precursor substances and secretes cytokines and nitric oxide. A number of enzymes such as - glutamyl transpeptidase, alkaline phosphatase, and aromatic acid decarboxylase present on the endothelium help to metabolize drugs and nutrient material. That the endothelial cells are metabolically active to carry out such function is suggested by the high mitochondrial density.

> **Table II: Conditions affecting BBB permeability**
>
> 1. Ageing
> 2. Hypoxia/ischemia
> 3. Kernicterus
> 4. Alcoholism
> 5. Inflammatoryconditions:
>
> HIV; Encephalitis; Alzheimer's Disease; Parkinsonism; Multiple sclerosis;
>
> 6. Brain tumours

The BBB is deficient in certain areas of the brain (the circumventricular orgnans). These are i. the posterior pituitary; ii. the median eminence of the hypothalamus; iii. area postrema of the supraoptic crest; iv. organum vasculosum of the lamina terminalis; v. the subfornical organ. These areas are covered by fenestrated capillaries, and hence do not constitute a part of the BBB. Because of the easy permeability, they act as a conduit for secretions by neurons Some of these organs have receptors for various peptides and can act as chemoreceptors. For example, the area postrema acts as a chemoreceptor trigger zone. Circulating substances may react with the receptors and initiate a reaction by the brain without the need for crossing the BBB. The circumventricular organs may thus be designated as windows through which the brain samples the environment immediately outside it.

A **breakdown of the BBB** may occurin a variety of conditions listed in **Table 2**

The basis of damage to the BBB in most instances is the disruption of the tight junctions.

Brain ischemia/hypoxia cause a sudden decrease in supply of essential nutrients. The resultant release of substances such as nitric oxide is then responsible for breakdown of the barrier. In the presence of inflammation, microglia may release a variety of cytokines, chemokines and reactive oxygen species which in turn affect the tight junctions adversely. In Alzheimer's disease, the β amyloid protein probably inter acts with the astrocytes and microglia to produces inflammatory substances such as tumor necrosis factor α (TNFα), and interleukins which damage the tight junction. It is also possible that thente β amyloid protein may directly damage the barrier. Brain tumours may also affect the barrier by releasing cytokines and other inflammatory factors.

Kernicterus is a devastating clinical entity which may affect a new born and cause wide spread CNS damage. Occasionally it may also affect adults. This condition may happen if i. the unconjugated serum bilirubin concentration increases to an extent that it can not be mopped by albumin; ii. the blood albumin is too low to combine with whatever bilirubin is being released by red blood cell breakdown as in premature babies; iii. if bilirubin is displaced from its binding with albumin by other substances which bind competitively with albumin. The exact mechanism of how neuronal damage occurs once the unconjugated bilirubin enters the brain is not clear. Various possibilities have been considered. The bilirubin may interact with neuronal mitochondrial and or the endoplasmic reticular membranes. This may lead to excessive availability of the excitatory amino acid glutamine with over activation of its receptors which leads to an influx of calcium in to the neurons. Once the Ca^{++} influx occurs, enzymatic damage and apoptosis may ensue.

Alcoholism disrupts the BBB by affecting various transport mechanisms. One of the major ones is the adverse effect on the Methionine Encephalin (ME) system. The BBB has a transport mechanism for controlling the ME level. A damage to the barrier may then disrupt the ME balance which in turn may induce alcohol related seizures. Thiamine deficiency induced by chronic alcoholism causes Wernicke-Korsakoff's syndrome which in turn has been associated with the disruption of the BBB.

Pathophysiology of Stroke (Cerebro vascular accident-CVA)

Stroke is an acute/subacute vascular event in which neurological deficit develops rapidly, some times in a step wise fashion. The symptom/sign complex must persist for at least 24 hours to qualify as a stroke event. The neurological effects of the stroke may progress over a few days(the progressive stage) as more and more area may get included in the ischemic zone, and stabilize thereafter (completed stroke). It is imperative therefore that repeated clinical examination is done to track the progress of the ailment. If the symptoms and signs disappear within 24 hours of onset, the event is designated as a Transient Ischemic Attack (TIA). The definition is entirely a clinical one, and laboratory investigations carried out are meant only to support the clinical diagnosis. The basic pathophysiology is an ischemia which must last for at least a few seconds. The clinical manifestations occur rapidly because the neurons are unable to continue with their metabolism in the presence of the reduced blood flow.

Strokes are classified as thromboembolic or hemorrhagic (**Fig. 3**). Majority are ischemic (thromboembolic) and constitute 85% of the total while hemorrhagic constitute the remaining.

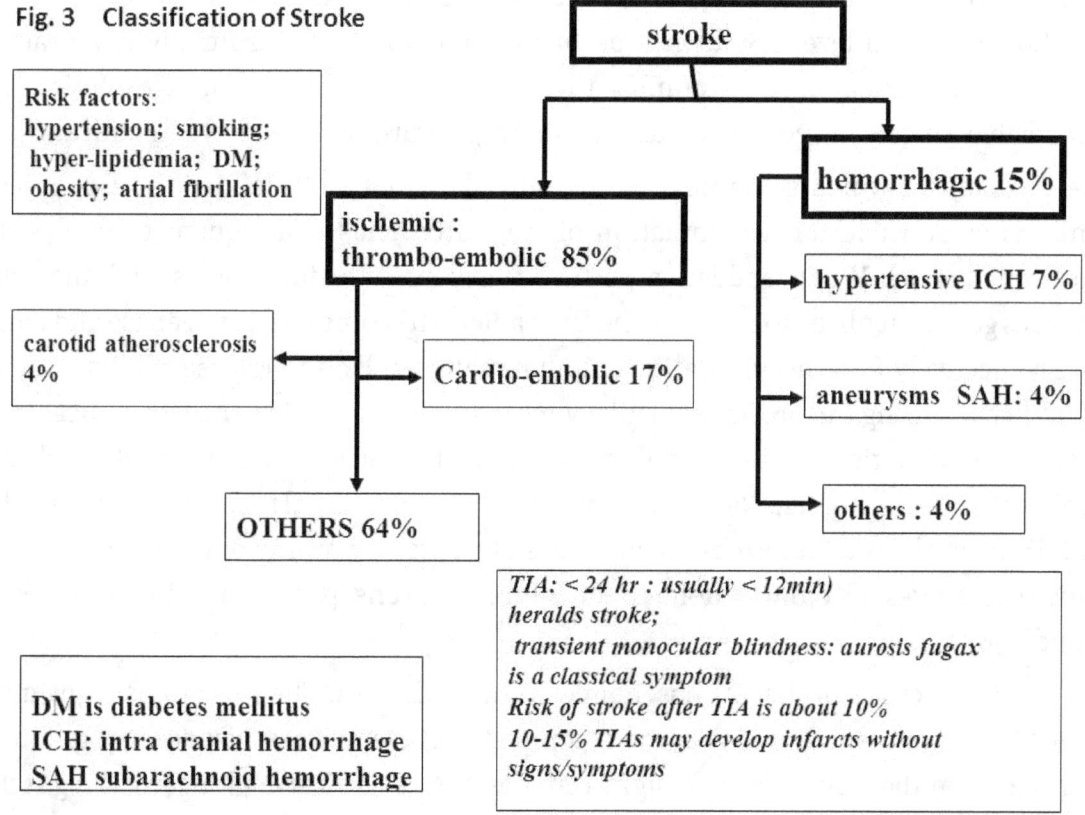

Fig. 3 Classification of Stroke

Etiology and Pathophysiology of Ischemic Stroke

The focal cerebral ischemia occurs because of thrombus and emboli formation in large arteries –most commonly in the carotid artery at its bifurcation. The vertebral, basilar and middle cerebral arteries can also involved. About 17% of the emboli originate from the heart, the commonest cause being atrial fibrillation followed by infarction in the apical region. The released emboli from whatever the source, flow up the internal carotid artery, and usually enter the middle cerebral artery to travel up the distribution of the middle cerebral artery. This happens because most of the blood from the internal carotid artery enters the brain tissue via the middle cerebral artery which is its main branch. The posterior (branch of the basilar artery) and the anterior cerebrals, get a relatively less share of the emboli. If the blood flow to the brain tissue reduces to <18 ml/100gms/min, brain death will occur in 4–10 min (infarction). If the flow is reduced to <20ml/100gm/minischemia without infarction is the result. But if the injury is prolonged for hours, it progresses to an infarction. The area surrounding the infarcted area is the ischemic penumbra, and can be demarcated using MRI. Rapidre-canalization treatment may help to reduce the extent of the penumbra, and the efficacy of the treatment may be assessed using repeated MRI. Following infarction the affected area becomes pale. The gray matter becomes congested and edematous because of dilated blood vessels and minute hemorrhages. If the offending embolus breaks away, sudden recirculation of the area may cause edema and hemorrhage (Reperfusion injury). Thrombolytic therapy instituted for embolic stroke can also produce reperfusion injury for the same reason.

The possible sequence of events that follows ischemia of brain tissue is illustrated in **Fig.4.** The proteolysis is brought about by the generation of free oxygen radicals subsequent to high intra cellular calcium levels. Sudden reperfusion which follows thrombolytic therapy does help to reduce the extent of the penumbra thus limiting the ischemic damage. But it may also be responsible for causing cellular injury, again by generation of free oxygen radicals.

The mechanism which generates free O_2 radicals is not well defined. It is possible that ischemia is responsible for accumulation of free fatty acids (consequent to lipolysis), and adenine nucleotides When sudden reperfusion occurs, free fatty acids and the adenine nucleotides get metabolized to generate the free radicals by using cyclooxygenase and xanthine oxidase respectively. Local accumulation of inflammatory cells, and release of nitric oxide (NO) may also occur during the process. The likely mechanism involved in the free radical induced damage is the disruption of brain capillary endothelial integrity which then adversely affects electrolyte and fluid exchange. Recently it has been shown in animal models that brain ischemia secondary to stroke releases excessive amounts of excitatory amino acids such as glutamate and this accentuates the damage done to the injured neurons, particularly in the centre of the ischemic injury.

The most innocuous of the brain ischemic events is the TIA. The symptoms/signs may last about 5–15 minutes usually followed by a complete recovery. The episode may graduate on to a stroke when the symptoms and signs last > 24 hours, indicating longer lasting ischemic

Fig. 4   Pathophysiology of cellular damage in stroke

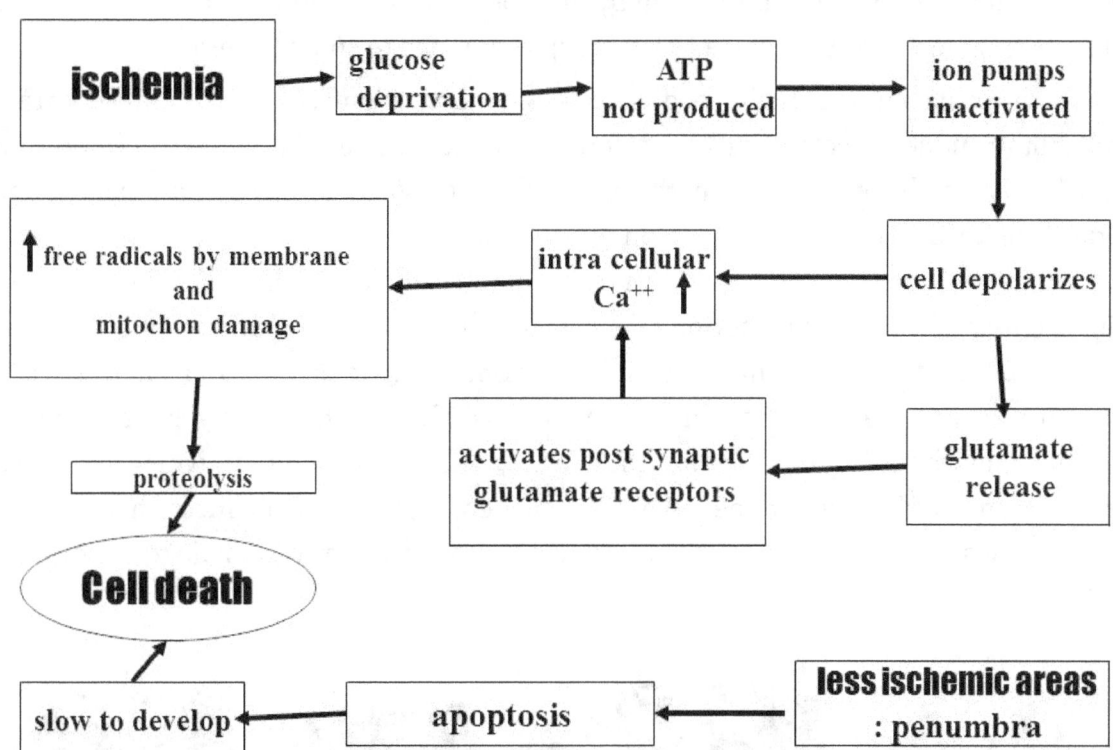

damage.50–70% of patients with stroke have had a previous episode of TIA. About 80% of the TIAs are embolic in nature (80%) and only about 5–10% of the are caused by heart emboli. The more common sources of emboli are i. the carotid artery and ii. the vertebro-basilar artery. The former usually results in a contra lateral hemi-paresis with or without a sensory involvement, a hemianopia; ordysathria if the dominant hemisphere is involved. The region of the brain commonly affected is the posterior limb of the internal capsule. All major sensory and motor tracts pass through this structure. The main reason for the involvement of this region is the peculiarity of its blood supply. Striate arteries which supply this area are end arteries.The vertebro-basilar insufficiency may manifest as vertigo; diplopia; visual blurring; facial nerve deficits; dysarthria; dysphonia; dysphagia, nausea, vomiting; ataxia; hemi or tetra paresis, sensory deficits; impairment of consciousness.

Pathophysiology of Acute Spinal Cord Injury

It is important at this stage for the medical student to appreciate the complexities of spinal cord injury- a topic that is seldom confronted by him/her in the early stages of the curriculum. The basic information may be of help in saving a person traumatized in an accident. The discussion on the spinal cord is hence restricted to just this facet. The student may do well to review the details of the sensory and motor tracts as they travel up and down the spinal cord and appreciate that the autonomic nervous system has its centres in the interomediolateral grey matter (C8/T1 to L3 for the sympathetic, and S2-S4 for the parasympathetic outflows). It is beyond the scope of this feature to describe details of interruptions in individual tracts.

Injury to the spinal cord may occur because of i. direct trauma (road accidents being by far the commonest reason); ii. compression by prolapsed intervertebral disk, hematoma/oedema, and bone fragments from fractured vertebrae; iii. damage to spinal arteries.

In the current context of high speed motor vehicular accidents, whip lash injuries to the spine result from a sudden deceleration or acceleration to the vehicle. With sudden deceleration such as in a head on collision, say in a car, the head follows Newton's 1st law of motion and continues to travel forwards. The cervical spine on which it is sitting is however a part of the rest of the body bound to the vehicle by the safety harness of the vehicle making the body and the vehicle a single entity. The force distribution over this part of the body –vehicle combination helps to reduce its per unit area concentration and thus offers relative protection from the crash force and also restrains the body from moving forwards. The weight of the head then makes the cervical spine which is by far the most mobile part of the spinal column whiplash forwards in hyper flexion. Similarly, when the impact is from behind, the opposite mechanism comes in to play, causing an hyper extension of the cervical spine. An attempt is made to illustrate this in **Fig. 5.**

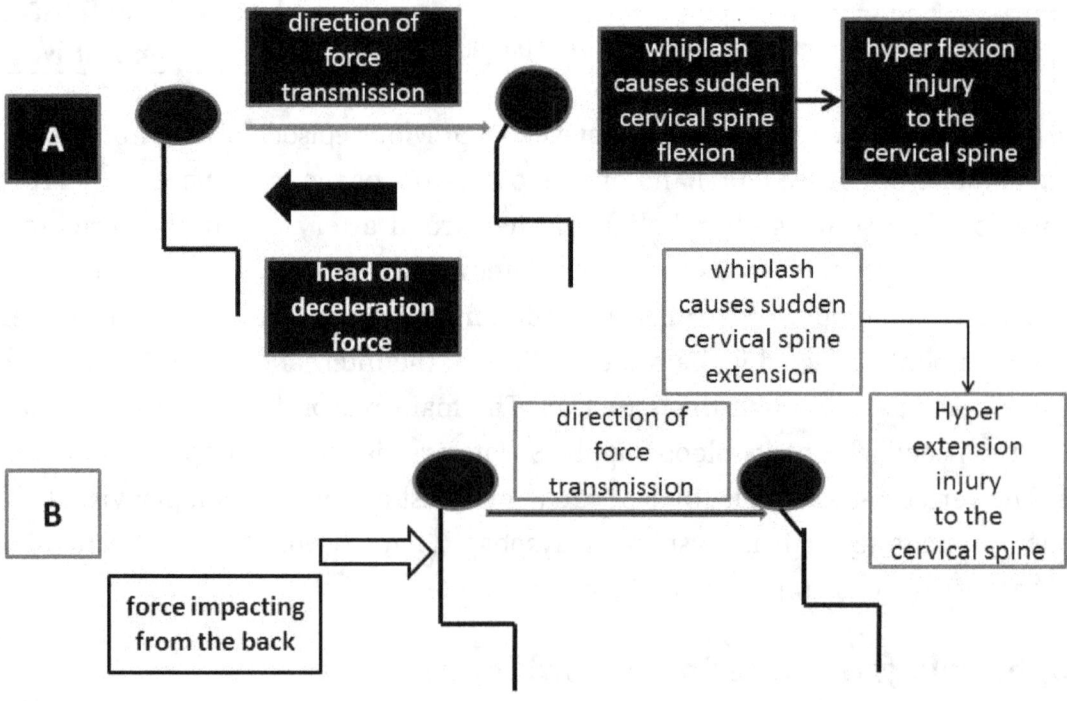

Fig. 5. Mechanism of spinal injury
A. sudden deceleration by a head on collision
B. sudden acceleration by impact from the back

Modern cars are fitted with head rests which prevent uncontrolled hyper extension. It is important however that the occupants adjust the head rest as given in the vehicle manual to ensure maximum safety. Unfortunately this issue is more often than not ignored. **(What position should you place a road side accident victim who may have had a cervical spinal injury and what supportive equipment will you use to transport him?).**

Pilots of high performance fighter aeroplanes are subjected to extremely high + Gz forces. During such a manoeuvre say 8 times the force of gravity, the weight of the body increases to 8 times normal. If the head normally weighs about 2–3 kilograms, within a second or two it weighs about16 kg and snaps forward resulting in flexion whiplash injury to the cervical spine.

Readers may recall that the human spine in the erect posture has natural curvatures. The cervical and lumbar spine have a natural lordosis (forward curvature) while the thoracic and sacral regions have a slight kyphosis (backwards curvature). The aim of these curvatures is to enable the spine to adjust to mechanical stresses. If an x-ray of the spine is taken in a motor cyclist immediately after a ride, he will be surprised to learn that the cervical spine has lost its lordosis though only temporarily. This makes the spine more susceptible to flexion injury. When confronted with sudden flexion in the cervical spine the initially loses its lordosis. This is a protective mechanism which attenuates the disturbing force to an extent. After the initial straightening, it bends forwards. It may be imagined that if this initial adjustment of spinal mechanics was not naturally available, the sudden flexion of the spine would make the local soft tissues more injury prone.

Table III: Physiology of the Autonomic Nervous System

1. It is an involuntary system
2. It co-ordinates and maintains homeostasis
3. It has a pre-ganglionic neuron inside the CNS (brain/spinal cord); and post ganglionic neuron outside the brain/ SC
4. Most organs are innervated by both arms of the ANS, and the effects may be excitation and /or inhibition. Kidneys,most blood vessels, and sweat glands are exceptions (only sympathetic).
5. The pre-ganglionic neurotransmitter for both sympathetic and parasympatheticarms is acetylcholine.
6. As a general rule, post ganglionic fibres are unmyelinated.
7. Post ganglionic neurotransmitter for parasympathetic is acetylcholine. For sympathetic it is nor epinephrine (exception sweat glands-acetylcholine)
8. Its physiological effects may be modulated by using either agonists or antagonists ofthe post ganglionic receptors

The degree of trauma may range from a neck muscle strain, injury to the supporting cervical spine ligaments, cervical vertebrae and damage to the spinal cord though the latter is a rare occurrence as the pilots are trained to handle such abrupt manoeuvres. The spine may also suffer vertical compression by a weight falling on the head or a diver hitting the bottom of a swimming pool head first. Here again the damage may extend from injury to the soft tissue and vertebrae which may get transmitted to the spinal cord. Rotational injuries to the spinal column my also occur.

Damage to the spinal cord is done by pressure produced by internal bleeding, oedema secondary to release of inflammatory substances, and ischemia. There could also be a transection. Functional liability depends upon the region of the spinal cord involved. For example, cervical spine damage may paralyse the diaphragm. Recovery is determined by the extent of damage to the neurons and tracts in the cord.

Autonomic Nervous System: Hyper and Hypo reflexia

The basic physiological principles under which the ANS operates are summarized in Table III. Autonomic hyper reflexia (Dysreflexia) is a potential life threatening conditioning which complicate transaction injury of the spinal cord at T6level or above any time after the spinal shock has resolved. The major issue is a very large increase in blood pressure which may be attributed to the following mechanism. Any strong sensory input below the level of injury (T6 and above) by harsh stroking of the skin, distended bladder (most common cause), distended rectum, is conducted by the sensory nerves in to the splanchnic nerve (T5 to T9). This input causes a powerful reflex vasoconstriction of the splanchnic vascular bed resulting in an abrupt and large increase in the blood pressure. Simultaneously, the nerve stimulation excites the adrenal medulla which releases large amounts of catecholamines. Apart from stimulating the splanchnic bed vasoconstriction, the catecholamine stimulate the myocardium by occupying adrenergic receptors and contribute further to the BP rise. It is for this reason that the injury level at or above T6 is pathophysiologically relevant. The steep rise in the blood pressure stimulates the carotid baro receptors to produce some reflex bradycardia via the intact vagus nerve. However, the compensatory mechanism is only partially effective because the sympathetic stimulation of the cardiovascular adrenergic receptors by the circulating catecholamines overwhelms the reflex bradycardia. The expected reflex vasodilatation by withdrawal of sympathetic tone (Marey's law) cannot occur as the descending inhibitory neural pathway is interrupted by the spinal cord injury. It has been postulated that the spinal cord injury is responsible for the increase insensitivity of the alpha adrenergic receptors in peripheral circulation below the injury level. This enhances the vasoconstrictor effect of the catecholamines. A role for serotonin has also been suggested. On spinal transection, the inhibitory influence of bulbno-spinal pathways on the interomediolateral serotonergic neurons is removed. The increased serotonin may in turn add to the vasoconstrictor effect. What finally contributes maximally to the crisis in human patients is not yet clear. During the crisis, typical signs of sympathetic over activity (apart from the high blood pressure) are seen in areas above the level of the section while below the section the skin is relatively pale and cool with reduced sweating. The severe hypertension if left untreated may cause serious problems such as myocardial infarction, pulmonary oedema, retinal and cerebral bleeds, seizures, acute renal damage, and even death though this is rare. Prompt treatment rapidly resolves the crisis (Q. What immediate treatment should be instituted in such a crisis? What preventive measures should be taken *by the nursing staff?)*.

The typical example of autonomic **hypotonia** commonly met with in clinical practice is orthostatic (postural) hypotension and is the most innocuous of the lot. Normally about 600 ml

Table IV. Response of normal subjects to orthostatic stress given as 70° head up tilt. + is increase; - is decrease. HR is hear rate; SBP is systolic blood pressure; DBP is diastolic BP; PP is pulse pressure. The cardiac out put is expressed as the cardiac index (l/min/m² body surface area). Data from Dikshit et al 1986. The figures are average from 168 tilts

HR/min	+ 18
SBP mmHg	- 1
DBP mmHg	+ 9
PP mmHg	- 12
CI lpm	-0.6

of blood is transferred from the chest cavity in to the lower limbs when a person stands up from a supine posture. This deactivates the arterial baroreceptors resulting in a withdrawal of the natural inhibitory tone of the cardiovascular centrein the medulla with activation of the sympathetic system. There is an increase in diastolic pressure and the heart rate with minimal changes in the systolic pressure (Table IV).The cardiac output in the standing posture is about 18% lower than in the supine posture. Under adverse conditions (Table V) the cardiovascular reflex mechanism is unable to cope despite the attempts made by a sympathetic system in overdrive. The powerful inotropism possibly activates left ventricular receptors which stimulate strongly the cardio-inhibitory centre via afferents C fibres. The vagal activation produces a bradycardia, while the sympathetic activity is suddenly withdrawn producing a wide spread vasodilatation. The perfusion to the brain is diminished followed by giddiness, sweating, nausea which may progress to a temporary loss of consciousness. The possible mechanism involved when the syncope is brought about because of loss of circulating fluid volume is depicted in figure. 6. Pain, heat stress produce a vasodilatation which may exceed its requirement and precipitate the vaso vagal syncope. Spaceflight deconditions the baroreceptors mechanism to postural stress and is one of the "true" autonomic hypotonia situations. The functionality of this mechanism depends upon the natural gravitational force to which the human body is subjected to on earth. During space flight when the gravitational force is zero, the arterial baroreceptor reflex "forgets" how to operate when back to earth. The dysfunction reverts to normal after a few days. Poor myocardial contractility may result in poor cerebral perfusion and loss of consciousness. It is important to note that myocardial disease may be a cause of syncope in the elderly who may be suffering from ischemic heart disease. This however cannot be classified as a typical autonomic hypotonia.

The person who has suffered syncope recovers back to normal in a few minutes. The possibility of injury as the individual collapses must be kept in mind especially in the elderly. Anxious bystanders who witness such an episode should not hasten to make the person sit up immediately after the episode. It is best to let the individual recover consciousness on his/her own. *(Why?)*

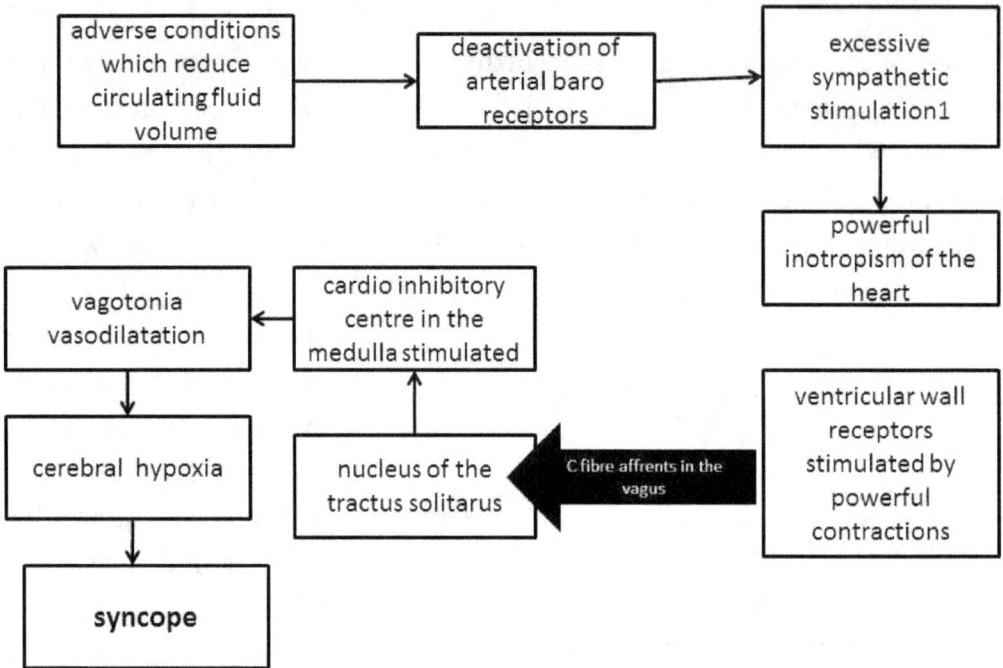

Fig. 6. Mechanism of orthostatic hypotension leading to syncope

The most serious condition of autonomic hypotonia of great severity is the Shy-Drager syndrome (Multiple System Atrophy). This is a progressive degenerative disease in which the degree of orthostatic hypotension increases in severity over time, finally reaching a stage when even an attempt to raise the head results in a loss of consciousness. There is loss of bladder and bowel control and absence of sweating. The condition is fatal.

Pathophysiology of Behavioral Abnormalities

Abnormal behavior ("Mental illness") in a person is often stigmatized, especially in our country, and the patient is ridiculed and on occasions maltreated and ostracized. At this phase of their studies, medical students should become aware of the fact that the so called "mental" illness is like any other disease that occurs because of physiological derangements, in this instance in the brain (neurotransmitters dysfunction) and has to be tackled accordingly. Discussed below is the pathophysiology of some of the mood disorders that a medical practitioner may meet with in practice.

Mood Disorders: Depression and Mania

Mood "Is a sustained emotional state." This state may alter briefly in daily life as joy, euphoria, sadness, depression, anger, anxiety (Affective states). If such a state persists, it becomes an Affective or Mood disorder. Depression, and extreme euphoria (Mania) are examples. Depression is the most common mood disorder (Unipolar). Sometimes patients may go from depression to extreme euphoria (mania). This is Bipolar disorder. The brain areas associated with of behaviour and cognition regulation lie mainly in the frontal cortex (the orbitofrontal cortex, dorsomedial prefrontal cortex, anterior cingulate gyrus, dorsolateral prefrontal cortex and the ventrolateral prefrontal cortex. In general, they are all likely to be involved in the pathophysiology of all mood disorders, be it depression or anxiety.

Depression Since the 1950s, it has been postulated that a deficiency of monoamines (Norepinephrine and serotonin) is the main causative factor of depressive illness. Genetic predisposition and early childhood adverse events have been thought to be involved. Normally, forebrain serotonergic and noradrenergic systems interact via their connections to the midbrain nuclei (the locus ceruleus, the parabrachial nucleus, and the raphe nuclei), the amygdala and the hippocampus to regulate behavior (Fig. 7). A large number of the mid brain nuclei synthesize and secrete serotonin which then reaches serotonin ($5HT_{1A}$ and $5HT_{2A}$) receptors spread over the forebrain. Some of the monoamines which are not utilized are reabsorbed in the pre-synaptic membrane and metabolized by the Mono Amine Oxidase (MAO) in the pre-synaptic membrane. It has been postulated that a malfunction of the post synaptic receptors where in there is an increase in their binding to the serotonin molecule, with a simultaneous reduction in binding to serotonin transporter may be responsible for the overall reduction in the availability of serotonin in patients with depression. Noradrenergic receptor dysfunction in the mid brain has also been suggested. Corticotropin Release Factor levels have been found to be elevated in the CSF of depressed patients while this was not observed in normal males and females, nor in schizophrenics. Structural changes such as a reduced forebrain volume and a difference in the volume of the temporal lobes have been detected in patients of depression by the use of imaging techniques. Functionally, positron emission tomography (PET) demonstrates reduced metabolic activity in the dorsal and dorsomedial prefrontal cortex. As against this, the ventro lateral ventro medial prefrontal areas show increased activity. The amygdale which is connected to these regions is also affected. **(Fig. 7).** Similar findings have been recorded in patients with bipolar depression. Never the less the exact mechanisms responsible for the pathophysiology of depression are not clear.

Fig. 7. Brain areas involved in the pathogenesis of stress and depression. The blue arrows are the serotogenic and adrenergic outflows from midbrain and amygdala whose neurons are activated by hypthalamic CRF

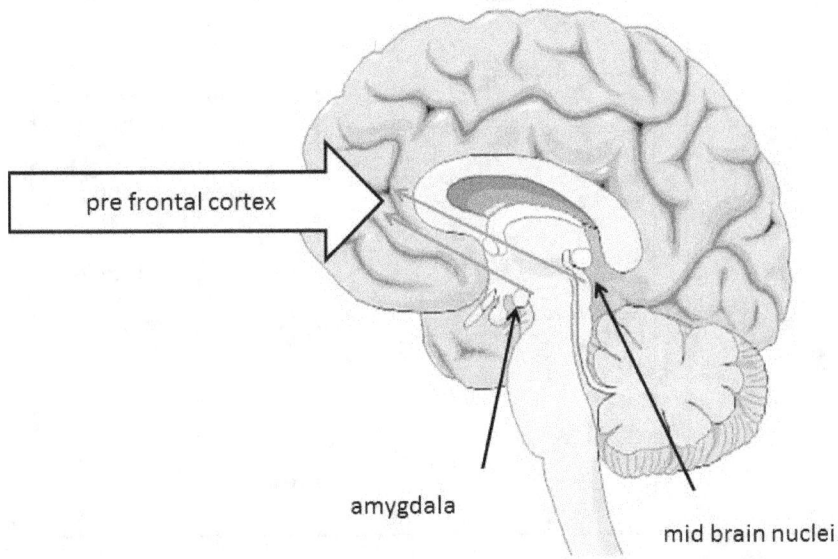

pre frontal cortex

amygdala

mid brain nuclei

The common symptoms in a depressed person are insomnia, general lack of enthusiasm in life, lack of concentration, irritability, loss of appetite and a dip in overall performance. At this stage of their studies, the young medicos could do well by spotting such symptoms amongst those in their close proximity and caution friends/relatives with the advice that expert medical help should be sought at the earliest. Talk of suicide by depressed person must be reported to those in authority on priority.

Considering the pathophysiology of depression, treatment revolves around the following principles: i. inhibition of reuptake of mono amines by the pre synaptic neuron (Selective Serotonin Reuptake Inhibitors - SSRIs) for 5HT or Tricyclic anti-depressants which are concerned mainly with noradrenaline). ii. inhibition of the activity of the enzyme that metabolizes noradrenalin and serotonin (5HT) at the pre-synaptic membrane-the Mono Amine Oxidase inhibitors. In both situations, the locally available concentrations of the monoamines increase within a day. However it is not clear as to why the onset of clinical improvement takes about 6–8 weeks. The most effective mode of treatment seems to be the use of serotonin reuptake inhibitors (SSRIs).

In **Mania,** there is an increase in the glutamate/glutamine levels in the dorsolateral prefrontal cortex. Imaging techniques corroborate this by showing increased metabolic activity in the region especially of the left side, as against hypo activity in patients with depression. With treatment the activity in these regions normalizes. The principles of treatment of mania revolve around use of drugs which reduce neuronal hyperactivity and receptor up regulation. The drugs involved are lithium salts, anti convulsive agents such as valproic acid and anti-psychotics.

Anxiety Disorders

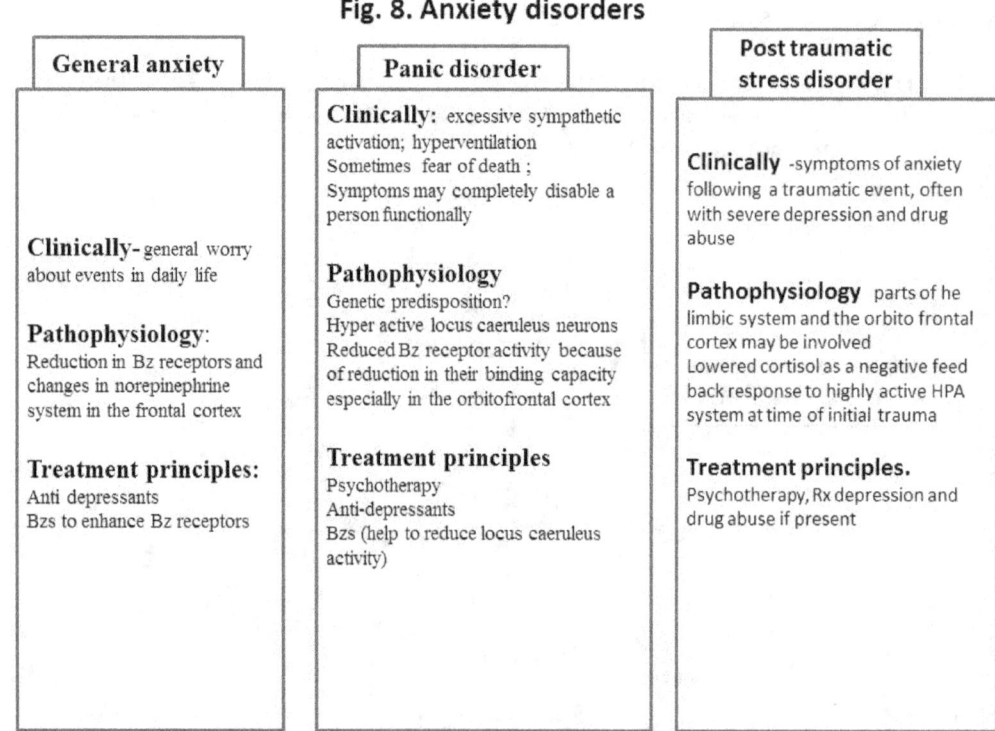

Fig. 8. Anxiety disorders

Anxiety is an emotion which is a natural reaction to an adverse situation resulting in behavioural and autonomic responses which help an individual to cope with the given situation. The brain areas involved are the locus ceruleus, amygdala, hippocampus, the hypothalmico-hypohyseal axis and the orbito frontal cortex. Functionally there is a GBA- Benzodiazepine (Bz) receptor system in the orbito frontal region. Activation of the Bz receptors which are located close to GABA receptors increases the responsiveness of GABA receptors. During anxiety, the GABA-benzodiazepine (Bz) receptor system shows diminished binding capacity. When the emotion (of anxiety) persists and interferes with daily life, it becomes pathological. Anxiety may present as a general anxiety syndrome, a panic disorder or a post traumatic stress disorder **(Fig. 8)**.

A patient of General Anxiety is constantly "worried" about some issue or the other, and is often tense and depressed. Panic attacks on the other hand are a sudden manifestation of a worrying issue as excessive sympathetic activity (tachycardia and palpitations, sweating, tremors and hyperventilation). These symptoms may cause a temporary functional paralysis in the sufferer. Post traumatic stress disorder is secondary to a highly personalized trauma which an individual may have suffered-whether physical or psychological. It was originally described in combat personnel (shell shock). Typical and severe manifestations of anxiety follow such a traumatic event after a variable latent period, and is by far the most difficult of the anxiety states. Pathophysiologically, the brain areas involved are common to both anxiety and depression, and hence anti depressants form an important part of management of anxiety states, particularly so in the posttraumatic stress disorder where morbid depression may co-exist with anxiety. Benzodiazepine (Bz) receptors present in the pre frontal cortex and the mid brain nuclei are thought to activate the inhibitory GABAnergic neurons and hence the use of this class of drugs in the treatment of anxiety is beneficial.

Schizophrenia (SCZ) is a thought disorder which affects about 1% of the adult population worldwide. It strikes in young adult hood and pervades throughout the most productive phase (work wise) of an individual. It has a devastating effect on the overall wellbeing of the patient and his/her family. It is probably the worst kind of psychiatric disability to affect mankind. Clinically, a patient may have **positive** and **negative** symptoms. The former consist of hallucinations, delusions, bizarre aggressive, abusive, illogical behaviour, and incoherent speech. The latter which is the hall mark of the disease is exhibited as asocial behaviour, with lack of social attention and general apathy to the surroundings. This and the lack of motivation in life with thought blocks, poverty of speech and over all non responsivity makes such a patient extremely difficult to deal with. It is also this aspect that makes the patient professionally incapable during the prime of life.

Pathophysiologically, neuro-developmental issues during the second trimester may be involved. During this phase cortical neurons develop and migrate to their assigned areas to make appropriate neural connections. Viral infection in the mother at this stage may interfere with this development. Prenatal dietary deficiencies or birth trauma could be the other possibilities. Neuro-anatomically, there is a widening of the 3rd and lateral ventricles as also a reduction in the volume of the frontal cortex, the limbic system, temporal lobe and the thalamus. These deficiencies have been confirmed by MRI in live patients, and by pathology during post mortems of patients of SCZ. The functional derangements are neurotransmitter oriented. The dorsolateral prefrontal cortex contains a large percentage of neurons which release the inhibitory neurotransmitter GABA. These neurons which develop most during adolescence when SCZ strikes, are encroached upon heavily by dopaminergic axons. They also interact with neurons which release glutamate. The latter then inter acts with N Methyl D-aspartate (NMDA) receptors. A dysfunction of GABAergic neurons as well as NMDA receptors has been hypothesized in the pathophysiology of this disorder. The fact that i. symptoms similar to those in SCZ appear with the intake of cocaine, excess of levo dopa, and amphetamine and ii. dopamine D_2 receptor blockers are useful in the treatment of SCZ has given credence to the Dopamine hypothesis of the disorder. Drugs which regularize $5HT_2$ and dopamine D_2 receptor activity have been used in SCZ. This suggests a disruption of 5HT and Dopamine interaction in the frontal cortex in SCZ.

Drugs used in the treatment of SCZ are the typical anti-psychotics which act blocking D_2 receptors (Haloperidol; Chlorpromazine), and the atypical anti psychotics (Clozapine) which may involve both 5HT2 and Dopamine receptor inter actions. The latter have been known to improve the negative symptoms of SCZ patients. (***What side effects can the use of typical anti psychotics have?***). At this point of time it is not the purpose of this discourse to make the student proficient in diagnosing this disorder and its treatment. It is only to make him/her aware of the seriousness of the condition and perhaps recognize its early symptoms so that those in authority may be warned in time so that a precious life may be salvaged to the extent possible.

References

1. Adiga U, Vickneshwaran V, Sen S. Electrolyte derangements in traumatic brain injury. Basic Research Journal of Medicine and Clinical Sciences ISSN 2315–6864 Vol. 1(2) pp. 15–18 September 2012.

2. Banks WA. Physiology and pathology of the blood-brain barrier: implications for microbial pathogenesis, drug delivery and neurodegenerative disorders. J.NeuroVirology. 1999;5:538 – 555.

3. Behan LA, Phillips J, Thompson CJ, Agha A. Neuroendocrine disorders after traumatic brain injury. J Neurol.Neurosurg Psychiatry. 2008;79: 753–759.

4. Cerebrovascular Diseases. Chapt. 34in Adams and Victor's Principles of Neurology Ed. Victor M, Ropper AH. 7[th]Edn. Mc Graw Hill London; 2001; pp 821–883.

5. Cookson J. Use of anti psychotic drugs and lithium in mania. Br. J Psych. 2001; 178: s 148 –s156.

6. Davalos A, Shuaib A, Gunnar Wahgren N. Neurotransmitters and pathophysiology of stroke: Evidence for the release of glutamate and other transmitters/mediators in humans and animals. J. Stroke and Neurovascular Diseases. 2000; 9; Part B 2–8.

7. Delgado PL, Moreno FA. Role of norepinephrine in depression. J Clin Psychiatry. 2000;61 Suppl 1:5–12.

8. Dikshit MB, Banerjee PK, Rao PLN. Orthostatic tolerance of normal Indians and of those with suspected abnormal cardiovascular reflex status. Aviat. Space and Environ. Med.: 1986; 57:168–173.

9. Dikshit MB. Syncope: An enigma in aviation physiology. Peripheral Elements of the Nervous System. Ed. Fahim M. Computronix India. New Delhi 1996; pp 59–72

10. Dunn LT. Raised intracranial pressure. J Neurol Neurosurg Psychiatry 2002;73(Suppl I): i23–i27.

11. Einaudi S, Bondone C. The effects of head trauma on hypothalamic–pituitary function in children and adolescents. Curr. Opin. Pediatr. 2007.19:465–470.

12. Fix JD. Neuroanatomy. 3rd Edn. Lippincot Williams and Wilkins. London; 2002.

13. Ganong WF. Review of Medical Physiology. McGraw Hill, New Delhi. 22nd Edn. 2005.

14. Gray's Anatomy: The Anatomical Basis of Medical Practice; 39th Edn. Ed. Standring S; Elsevier, Churchill Livingstone, London; 2005.

15. Heim C, Nemeroff C. Impact of early adverse experiences on brain systems involved in the pathophysiology of anxiety and affective disorders. Biol. Psychiatry. 1999; 46: 1509–1522.

16. Hirschfeld RM History and evolution of the monoamine hypothesis of depression. J. Clin. Psych. 2000; 61; Suppl. 4–6.

17. Katzman M, Jacobs L. Venlafaxine in the treatment of panic disorder. Neuropsych. Dis. Treat. 2007; 3: 59–67.

18. Kelley R, Kovac A. Horizontal gaze paresis in hemispheric stroke. Stroke. 1986; 17: 1030–32.

19. Kerner B. Glutamate neurotransmission in psychotic disorders and substance abuse. The Open Psychiatry Journal. 2009; 3, 1–8.

20. King NS. Post Traumatic syndrome: clarity amid the controversy. Editorial. British J. Psychiatry 2003; 183: 276–278.

21. Lewis DA, Pierri JN, Volk DW, Melchitzky DS, Woo TU. Altered GABA neurotransmission and prefrontal cortical dysfunction in schizophrenia. Biol. Psychiatry 1999; 46: 616–626.

22. McPhee SJ, Hammer GD. Pathophysiology of Disease. An Introduction to Clinical Medicine. 6th Edn. Mc Graw Hill; New Delhi. 2010.

23. Micheal N, Erfurth A, Ohrmann P et al. Acute mania is accompanied by elevated glutamate/glutamine levels within the left dorsolateral prefrontal cortex. Psychopharmacology.2003; 168: 344–346.

24. Nutt DJ. The neuropharmacology of serotonin and noradrenaline in depression. Int.Clin. Psychopharmacology. 2002; Jun, 17, Suppl. 1: S1-S12.

25. Owens MJ, Nemeroff CB. Role of serotonin in the pathophysiology of depression: focus on the serotonin transporter. Clin. Chem. 1994; 40: 288–295

26. PersidskyY, Ramirez SH, Haorah J, Kanmogne GD. Blood–brain Barrier: structural components and function under physiologic and pathologic conditions. J Neuroimmune Pharmacol. 2006; 1: 223–236.

27. Phillips M, Ladouceur CD, Drevets W. A neural model of voluntary and automatic emotion regulation: implications for understanding the pathophysiology and neurodevelopment of bipolar disorder. Molecular Psychiatry. 2008;13:833–857.

28. Stobart J, Anderson C. Multifunctional role of astrocytes as gatekeepers of neuronal energy supply. Frontiers in Cellular Nueoscience. 2013; 17: 1–20.

29. Text Book of Medicine. 2nd Edn. Ed. Souhami RL, Moxton J. Churchill Livingstone (London) 1994; pp 905–907.

30. Traytsman R, Kirsch J, Koehler R. Oxygen radical mechanisms of brain injury following ischemia and reperfusion. J Appl Physiol.1991; 71:(4) 1185–1195.

31. Volk DW, Lewis DA. Impaired prefrontal inhibition in schizophrenia: relevance for cognitive dysfunction. Physiol. Behav. 2002; 77: 501–505.

32. Wade S, Smith S, Johnston J, Esaton D. Cerebrovascular Diseases in Harrison's Principles of Internal Medicine. Edt. Kasper DL Braunwald E, Fauci A, Hauser S, Lomngo D, Jameson JL. 16th edn. 2005;pp. 2372–2393.

33. Watchko JF. Kernicterus and molecular mechanisms of bilirubin induced CNS injury in new borns. Neuromol. Med. 2006; 8:513–529.

CHAPTER - 13

Pathophysiology of Common Thermoregulatory Disorders

Plan

1. Generalintroduction
2. Normal body temperature and its regulation
3. Physiological effects of heat and cold stress
4. Heat induced illnesses
   i. Heat exhaustion
   ii. Heat hyperpyrexia
   iii. Heat stroke
4. Acclimatization to heat stress
5. Pathophysiology of fever
6. Hypothermia

General Introduction

The Indian subcontinent has a tropical climate with the sun shining down mercilessly most of the year round. Hence it is the heat stress that is of greater concern to our population. Only in the northern parts severe winter conditions may adversely affect health of the population. As such given the poor socio-economic status most of the population thrive under makes them overtly susceptible to these environmental hazards. Our Armed Forces operate under severe weather conditions to remain battle fit. Unfortunately our physiology text books do not cover adequately this extremely pertinent issue of environmental hazards. The young medico needs to be conditioned to this real live problem well in time to make him/her a well informed medical practitioner.

Normal Body Temperature and Its Regulation

The human body generates its own body heat and endeavours to maintain it's constancy under a very wide range of environmental conditions. It is interesting to note that the average deep body temperature (Core temperature or T_{core}) is about 37^0C which is quite close to the T_{core} of 42^0C which could be rapidly fatal in humans. Hence the mechanisms which help to dissipate heat from the body need to be more efficient because of the low margin of permissible error as compared with mechanisms which help to increase the T_{core} if it falls below the normal. Body temperature $<27^0C$ may be lethal but here the margin between the normal and the dangerous lower limit is as much as 10^0C.

Normal body temperature measured orally varies widely in normal individuals from a low of 35.6°C to a peak which is slightly more than38°C. It is at its lowest in the early part of the morning and highest in the late afternoon/early evening. Arbitrarily it is generally taken as 37°C the so called "Set point," more recently designated as the "Balance" point. The variability in the elderly is similar to that of young adults. Temperature regulatory mechanisms in the elderly function less efficiently when confronted with extreme environmental temperatures. For example, the ability to produce sweat, and respond with suitable cardiovascular adjustments to heat stress are limited making this group easily susceptible to heat induced illness. Also the body temperature response to infection in the elderly may be muted. This may misguide the treating physician while assessing the severity of infection. In children the range of variation in daily temperature settles down to adult values in early childhood while it may be high in children below the age of 1–2 years. Very young children too have immature temperature regulating mechanisms. Some authors have opined that an early morning value of >37.2°C and an evening value of > 37.7°C should be taken as abnormal (fever).

Hypothalamus and Body Temperature Disturbances

The hypothalamus is a tiny collection of nuclei deeply embedded in the diencephalon. It is virtually an "Intelligence Head Quarter" for all goings on inside and outside the body. It performs a myriad functions **(Table I)**. Most of these have been brought out in various chapters in this book (Pathophysiology of Stress, and Obesity). In this chapter, its role in body temperature regulation and various heat illnesses have been emphasized.

Table I: Functions of the hypothalamus

A. Endocrine

i. Hormonal control of the anterior pituitary
ii. Secretion of vasopressin and oxytocin
iii. Neurotransmitter functions of hypophysiotropic homones in the brain and some other tissues

B. Non Endocrine

i. Body temperaturecontrol
ii. Feeding/drinking
iii. Sleep
iv. Sexual functions
v. Emotions
vi. Autonomic control
vii. Circadian rhythm regulation

The integration of thermal regulation is probably done at the level of the pre-optic region of the anterior hypothalamus for both heat and cold. The sensory input to the hypothalamus is provided by warm and cold thermal receptors present both centrally (hypothalamus, great veins, deep tissues) as well as in the periphery (skin, oral mucosa and the uro-genital tract). In the former it is the warm receptors that predominate while the cold receptors situated just below the epidermis are more active peripherally. The warm receptors in the skin are located deeper below the dermis. More recently it has been concluded that when the warm receptors are activated, they suppress the activity of cold receptors. For activation of autonomic responses to cold exposure, the warm sensitive neurons need to be suppressed. The transduction of various ranges of peripheral thermal sensations are now known to be made through proteinous cation channels (transient receptor potential (TRP) channels) which are activated at specific ranges of temperatures. These channels are located on the cell membranes of various cells including the peripheral thermal receptors. Opening of these channels may release secretions which in turn stimulate the sensory nerve endings.TRPV3 and V4 function best over a range of 30°C to 50° C, peaking at about 42°C while the range of temperature between about 7°C to 40°C (peak at about 25°C) is best suited to activate TRPM8. Extreme cold activates TRPA1 to produce severe pain while temperatures exceeding 45°C associated with pain open TRPV 1 andV2channels. The activity of these channels at various temperature ranges overlaps considerably. As yet there is no evidence that the TRP channels are involved in the function of the central thermal receptors. A working model for control of body temperature is depicted in **Fig. 1.**More recently it has been shown in rats that behaviour on exposure to thermal stress (hot or old) is controlled by a set of hypothalamic neurons which function independently of neurons in the pre-optic hypothalamus which control sweating and cardiovascular effects of thermal exposure. Damage to the hypothalamus results in dysfunction of thermoregulatory mechanism.

The average skin (shell) temperature in the human being is 33°C, much lower than the core (body) temperature which is about 37°C, generating a gradient from the core to the skin of about 4°C **(Fig. 2)**. This establishes a passage for excess heat accumulation to flow out of the body in to the environment in order to maintain normal body temperature. The shell provides the physiological insulation and consists of a variable component (skin blood flow and sweat glands). A temporary decrease in insulation may be brought about by vasodilatation of skin blood vessels and sweating. The extent of the former is large. Skin blood flow at rest (0.5 l/min) may increase to as much as 8 l/min on exposure to severe hot environment and exercise in the heat. Both these effects help to increase the core to skin gradient, facilitating loss of heat when required. Alternatively the insulation increases if vasoconstriction occurs on exposure to a cold environment. The fat in the dermis of the skin is a more permanent source of insulation. It has been noted that in cross sectional CT scans of the trunk of residents of extremely cold areas, the percentage contributed by fat to the overall circumference is quite high.

The non hairy (glaborous) skin of the palms, soles of feet and lips is innervated by sympathetic vasoconstrictor nerves. Exposure to cold increases sympathetic tone to these vessels and reduces heat loss by reducing surface area for exposure of blood to the environment. Withdrawal of sympathetic tone produces vasodilatation which helps to cool blood by convection. The non-glaborous skin on the other hand is innervated by both sympathetic adrenergic as well as cholinergic nerves. Activation of sympathetic cholinergic fibres with heating of the skin dilate the blood vessels for facilitating heat loss when just the passive withdrawal of vasoconstrictor tone leading to vasodilatation is not adequate to lose excess heat to the surroundings. Neuro/ chemo transmitters involved in active vasodilatation in the skin include acetylcholine (most important), nitric oxide (NO), prostaglandins, substance P and histamine. The insulation offered by the shell increases if vasoconstriction occurs on exposure to a cold environment.

The blood vessels in the fingers when exposed to extreme cold undergo consecutive phases of vasoconstriction followed by vasodilatation. This is also called hunting reaction of Lewis. Severe cold exposure induces α adrenergic induced arterial vasoconstriction in the fingers. This rapidly brings down the skin temperature. In a few minutes time, the α adrenergic receptors become insensitive to nor epinephrine. This brings about a local vasodilatation and the blood flow to the fingers increases warming of the skin. The cycle repeat itself as the warmed arteries again become responsive to nor epinephrine. The reaction is well established in people living in cold environment such as the arctic and may be an indication of acclimatization to severe cold. It also occurs in residents of tropical regions operating in cold environment such as Indian Army soldiers at high altitude during the winter months. Whether it protects them from cold injuries is a matter of debate. The fat in the dermis of the skin is a more permanent source of insulation.

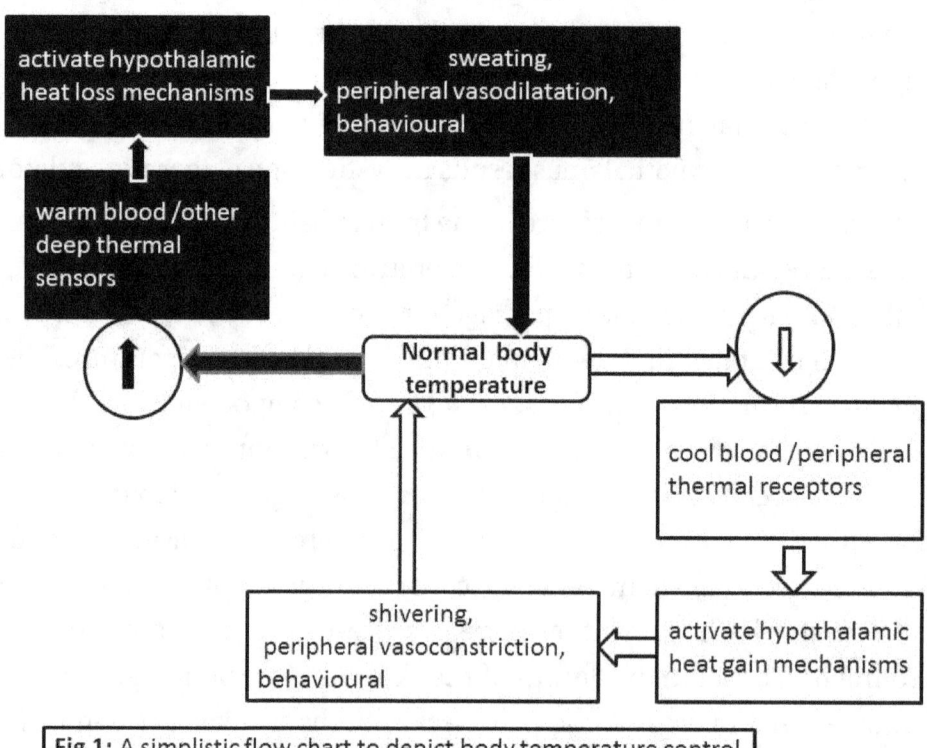

Fig.1: A simplistic flow chart to depict body temperature control mechanisms

Evaporative cooling of sweat is by far the most important physiological method of dissipating accumulated heat to the hot environment, particularly during exercise in hot environment. In hot, humid atmosphere plenty of sweating occurs but as evaporation of this sweat cannot occur effectively, cooling of the skin is not effective. Hence a hot humid environment is a more serious proposition. Sweat is produced by eccrine sweat glands distributed all over the skin. The secreted sweat is hypotonic because of re-absorption of NaCl. In heat acclimatized subjects (Indians for example) the salt reabsorption is more than in non acclimatized subjects. However when sweating occurs profusely, large amounts of electrolytes are lost resulting in electrolyte disturbances with dehydration. Sweating is initiated centrally by the anterior hypothalamus via sympathetic cholinergic nerves to the sweat glands. Increase in sweating occurs initially because of recruitment of a greater number of sweat glands followed by an increased rate of sweating. Excessive fluid loss during heavy exercise in high temperature, use of anti-cholinergic medication which block post ganglionic muscarinic receptors are factors that can adversely affect sweating and with it, heat loss to the environment. Hypothalamic damage may produce cessation of sweating.

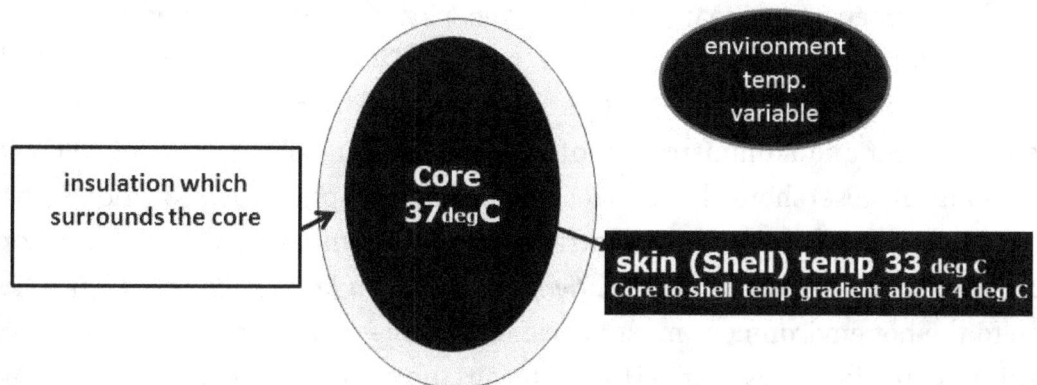

Fig. 2. Concept of inter dependence of core & skin (shell) temperature

Disorders of Thermal Regulation

The pathogenesis of increased body temperature in human beings is of two types. The body temperature increases because the so called "Set point" has been set to a higher level by action of pyrogens which affect the hypothalamus. This is broadly classified as fever. The other variant is caused by exposure to high environmental heat stress. It may vary from heat exhaustion to the very severe form- heat stroke which has high mortality and morbidity.

Pathophysiology of Environmental Heat Induced Illness.

The core to shell gradient **(Fig.2)** is of paramount importance in maintaining the core temperature when an individual is confronted with high environmental temperatures. Under the circumstances, the skin temperature rises as it absorbs the environmental heat mainly by radiation and convection. This reduces the core to shell gradient, resulting in an increase in core temperature. As a response to this, reflex sweating and vasodilatation are induced by the hypothalamus. The sweating helps to reduce skin temperature by evaporative cooling, while vasodilatation increases the surface area of blood vessels which in turn helps the blood flowing through the skin to cool. Both factors enable the excess accumulated heat to be dissipated. More the sweating, better is the cooling. However under a given set of circumstances, the physiological mechanism described above may be exhausted or prove inadequate.

Heat Exhaustion which may occur could be **exertional** as in military personnel operating in hot environment, during exercise in very hot environment, and **non exertional**. The latter is seen in infants and the elderly, in malnourished individuals and in those with various pre-existing ailments when they are exposed to high environmental temperature. The "set point" is not altered. The sweating causes substantial fluid and electrolyte loss. If this is not replaced at regular intervals, dehydration sets in **(Fig. 3)**.The fluid loss leads to a fall in the cardiac output and the blood pressure with are flex tachycardia. This may precipitate a vaso-vagal syncope, and in severe cases a cardiovascular collapse. The urine is concentrated and high coloured. All these are evidence of a physiological strain. The electrolyte depletion (Na^+ and K^+) through the sweat and the low cardiac output are mainly responsible for the fatigue and muscle weakness. The loss of electrolytes may cause painful repeated spasms of various limb and abdominal muscles. These sometimes appear at a later stage, particularly if the fluid replacement has not included adequate quantity of salts (Na^+ and K^+). Sweating helps to maintain the core temperature within normal limits. At this stage a rapid evacuation of the individual to a cooler environment and administration of oral rehydration solution (ORS)(which contains electrolytes and glucose) should be resorted to. The latter helps rapid absorption of water and electrolytes in the small intestines) or intravenous fluids will rapidly reverse the condition. *(What could be the disadvantage to the patient who drinks a large volume of plain water?)* Re-exposure to the hot environment must be avoided. If due care is not taken, the dehydration may result in depletion of sweating. Under the circumstances, the core to skin temperature gradient maybe disturbed and a rise in core temperature may occur (heat hyperpyrexia). The body temperature is usually <40^0C in this condition if at all it does rise. However this could progress on to heat stroke if suitable treatment is not given too the patient. The principles of treatment remain the same. Whole body coolinghelps if the body temperature does not reduce after the recommended first aid treatment. Using ant-pyretic medication does not help *(Why?)*.

The elderly and the very young are more susceptible to developing heat induced illness. The water content of the elderly is about 50% of total body weight. Hence the fluid loss they suffer on heat exposure will rapidly deplete their body water reserves when copious sweating occurs. This results in rapid, severe dehydration. They may be on various medications which may interfere with the requisite compensatory mechanisms. The ability of tissues in the elderly to generate heat shock proteins which offer cytoprotection during various stresses is likely to be compromised. Drinking behaviour in the elderly is often deranged and this factor may get further confounded if their debilitated health and sarcopenia restrict movements which are required to take them to the source of food and water or to that matter move away from the source of environmental heat stress.

Children have a relatively large skin surface area as compared with their body mass. This has two disadvantages when confronted with environmental heat stress. The area of exposure available to convective and radiant heat gain from the environment is more, and sweating because it spreads over a large body surface area causes rapid fluid depletion. Added to this is the high level of physical activity which children may continue with in hot weather. Children often suffer from gastrointestinal disease during the summer. The ensuing diarrhoea and vomiting rapidly dehydrates them exposing them to severe effects of heat induced illness. Care must be exercised by adequate rehydration, and by judicious use of anti-cholinergic medication if indicated *(Why?)*.

Fig. 3. Flow chart: Pathogenesis of heat induced illnesses

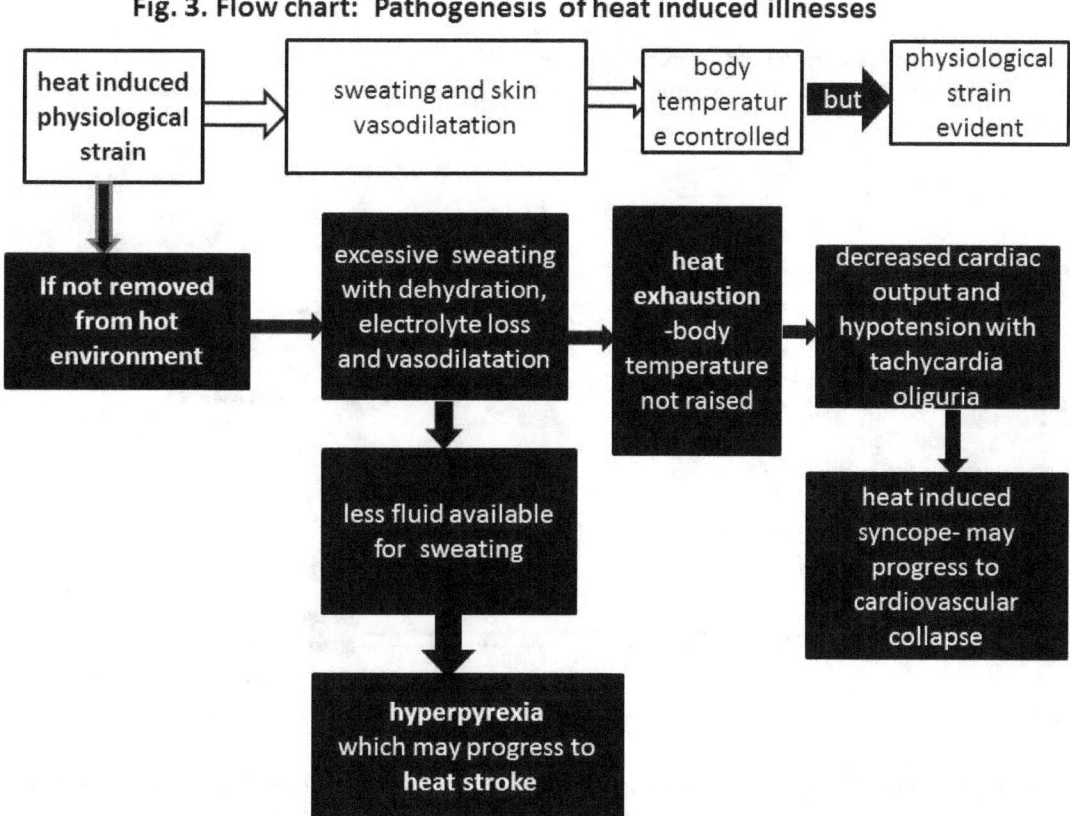

When exposed to severe hot environment, the skin absorbs the external heat because of conduction convection and radiation. The core to skin gradient reduces, and loss of heat which is dependent upon the normal gradient cannot occur easily. As a result, the heat accumulates in the body and the core temperature rises. The reflex sweating (which cools the skin) and vasodilatation (which increases the surface area over which blood flows) help to try and keep the skin cool thereby attempting to restore the core to skin gradient. The overall adjustment tends to limit the rise of the core temperature. In very severe hot weather conditions, despite this compensatory mechanism, the skin temperature keeps increasing until it nears or equalizes the rising core temperature. This is illustrated in **Fig.4** which is constructed on the basis of data from Dikshit 1980 and Iyer et al 1983.The point of this confluence signals the tolerance limit of exposure to severe acute heat stress. Physiologically this is an extremely stressful condition and produces loss of concentration, irritability, drastic reduction in performance, and severe cardiovascular strain. It may rapidly progress to a collapse. Such situations have been known to arise in high speed low level flying in fighter aeroplanes and in battle tanks during the summer months. It may also be seen in military personnel engaged in strenuous physical work and during exercise in hot environment. This may happen in patients of heat stroke when the sweating ceases and the skin temperature increases rapidly. It must be appreciated that the "Set point" is not disturbed during such derangements as in the pathogenesis of fever. Whole body "Pre-cooling" by circulating cold water through a cooling suit has been known to extend the time at which confluence of the core and skin temperatures occurs, there by increasing tolerance to Severe Acute Heat Stress.

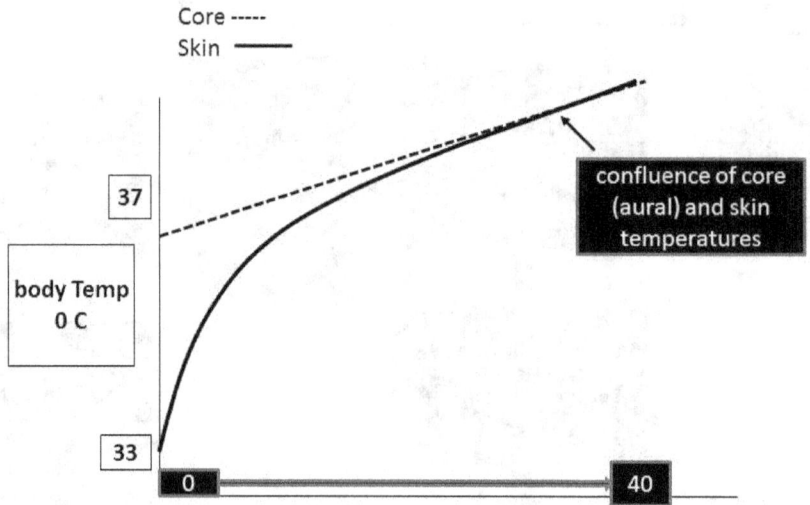

Fig. 4. This constructed graph depicts likely responses of core and skin temperatures to high environmental temperatures. The same may happen in conditions such as heat stroke when sweating ceases and the skin is no longer cooled.

Time of exposure in minutes in a "Hot cockpit". Environment setting: Temperature 50 deg C with relative humidity 50%

Heat stroke is by far the most serious and dangerous consequence of heat exposure. It presents as body temperature in excess of 40°C with derangement in brain function of various degrees. It is often accompanied by a cessation of sweating (in most cases) resulting in a dry hot skin, and severe cardiovascular strain. The condition has high mortality because of multiple organ system failure. Patients who recover from it may suffer premature death because of renal, cardiac, brain or other systemic dysfunction. It may be noted that cessation of sweating and a dry skin may not be present in a patient of heat stroke.

Heat stoke may be a consequence of heat exhaustion/hyperpyrexia if the sufferer is not treated adequately. Heavy physical activity undertaken in a hot environment is a deadly combination which invites this malady. High humidity makes the situation even more dangerous because evaporative cooling of the sweat is not effective. The metabolic heat generated during exercise is normally dissipated via the skin by sweating and vasodilatation. But when the skin itself is absorbing heat rapidly from the environment this heat gets added on to the overall heat load. Under the circumstances, body temperature mechanisms are unable to cope and body core temperature increases to dangerous levels to precipitate heat stroke. Military commanders in the field, and of military training centres, coaches of high performance athletes are always aware of the problem of exertional hear stroke so that it may be prevented. It may be interesting to hypothesize that hard physical work carried out in extremely low temperatures (-20°Cto -50°C) may also invite this malady *(Can you explain this possibility?)*.

In the elderly sweating and cardiovascular regulatory mechanism may as such be less effective. This added to existing debilitating conditions makes the elderly highly susceptible to non exertional heat stroke. The very young too are more prone to non exertional type of heat stroke if they are poorly hydrated, suffer from gastrointestinal disease which promotes dehydration, and generalized mal nutrition. These issues play a significant role in our country because of the poor socio economic conditions. Clothing which is impermeable to evaporative cooling by sweating may be a precipitating factor. This variety of heat stroke (non exertional) may occur in an epidemic form when an area is affected by a "Heat wave" during the summer months. The local government usually warns of an impending heat wave by public announcements in the media.

Pathophysiology of heat stroke is complex. A vicious cycle of events culminates in to multiple organ system failure which accounts for the high morbidity and mortality. High core temperature may directly damage brain cells, the blood brain barrier, liver cells, muscle cells, renal tissue and the capillary endothelium. The generation of protective heat shock proteins particularly in the liver, intestines and kidneys may prove inadequate to prevent cytotoxicity. The cardiovascular system is maximally strained because of the high temperature. Redistribution of blood to the skin as a heat dissipating mechanism probably compromises gut blood flow to the extent of causing cell wall hypoxia which damages the cell membranes, and induces leakiness through which gut bacteria enter circulation to release toxins which may produce a severe systemic inflammatory response. A number of cytokines like the interleukin series, TNF α and β are thought to be responsible for the systemic inflammation. The kidneys suffer early damage and acute renal failure is often an early manifestation. The damaged blood brain barrier allows leakage of cytotoxic proteins and inflammatory cytokines to enter the brain tissue. The deleterious effects can be wide ranging from brain edema and brain cell damage in various areas, manifesting as a range of symptoms from mental confusion, delirium to frank coma and death. What is difficult to explain is the fact that acetaminophen (paracetamol) is not effective in controlling the hyperthermia of heat stroke even though "Fever producing cytokines" are secreted during the systemic inflammatory response.

Extremely high body temperature damages capillary endothelium leading to platelet aggregation and micro emboli formation. The emboli may hen disperse to various tissues and produce ischemic damage. It may also activate coagulation proteins. The combination may progress on to the syndrome of Intravascular coagulopathy. Muscle cell damage may release muscle protein and myoglobinin to the circulation (Rhabdomyolysis)which may clog the renal tubules to precipitate acute renal failure. If this happens, the urine colour turns brownish red. Because of the muscle cell damage large amounts of K^+ is released in blood. Other mechanisms for acute renal failure could be i. shock because of dehydration which may have preceded the condition; ii. direct renal cell damage because of the hyperthermia.

Major electrolyte disturbances accompany heat stroke. Initially the sweating with dehydration are responsible for hypokalemia and hyponatremia. This may rapidly change to hyperkalemia if acute renal failure sets in. The hyperkalemia of Rhabdomyolysis is very severe and is accompanied by hyperphosphatemia and hypocalcaemia.

Liver commonly suffers damage because of the hyperthermia. This is reflected by raised blood amino transaminases (Aspartate and Alanine)and bilirubin, with hypoglycemia. However the liver has the capacity to recover from the trauma, and liver failure as a cause of death by heat stroke is not very common. Use of anti-pyretic medication such as paracetamol is as such not indicated in heat stroke. In fact it may accelerate liver damage.

Lung damage may lead to acute respiratory distress syndrome (ARDS). This may be secondary to direct damage by the heat stress, or to shock, or precipitated by the systemic inflammatory response.

Body temperature > 44 to 45^{0}C results in protein degeneration and cellular death.

The clinical manifestations of this pathophysiology are wide ranging. The body temperature is very high; there are various central nervous system manifestations; there is often a cessation of sweating manifesting as hot dry skin and severe cardiovascular strain. The latter is indicated by tachycardia, and low blood pressure which may progress on to cardiovascular failure; ecg abnormalities caused by electrolyte imbalance (mainly K$^+$) are seen. Disturbances of the nervous system include delirium, mental confusion, progressing on to coma which may be secondary to cerebral edema and, or ischemia. Renal failure is a common systemic involvement with oliguria, increase in serum creatinine and urea. Brownish red urine heralds the occurrence of Rhabdomyolysis. The damaged muscles are tender and fluid filled. Small hemorrhages from multiple sites such as points of insertion of needles for intra venous fluids, gum bleeds etc indicate the onset of intravascular coagulopathy. Breathlessness, cyanosis and raised jugular venous pressure suggest onset of pulmonary edema because of ARDS. Once systemic inflammatory response sets in, multiple organ system failure is not far behind.

Principles of management include rapid reduction of the body temperature to 39^{0}C by whole body cooling in ice packs or continuous sprinkling of the body by water at room temperature under a fan. Reducing the temperature to below39^{0}C is likely to excite heat gain mechanisms such as vasoconstriction and shivering which are not desirable at this point in time. Intra venous fluids must be carefully regulated in order to avoid circulatory overload which if it occurs may precipitate a heart failure if the heart has been damaged, or pulmonary edema if ARDS is impending. Monitoring of the blood pressure, heart rate, respiratory system (ARDS/ Pulmonary edema), oliguria (acute renal failure) and most importantly signs of the central nervous system involvement must be done on a regular basis. Development of multiple organ failure secondary to systemic inflammatory response often results in a fatality.

(Which common blood parameters will help you to monitor the clinical status of a patient of heat stroke?)

Acclimatization to Heat Stress

We as Indians are well conditioned to living in hot climates but even then when the weather changes from the cooler months to summer, or when citizens of our relatively cool areas move to the northern plains of our country during the summer, we initially feel some adverse effects on our health but recover quickly as we progress into the summer. Individuals who are permanent inhabitants of cool climates (westerners) suffer from a greater degree of discomfort if they travel to India during the summer. The commonly felt effects are easy fatigability; loss of appetite;

Table II. Effects of acclimatization to heat

1. Increase in appetite and general well being
2. Production of more volume of light coloured urine
3. Lowering of resting and exercise heart rates
4. Improved exercise capacity
5. Increase in the amount of sweating with sweat which is more hypotonic

restlessness. The core temperature is slightly raised (though still within the upper limit of normal). The daily urine output is reduced with high coloured urine. The electrolyte content of the sweat is high and effects of Na^+ and K^+ depletion are seen as muscle weakness and cramps. *(Increased blood concentration of which hormone explains the low volume and high coloured urine in such people?)*. The resting and exercise heart rates are high, and exercise capacity is reduced. After about a week to 10 days, the overall situation improves as heat acclimatization sets in **(Table II)**. This continues for about a month or so before it achieves peak levels.

Acclimatization probably occurs at the level of the hypothalamus. Sweating is initiated at a lower core temperature and the volume secreted increases because of increased sensitivity of sweat glands to neuro –endocrine factors. The electrolyte content of the sweat (Na^+ in particular) decreases because of increased reabsorption of Na^+ under the influence of aldosterone. Despite increased sweat loss, there is an expansion of plasma volume with a decrease in cardiovascular strain.. The increase in daily urine output which now becomes more dilute is secondary to the increase in glomerular filtration rate because of restoration of circulating fluid volume. The rennin–angiotensin-aldosterone system (RAS) plays an important role here. Regulated exercise particularly in dry heat during the period of acclimatization facilitates the process. Heat acclimatization is lost if the person leaves the hot environment. This is applicable to tropical subjects too. However as those of us who live in the tropics are genetically "adapted" to a hot environment, our re-acclimatization to a hot environment is swifter and better. It is interesting to note that body physiology is designed to acclimatize to heat stress, but for all practical purposes not to cold stress.

Pathophysiology of Fever

Fever is a classical response to infection introduced into the body. The temperature rarely exceeds 42^0C. However patients of some infections such as Typhus may develop fever $> 42^0C$. Certain pathological situations such as cancer, myocardial infarction, injuries may produce fever with a pathogenesis similar to that of infections. The basis of fever production is the release of pyrogens. This may happen endogenously or as a result of exogenous materials such as bacterial toxins. For generation of endogenous pyrogens, the phagocytic activity by neutrophils and macrophage releases a number of cytokines, the premier amongst them being Interleukins 1 and 6. There are also some others as given in **Fig, 5.** These mediators enter the brain via the organum vasculosum lamina terminalis (OVLT or the circumventricular organs) where the blood brain barrier is relatively permeable to such material. The mediators are thought to occupy receptors on the endothelial surfaces of brain capillaries as also microglia to generate prostaglandin E2 (PGE$_2$) which then reduces the activity of heat loss neurons in the pre-optic anterior hypothalamic neurons raising the bar of the so called "set point" to a higher level. In response to this, heat gain mechanisms are activated to meet the new higher setting and fever is the outcome. Anti-pyretic medication is helpful in bringing down the temperature as it inhibits the formation and release ofPGE$_2$. Exogenous pyrogens situated on the walls particularly of gram –ve bacteria, get attached to a liposaccharide binding protein. The complex then is picked by CD14receptors of macrophages. The interaction releases the same interleukins as in the endogenous pathway to produce fever. Cancerous tissue is known to produce Tumour Necrosis Factor (TNF) α which may be responsible with other interleukins for the fever often seen in cancer patients. Tissue trauma may result in fever because of the release of inflammatory cytokines secondary to a response by the immune mechanisms. A typical example is mild fever which may appear after a myocardial infarction. The fever in many situations may be self limiting as the release of the pyrogens may subside without any medical intervention. At other times antipyretic medication may be required until the causative pathology is treated specifically with antibiotics and other indicated medication. As to whether fever is physiologically useful for the body has been debated. It definitely acts as a warning that something awry is going on in the body systems. The increase in body temperature may make life difficult for metabolism of bacteria that are causing the infection. It may also be helpful in enhancing body immune activity to counter the exogenous threat.

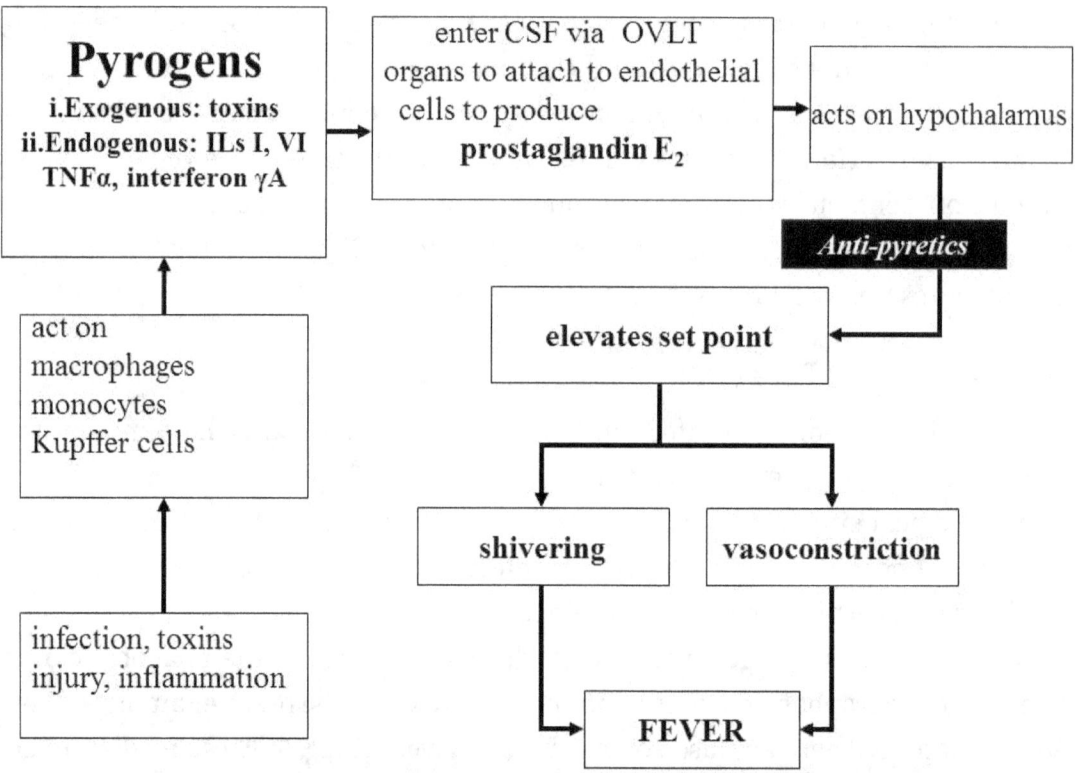

Fig. 5. Pathogenesis of fever. OVLT is organum vasculosum lamina terminalis (Circum ventricular organs)

Hypothermia

When core temperature drops to <35°C, **hypothermia** is said to occur. Temperature<27°C is usually not compatible with life. Hypothermia is generally accidental. Only under specific situations it may be deliberately induced. We as Indians are less exposed to a cold environment. In our northern states however, severe cold waves are known to produce hypothermia casualties in the destitute, and in the very old. The elderly are particularly susceptible because their thermoregulatory system is unable to meet with the demands imposed by this stress. Added to this other factors such as poor nutrition, inability to move freely when required to clothe themselves adequately, pre-existing disease such as stroke, chronic infections also play a role. This discussion pertains only to environmental hypothermia (Primary hypothermia). Secondary hypothermia caused by Endocrine and neurologiccauses and sepsis etc is not discussed.

Autonomic vasoconstriction (noradrenergic) and shivering induced by the hypothalamus is aimed at increasing metabolically generated body heat and peripheral insulation to conserve body heat. Endocrine responses involving thyroxin are triggered at a later stage. Non-shivering thermogenesis by Brown Adipose Tissue (BAT) is carried out in neonates and young infants, and perhaps in some adults (Korean female pearl divers (Amas), Japanese fishermen). Some Nagna (naked) Sadhus (Hermits) have been seen in the Himalayas beyond the snow line. They may also be using BAT to protect themselves against severe cold. Behavioral responses such as wearing of warm clothing etc are more effective in long term protection against cold as compared with autonomic and endocrine responses. Acclimatization to cold in humans is not well documented. Lewis's hunting phenomenon (described above) may be an example of acclimatization to cold.

Pathophysiologically the cardiovascular (CVS)and nervous systems (CNS) are maximally affected by hypothermia.

Initially the cardio vascular system responds proactively to the hypothalamus induced sympathetic activation with a tachycardia, peripheral vasoconstriction and an increase in blood pressure. As the hypothermia advances, the peripheral vasoconstriction becomes confounded with an increase in blood viscosity because of cooling of the blood. Both factors together lead to a compromise of tissue blood flow and may account for the ischemic changes which are recorded on the ecg. The tissue hypoxia causes acidosis which accentuates myocardial depression. There is a drop in cardiac output with its adverse implications on perfusion of various tissues. Hypothermia makes the cardiovascular tissue refractory to catecholamines. In later stages the peripheral vasoconstriction that was sympathetically mediated is no longer effective as the adrenergic receptors become insensitive to catecholamines. This worsens the cardiac out put situation. When the body temperature drops even further severe bradycardia sets in because of direct depression of the SA node. Intravenous atropine does not reverse the bradycardia as the SA node depression is not brought about by an increase in vagal activity. The bradycardia may progress to a ventricular asystole. Re-warming brings forth other complications involving the heart and circulation. The myocardium now becomes responsive to circulating catecholamines increasing its irritability. Arrhythmias may set in with ventricular fibrillation as the most serious consequence. Body physiology attempts to increase myocardial contractility may further increase the existent hypoxic status of the heart which gets reflected as myocardial ischemia on the ecg. Cardiovascular reflex adjustments are disturbed as hypothermia slows down nerve conduction. Such patients can easily develop orthostatic hypotension if they are allowed to stand up or in more severe state, even sit up. This issue is of importance while nursing a hypothermia patient. An interesting aspect of use of ecg in monitoring patients of hypothermia is that hypo or hyperkalemia are not easily reflected on the ecg because of the reduced sensitivity of the myocardium.

Hypothermia progressively depresses the CNS, decreasing its metabolism in a linear fashion as the core temperature drops. At core temperatures less than 33°C, brain electrical activity becomes abnormal; between 19°C and 20°C, an electroencephalogram (EEG) may be un-recordable (as in brain death). Oxygen consumption by tissues is reduced at lower temperatures probably because hypothermia reduces metabolic rate. It may also happen because the haemoglobin tends to keep attached to oxygen (Oxy-haemoglobin dissociation curve shifts leftwards). Whatever the reason, this may be a saving grace as far as damage to brain cells is concerned as long as the hypothermia does not progress to an un-retrievable situation. (This principle is used at times in treating patients with brain damage or heart damage and is known as therapeutic controlled hypothermia). Progressive decline in brain function clinically presents as mental confusion, irrational behaviour, coma and finally death. Nerve conduction velocity is diminished. All reflex activity is there for slowed down. The usual spinal reflexes are sluggish and at later stage, difficult to elicit.

Haematological changes in hypothermia further complicate the pathophysiology. There is an increase in blood viscosity which makes the blood flow sluggish. In addition the endothelial damage incurred allows fluids to flow out in to the tissues. The resulting hemo concentration accentuates the restriction of blood flow adding to the tissue hypoxia. Micro thrombi are formed because of enhanced platelet aggregation. There is an overall suppression of activation of the coagulation proteins, making the patient susceptible to internal haemorrhages. There is a predisposition to infections as the leukocyte activity is diminished. The leftward shift of the oxy-haemoglobin dissociation curve has already been mentioned. It is there for apparent that a combination of cardiovascular changes, vasoconstriction being a major one, combined with haematological changes sets up a pathophysiological sequence which results in poor perfusion leading to tissue hypoxia. The initial saving grace as mentioned in the earlier paragraphs is the reduced cell metabolism particularly in the brain that keeps at bay the more serious effects of perfusion deficit and tissue hypoxia.

The term "Frozen stiff" with cold used popularly is in fact a sinister implication for a patient in severe hypothermia. The increased viscosity of the joint fluids makes joint movements progressively difficult thus limiting mobility. Added to this is the poor contractility of the striated muscle because of cooling and impaired transmission of nerve impulse at the neuromuscular junction. So even if a person has the will to move out of the danger of the hypothermic environment, mobility restrictions make it difficult for him/her to do so. This is of particular relevance to the elderly who as such have mobility problems because of chronic joint disease especially of the lower limbs, sarcopenia and general debility.

Respiration is depressed at low body temperature. Because of lowered metabolism, the activity of the cilia in the respiratory tract is depressed. This makes the patient susceptible to airway infection. The depression in respiration in deep hypothermia invites the onset of hypercarbia and hypoxia. The degree of hypercarbia may be somewhat decelerated as CO_2 production is dampened by the reduced tissue metabolism. Because of respiratory depression the role of CO_2 in controlling respiratory activity is less important than the direct stimulatory effects of the co-existing hypoxia. It is interesting to note that it is difficult to obtain an accurate measurement of arterial blood gases because the gas analysers automatically measure the gas tensions at 37^0 C while the solubility of respiratory gases at low body temperatures is altered thus vitiating the recording of the prevalent tension of dissolved blood gases.

Diuresis on exposure to cold is well known. This sets in even before frank hypothermia sets in and has been attributed to a vasoconstriction mainly of the efferent arterioles which in turn raises the afferent-efferent arteriolar pressure difference and hence the glomerular filtration rate (GFR). Later with increasing fall in body temperature the GFR actually begins to come down *(Can you explain this?)*.Suppression of tubular function adds on to the respiratory acidosis because of poor secretion of H^+.

By far the most dominant electrolyte disturbance in hypothermia is linked to K^+. This is caused by disturbances in Na/K ATPase pump activity and cell membrane permeability changes. The ecg does not reflect the K^+ induced changes easily as the myocardium itself is under severe duress because of the cold temperature, poor perfusion and acidosis. The major pH disturbance is a metabolic acidosis. *(Can you work out why?)*.

Some contradictory effects are seen in the gastrointestinal tract. These vary from increased acid secretion in the stomach on one hand to intestinal obstruction on the other

The student should be able to work out the various clinical manifestations of a patient of hypothermia. Cardiovascular and CNS monitoring is of paramount importance. It is important to realize that here is often a disconnect between the clinical signs and symptoms and the pathophysiology that is going on. Principles of management include continuous monitoring of rectal and where ever possible, oesophageal temperatures; controlled slow re-warming, intra venous fluid management; looking out for cardiovascular and renal complications and control of infection if present. Assisted ventilation may be required. Cardio-respiratory failure and septicaemia are often the terminal complications.

References

1. Boulant J. Role of pre-optic anterior hypothalamus in thermoregulation and fever. Clin. Infectious Dis. 2000; 30:s157–S161.

2. Danzl D. Heat related illnesses. Part 19,Chapt 479e; pp e1-e3 in Harrison's Principles of InternalMedicine. 19th Edn. Edt. Kasper D, Saucy A et al. McGraw Hill London. 2015.

3. Dikshit MB. Heat problems in high sped low level flying. Aviat. Med. (Journal of the Aero. Med. Society of India). 1980; 24:31–36.

4. Greenleaf J, Kaciuba-Uscilko H. Acclimatization to heat in humans. NSA Technical report 101011; 1989.

5. Iyer EM, Dikshit MB, Banerjee PK and Suryanarayana S.100% oxygen breathing during acute stress: effect on sweat composition. Aviat. Space and Environ. Med.1983; 54: 232–235.

6. Leon L, Helwig B. Heat stroke: Role of the systemic inflammatory response. J. Appl. Physiol. 2010; 109: 1980–1988.

7. Mallet M. Pathophysiology of accidental hypothermia. Quartly J Med. 2002; 95: 775–785

8. Romanovsky A. Thermoregulation: some concepts have changed. Functional architecture of the thermoregulatory system. Am. J Physiol. (Reg., Intg. And Comp. Physiol.) 2007; 292: R37-R 46.

9. Shibasaki M, Wilson T, Crandall C. Neural control and mechanisms of eccrine sweating during heat stress and exercise. J. Applied Physiol. 2006: 100: 1692–1701.

10. Tansey E, Johnson C. Recent advances in thermoregulation. Adv Physiol. Edu. 2015; 39: 139–148.

Short Answer Questions and Their Solutions

Chapter 1: Cellular Adaptations to Changed Environment

Q.1. A person donates a part of his liver to his sibling. What will happen to this person's liver in 6 months time?

Ans. The liver will have regenerated because it undergoes hyperplasia

Q.2. What makes the uterus regress to its near original size after pregnancy is over?

Ans. The hormonal stimulus is withdrawn, and the autodigestive enzymes break down the now unwanted tissue.

Q.3. Theoretically, can a tumour be treated by producing atrophy in it? If so how?

Ans. Yes. By shutting off its blood supply and producing ischemic atrophy.

Q.4. How may excess growth hormone lead to a frank diabetes mellitus.

Ans. This produces persistent hyperglycemia which in turn leads to an exhaustion atrophy of insulin producing beta cells in the Islets of Langerhans.

Q.5. In a box in the flow chart in Fig. 3, it is indicated that water retention takes place in the cell, resulting in its swelling. How does this happen?

Ans. The increase in intracellular Na^+ pulls in water by osmosis.

Q.6. In what manner may the free radicals interfere with ionic pumps?

Ans. Protein channels via which ions move may get denatured by free radicals.

Chapter 2: Fluid, Electrolyte and pH Disturbances

Q.1. What should be the property of the substance which may be used to measure TBW?

Ans. It should be able to distribute itself easily between intra and extra cellular compartments. Substances used are antipyrine, D_2O (deuterated water), and HTO (tritiated water)

Q.2. If this is true, how can one estimate the volume of water in these two spaces?

Ans. TBW and ECF can be measured. Then ICF = TBW- ECF. Similarly, interstitial fluid volume = ECF- plasma volume.

Q.3. Work out what will happen to cells surrounded by i. hypertonic ECF and ii. hypotonic ECF. What sort of solution would you like to use intravenously to treat edema of the brain and why? Find out the name of this solution used in practice.

Ans. In a hypotonic ECF, cells will imbibe water and swell until osmotic equilibrium is restored. The danger is that they may burst before this happens. The opposite is true in a hypertonic ECF and the cells will shrink (crenate) because the ECF will now pull out water out of them. A substance which will raise the osmolality is given intra venously to treat brain edema so as to pull excess water out of the brain cells. The solution usually used is mannitol.

Q.4. How do low pressure cardiopulmonary (atrial type B receptors) help in reducing the excess circulating fluid volume. What is the atrial natriuretic polypeptide, and what do you think would be the role of this hormone in this situation?

Ans. With increased filling of the atria, the type B volume receptors will be activated to inhibit ADH and thus increase water loss. Also, atrial distension will initiate secretion of atrial natriuretic polypeptide. This is secreted by atrial myocytes, and increases urine output as follows: i. dilates the afferent arterioles and increases GFR ii. reduces Na^+ re-absorption by the PCT and inhibits release of renin and aldosterone, and iii. Opposes arteriolar and venous dilatation by blocking vasoconstrictor actions of Angiotensin II, ADH and catecholamines.

Q.5. What is the reflex mechanism involved in activation of the sympathetic system. How do the kidneys help?

Ans. Reduced circulating fluid volume deactivates low pressure cardio-pulmonary receptors and arterial baroreceptors. Afferents travel up the vagus from the former and the glosso-pharyngeal from the latter to activate the reflex sympathetic drive. Circulating catecholamines also increase because of this. The kidneys aim at retaining water and salt

Q.6. What do you think is the "appropriate fluid" to be used in cases of a i. road accident; ii. burns; iii. Excessive sweat loss?

Ans. Whole blood for i.; plasma (mainly) for ii; and 0.9% saline for iii.

Q.7. Try and list out situations which may produce hypervolemia

Ans. Kidney diseases with reduced urine formation; congestive heart failure

Q.8. How does desmopressin help to distinguish between hypothalamic diabetes insipidus and nephrogenic diabetes insipidus?

Ans. In hypothalamic diabetes insipidus, this well reduce urine volume as the condition is caused by a deficiency of ADH. In nephrogenic diabetes insipidus, the defect is the collecting tubules and hence desmopressin cannot act to decrease the polyuria.

Q.9. If you measure plasma K^+ immediately after a patient of epilepsy has thrown violent convulsions, what will you find?

Ans. Plasma K^+ will be raised as the cation will have been released in to ECF during muscle contraction. The situation is easily reversed.

Q.10. Look up the pH values for urine, gastric juice and pancreatic juice, and reason out why they should be so?

Ans. Urine pH around 6 because kidneys secrete H^+ ions; gastric juice is between 2–4: highly acidic because of HCl in the stomach; pancreatic juice: 8–8.4 because most of this juice is constituted by $NaHCO_3$

Q.11. A healthy young man has the following arterial blood gas and pH picture: PaO2= 20 mmHg; PaCO2= 9 mmHg; pH =7.78. Explain how he could have developed this situation?

Ans. This is in fact the ABG picture of a young man at a simulated altitude equivalent to that of Mt Everest. The very high altitude induces severe arterial hypoxia (PaO2 = 20 mmHg) which stimulates peripheral chemoreceptors to increase ventilation considerably. This results in CO2 washout (hypocarbia; PaCO2= 9 mmHg), and the subsequent highly alkaline pH.

Chapter 3: Some Aspects of Renal Dysfunction

Q.1. Which type of animals do you think will have long loops of Henle. State a plausible hypothesis for your answer.

Ans. Animals like the camel, the desert rat, which live in the desert and need to conserve as much water as possible by concentrating the urine. In proportion to its size, the desert rat is supposed to have the longest loop of Henle

Q. 2. List sequentially the factors that determine the GFR.

Ans. Normally the GFR is governed by: i. blood flow into the glomerulus; ii. adequate glomerular capillary pressure generated as an adjustment between the resistance offered by afferent and the efferent arterioles; iii. Glomerular membrane permeability and iv. low hydrostatic pressure in the tubules.

Q.3. Define anuria. Would it be correct to say that reduced GFR as a result of poor perfusion of the kidneys is in fact an adaptive response?

Ans. Urine output <50 ml/day is anuria.

Yes. The GFR is reduced in response to lowered perfusion because the kidney wants to help by reducing further water loss, and thereby prevent deterioration in the situation. The problem arises because in spite of the good intentions, unknown to the kidneys, other changes occur which produce imbalance in the milieu interior.

Q.4. Why are the terminal parts of the medullary PCT and the thick ascending limb maximally affected by ischemic injury?

Ans. This is because these parts contain the maximum number of mitochondria and are heavily oxygen- dependent.

Q.5. Can the kidney of a patient who has recovered from pre-renal azotemia be successfully transplanted in to another human being who requires the transplant? Explain your answer

Ans. Yes because such a kidney has not suffered permanent tissue damage. However the perfusion available to the transplanted kidney must be adequate

Q.6. How may K^+ disturbances be detected during the evolution of ARF?

Ans. In the initial and maintenance phases, HYPERKALAEMIA (serum K^+> 5mmol/l) is the usual problem, and results from retention of the ion because of inhibition of the Na/K ATPase pump by acidosis. Intervention is needed if the serum K^+ exceeds 6 mmol/l for fear of dangerous cardiac arrhythmias. It manifests as muscle weakness and fatigue. Hyperkalemic

ECG changes are diagnostic (tall T waves and widening of PR interval and QRS complexes. In the recovery (diuretic) phase, HYPOKALEMIA occurs. The symptoms are essentially similar to those of hyperkalemia. ECG should be monitored to look out for ECG changes (normal rhythm; flattened T waves and appearance of U waves).Monitoring of serum K^+ clinches the diagnosis.

Q.7. Recapitulatewhat is the anion gap. How may its measurement help in ARF?

Ans. This is calculated as the sum total of plasma Na^+ - (Cl^- + $HCO3^-$). The normal value is usually about 12 + 4 mmol/l. Addition of any acid radical other than HCl will increase the anion gap. It is useful in detecting addition of H^+ ions from non-volatile sources such as proteins, and hence its estimation in ARF is useful.

Q.8. Keeping the consequences of ARF in view, what should be the approach to fluid management in such a patient?

Ans. In the early phases, there is an obvious need to restrict fluid intake. The replacement should include the daily urine output which is to be carefully measured. Add to this about 500 ml/day for insensible water loss and about 100–150 ml for the fluids lost in the GIT. Additions must be made for sweating, vomiting, diarrhea etc. In the recovery phase, careful monitoring of fluid loss must be done in order to preempt dehydration and with it, hypernatremia. K^+ supplements will prevent development of hypokalemia.

Q.9. What do you think happens if a person is left with only one kidney because the other has been removed surgically?

Ans. At the time of removal of one kidney, there is a fall in the GFR by 50% (obviously). But soon the remaining normal kidney nephrons start to function overtime (hyper filtration) and try and push up the GFR to as much as 80% of normal. This is achieved by adjustments of renal hemodynamics where in there is an increase in renal plasma flow and glomerular capillary pressure, as also glomerular hypertrophy. The changes may be visualized on histology. In human beings a solitary kidney after removal of the other kidney is known to function normally for about 20 yrs.

Q.10. What could be the best method of treating anemia of chronic renal failure?

Ans. Use of injectable erythropoietin

Q.11. Why do you think patients of nephritic syndrome are more susceptible to infections? Also, why may the chances of iron deficiency anemia, hypercoagulability and endocrine disorders increase?

Ans. Along with albumin, other proteins also leak out via the urine. These include gamma globulins which form the basis of antibodies. This makes such patients susceptible to infections. Iron binding proteins are in short supply and explain the iron deficiency anemia, while anti thrombin III is lost, and hepatic synthesis of fibrinogen is increased, encouraging hypercoagulability. Binding proteins for thyroid and sex hormones are also lost.

Q.12. How do you think are these reactions which increase ADH and inhibit atrial Natriuretic peptide triggered?

Ans. The loss of fluid into the interstitial space reduces circulating fluid volume. This reduces atrial filling and (remember the atrial volume receptors?) which in turn excites the release of ADH. Also the atrial cells which contain ANP are also deactivated, thus reducing circulating ANP.

Q13. Now that you are aware of the pathophysiology of Nephrotic Syndrome (NS), list the possible clinical complications of this disorder.

Ans. Hypo-proteinemia; coagulation disorders- mainly thrombotic tendencies, including stroke; normocytic normochromic anemia and iron deficiency anemia; atherosclerosis; hypovolemia, generalized anasarca (massive edema); repeated infections; calcitriol deficiency and hypocalcemia.

Chapter 4: Physiologic Disturbances in Cardiovascular Disease

Q.1. Look at the left atrial and left ventricular pressures in Fig. 3 B. Why should the diastolic pressure in the LV be so low?

Ans. In order for the blood to flow easily from the LA in to the LV as the transfer is dependent almost entirely on the pressure gradient between the two chambers.

Q.2. During exercise, when HR increases, why is the reduction in systolic time relatively less than that of the diastolic time?

Ans. This is to ensure that the stroke output remains unchanged as far as possible. But at heart rates near maximum, even the stroke output begins to reduce, and the continuous increase in CO then maintained by the increasing heart rate.

Q.3. How does Rate Pressure Product actan index of clinical status in a patient of ischemic heart disease.

Ans. Suppose a patient having ischemic heart disease starts to have anginal pain during a tread mill exercise test at SBP of 150 and HR of 120 (RPP = 180). An increase in RPP before ischemic pain appears on retest will indicate improvement while a decrease in RPP when chest pain first appears will mean deterioration.

Q.4. What do you think is the significance of development of smooth muscle in the large collaterals?

Ans. They are likely to be responsive to vasodilator substances such as NO. Against this, serotonin and vasopressin may constrict these vessels and exacerbate ischemia.

Q.5. What could be the link between severe anaemia and IHD?

Ans. Anemia by itself may not result in IHD. But when severe anemia is combined with atherosclerosis, in situations where demand increases as in exercise, it may lower threshold of IHD onset.

Q.6. Explain why low blood pressure in a patient with MI is considered as a serious situation?

Ans. This suggests that the myocardial damage is extensive enough to prevent the heart from pumping effectively. This is usually accompanied by a baro-reflex induced sympathetic tachycardia and other evidence of sympathetic activation such as sweating and cold clammy skin.

Q.7. Physiologically what will a patient who develops myocardial stunning present with?

Ans. The biggest problem will be low cardiac output and systolic BP with associated heart failure. This will need inotropic support for the duration of the stun effect.

Q.8.Work out how such a patient will present clinically?

Ans. Basically, the same as what is given in ans 7. The signs and symptoms will pertain to acute left ventricular failure with pulmonary edema.

Q.9. What are the mechanisms that help the body in getting rid of excess fluid?

Ans. i. Atrial volume receptors get stimulated and reflexly reduce ADH secretion ii. The stretch on the atria releases atrial natriuretic polypeptide. Both these mechanisms promote enhanced excretion of water and electrolytes by the kidneys and help to restore the extra cellular fluid volume to normal and therefore prevent hypertension.

Q.10. Why does the baroreceptor inhibitory reflex get deactivated during exercise?

Ans. During exercise, the BP needs to increase in order to increase muscle perfusion. If the Marey's law keeps active even during exercise, this required rise in BP will be inhibited, and the tachycardia which is also needed will be inadequate to generate the required cardiac output. The adjustment probably occurs at the hypothalamic level.

Q.11. Why only LVH?

Ans. That because it is the left ventricle which is the main pump that pushes the blood column against the offered resistance.

Q.12. Explain why IHD may complicate LVH caused by HBP?

Ans. The hypertrophy results in increased muscle mass with an increase in O_2 requirement. When there is coronary artery narrowing, coronary blood flow reduces. A relative discrepancy between demand and supply surfaces, producing the IHD.

Q. 13. Can you think of any examples at this stage?

Ans. Acute left ventricular failure (Pulmonary edema) in a patient with hypertensive heart disease.

Q.14. Why is the left ventricle considered the more important functionally?

Ans. It is the left ventricle that generates the power to pump out a bolus of blood per beat, giving it enough momentum to make it circulate throughout the system.

Q.15. Which of the causes listed in Table IV do you think by pass myocardial muscle damage as the primary cause of failure?

Ans. III and IV under "acute failure, and IV, V, VI in chronic failure.

Q.16. In Fig. 10, explain why the end systolic volume at "a" in abcd is greater than volume at A in ABCD?

Ans. This is because the lower power generation by the LV has not been able to clear the normal volume load (70ml to 140 ml in ABCD), leaving behind an extra volume at the end of systole.

Chapter 5: Pathophysiology of Shock and Burns–Multi Organ Dysfunction

Q.1. What causes of hypovolemia other than hemorrhage can you think of?

Ans: Diarrhea and vomiting; excessive sweating; burns (evaporative water loss); excessive urination as in patients with Diabetes Inspidus; and use of diuretics; fluid deprivation in hot environment

Q.2. Students should work out as to how an increase in fluid volume is organized by body mechanisms.

Ans. Decrease in perfusion to the kidneys excites the release of renin which in turn becomes responsible for activating the renin-angiotensin system into release of angiotensin II. The latter is a powerful vasoconstrictor, and also stimulates the adrenal cortex to release aldosterone, which albeit slowly, gets the kidneys to retain Na^+ and water.

Q.3. Acute renal failure (ARF) is a single organ failure syndrome which may occur as a complication of hemorrhagic shock. How may it change into a MODS?

Ans. ARF is a classical example of a single organ failure syndrome following shock. When managed enthusiastically, recovery is good. However if some of the problems that are generated during this syndrome are not tackled effectively, MODS may result. For example, hyperkalemia during the early (acute) phase of ARF if not detected and corrected in time leads to myocardial failure; the metabolic acidosis which is usually present if allowed to persist may help in aggravating uncontrolled inflammatory response. In this manner, ARF can develop into a MODS. This variety of MODS is a delayed or secondary MODS while ARF is termed a primary organ system dysfunction.

Q.4. What physiological indices do you think indicate that the cardiac output delivery to the tissues is improving?

Ans. increase in pulse pressure, increase in systolic BP, reduction in heart rate; a rise in central venous pressure; increase in urine output: these indicate the good that is being done. (On the other hand excess/harmful effects of over correction are indicated by onset of pulmonary edema).

Q.5. In a normal situation, will the tissue pressure help to increase the transmural pressure or decrease it?

Ans. Almost everywhere, the tissue pressure is thought to be sub-atmospheric (negative), and will therefore help to increase the transmural pressure and keep the blood vessel patent.

Q.6. Theoretically, do you think under such circumstances, vasodilators could be used to restore tissue perfusion?

Ans. Theoretically yes, but physiologically this may not be helpful because: i. What this may do is to worsen the volume available/volume capacity discrepancy because the lowered volume (which was initially responsible for the low ratio, has not been corrected, and to that is added an increase in volume capacity; ii. After producing a vasodilatation, it may become difficult to titrate vascular tone requirement in order to adjust perfusion by using vasoconstrictors. In both situations the pendulum may swing from a hypovolemic shock to a neurogenic variant, and complicate the clinical scenario. The closest one may come to the use of vasodilators is when β adrenergic agonists such as dobutamine, are used as cardio tonic agents to increase the cardiac output. Keeping in view the current understanding of pathophysiology of shock, the best bet is to use volume expanders and after that judiciously use vasoconstrictors with the over all goal of maintaining perfusion at the auto regulatory level

Q.7. Given the above, what physiological principle of treatment do you feel is indicated in a patient with burn injury?

Ans. Rapid infusion of fluids in order to i. limit the zone of stasis, and ii. To make available enough circulatory fluid volume so that cardiac out put is maintained as close to optimum as possible.

Q.8. Work out on your own as to why this is likely to happen.

Ans. Hemodilution is one of the reasons. This occurs with massive fluid infusion. This may also be responsible for reduction in other micronutrients such as Magnesium, phosphate and Calcium though ionized Calcium usually remains within physiological limits. The other reason for hypoalbuminemia is the extravasation through the capillaries because of the increased permeability.

Q.9. How does the appearance of laryngeal stridor indicate airway injury??

Ans. Because it means laryngeal edema has set in.

Q.10. How does acute renal failure occur in a patient with severe burns?

Ans. The hypovolemia that complicates burn injury as its immediate aftermath will lead to reduced renal blood flow, and glomerular filtration. Oliguria may progress to ARF, particularly in patients with major burns (>40% body surface area involved).This is a life threatening complication when it happens in a burn patient.

Q.11. Physiologically, what will be the disadvantages of such a metabolic state in a patient with burns?

Ans. Rapid tissue break down, with possibilities of hyperkalemia as a complication; rapid weight loss, increased susceptibility to infections, especially with abundant availability of a medium (glucose) for bacterial growth; increase in fluid requirements.

Chapter 6: Physiological Disturbances in Respiratory Disease

Q.1. If there is no evidence for direct sympathetic innervation of human airways, how does sympathetic system excitation produce bronchodilatation?

Ans. There are beta2 adrenergic receptors on airways smooth muscle. Circulating catecholamines, especially adrenaline, stimulates these receptors to produce bronchodilatation

Q.2. With reference to table I: Contributing situations which are involved in

Dyspnea.

Ans. a) obstruction to airflow: produces hypoxia because of ventilation /perfusion mismatch: also when severe, may distort respiratory muscle afferents

b) resistance to expansion of lung and chest wall: distortion of respiratory muscle afferents; hypoxia/hypercapnia which may supervene conditions producing this: eg. COPD

c) excessive respiratory drive: hypercapnia in COPD; to experience effect of hypercapnia as a cause of dyspnea: hold your breath as long as possible. A point in time will be reached when you cant do so any more: "break point." List symptoms that you feel. Also stimulation of J receptors by an increase in interstitial fluid pressure: severe exercise; pulmonary edema will result in dyspnea

d) increased sensitivity of respiratory neurons: one of the reasons as to why near maximal exercise produces dyspnea could be increased sensitivity of respiratory neurons to peripheral chemoreceptor input even when blood PO_2. may be within normal range. At high altitude, sensitivity of central chemoreceptors to CO_2 has been shown to increase and breath holding time is shortened: earlier onset of break point and dyspnea

Q.3. Patho-physiological basis for symptoms and signs of asthma

a) Tightness of chest and dyspnea: with air flow limitation because of obstruction both these symptoms will occur as a result of afferent inputs from respiratory muscles which have now to work overtime. Initially dyspnea may be on exertion because hypoxia occurs during exertion when airways are narrowed and ventilation is inadequate.

b) Cough: distortion of airways, their irritation by mucus secretion, will stimulate irritant lung receptors and produce reflex coughing. When mucus secretion increases cough will become productive. Mucus secretion occurs because sensory receptors in the airway are irritated by inflammatory substances, and a vago-vagal (cholinergic) effect which stimulates mucus secreting glands.

c) Accessory muscles of respiration are recruited in order to overcome the airway resistance

d) Wheezing occurs because air has to flow through partial obstruction of airways brought about by mucus plugs, bronchoconstriction and mucosal edema

e) Hyper resonant chest wall: the lungs contain more than normal volume of air because of trapping of air due airway narrowing. This produces the hyper-resonant note on percussion

Q. 4. Why is the appearance of cyanosis a dangerous sign?

Ans. This indicates severe hypoxia (possibly with carbon dioxide retention). The reduced Hb exceeds 5gm% and colors the skin and mucus membranes. That means the airway obstruction has become very severe.

Q.5. Physiological principles of management:

Ans. i. bronchodilators using beta 2 adrenergic stimulants and/or vagal inhibitors; ii. anti inflammatory agents to treat inflammation; iii. supplemental oxygen to relieve hypoxia

Q.6. Clinical findings which a patient with advanced COPD is likely to have?

Ans. Dyspnea which has been progressive because of worsening hypoxia; barrel shaped chest (the Antero- posterior diameter increases as the chest remains chronically hyper inflated); hyper-resonant chest because of air trapping; wheeze because of mucus secretion and inflammatory products in air way lumen; may have cyanosis as ventilation perfusion mismatch results in increasing hypoxia and hence cyanosis; ABGs classically PO_2 <60 mmHg, CO_2 > 45 mmHg; peripheral edema, palpable liver and increased jugular venous pressure if corpulmonale has set in.

Q.7. If the normal alveolar oxygen tension is about 104 mmHg, why does the PaO_2 remain at 95–98 mmHg?

Ans. This is because there is a physiological shunt of about 1–2% which does not allow complete oxygenation of mixed venous blood. This results in the average ventilation/perfusion ratio of 0.8 instead of an ideal 1. That means that ventilation of alveoli is not uniform even in normal people

Q. 8. If obstructive disease is a COPD, the TLC is likely to be higher than normal: why?

Ans. The lungs remain hyper inflated and thus the volume capacity of the lungs increases but the residual volume and thus the FRC increase. In spite of a raised TLC, the FVC and FEV1 are both lowered, more so the FEV1 because of the air trapping (obstruction)

Q.9. Patients of lung collapse tend to have lowered blood CO_2 tension?

Ans. Stimulation of irritant receptors. Also hypoxia stimulates respiration via peripheral chemoreceptors which in turn drives off the CO_2

Q.10. Suggest a method of overcoming this problem and improve blood oxygenation in a patient of hypoxic hypoxia?

Ans. Normally alveolar oxygen tension is about 100 mmHg. This may be increased by increasing the percentage of oxygen in the breathing mixture. For example, normal atmospheric air has a concentration of 20.93% of O_2. If instead of this, the patient is made to breath 50% Oxygen, then the PO_2 in the new mixture breathing mixture will go up (50% of 760 mmHg which is the atmospheric pressure at sea level, will be much more than 20.93% of 760 mmHg in atmospheric air). This raised tension will be transmitted down to the alveoli, thus increasing alveolar PO_2 which in turn will improve diffusion of the gas.

Q.11. In a patient with low DLCO, how will spirometry and arterial blood gases (ABGs) help to distinguish between COPD and interstitial lung disease such a interstitial fibrosis?

Ans. In both conditions FVC and FEV_1 will have reduced. However, in ILD, the ratio of FEV1/FVC will be normal or high, as against a low ratio in COPD. Also in fibrosis, all lung volumes with TLC will be reduced while in COPD, the TLCC, FRC and RV/TLC ratio will all be higher than normal. ABGs in COPD will have low PO_2, raised PCO_2 and low pH, while in fibrotic lung disease initially the PCO_2 will be lowered with an alkaline pH. Oxygen tension will be low.

Q.12. How could drowning result in a diffusion abnormality?

Ans. Inhaled water lines the alveoli and increases the diffusion distance for respiratory gases.

Q.13. Supplemental oxygen does not improve hypoxia produced by a vascular shunt but does so in ventilation perfusion mismatch. Why?

Ans. In a shunt the blood bypasses the exchange surface altogether while in a ventilation/ perfusion mismatch as result of lung disease, the supplemental oxygen, because of its high concentration, and therefore tension, diffuses through the respiratory membrane and improves the situation

Q.14. Usually, patients with chronic hypoventilation such as COPD do not suffer from anemia?

Ans. Because of low PaO_2, the kidneys increase production of erythropoietin which in turn is a powerful stimulator of erythropoieisis

Chapter 7: Pathophysiology of Some Gastrointestinal Disorders

Q.1. Use of which kind of medication could be guided by this type of innervation?

Ans. Sympathomimetics/parasympatholytics should logically be useful in the management of excessive motility of the GIT.

Q.2. How do you differentiate between hematemesis and hemoptysis?

Ans.i. colour: black in hematemesis and bright red in hemoptysis which is fresh bleeding from the respiratory tract); pH (acidic in this while its alkaline in hemoptysis).

Q.3. What will be the end result of such a phenomenon (reduction/loss of spontaneous electrical rhythm)?

Ans. The spike potentials can not be generated, and hence the intestinal motility will be hampered.

Q.4. Why may a patient in coma be at a risk?

Ans. The swallowing reflex is absent. This is likely to result in the food bolus tracking down the larynx and result in an asphyxia.

Q.5. Explain why drinking excessive tea/coffee often produces a transientgastro-esophageal reflux.

Ans. These beverages contain caffeine which inhibits phosphodiesterase. This in turn prolongs activity ofβ adrenergic hormones in circulation, and act as smooth muscle relaxants.

Q.6. What do you think is the clinical relevance of the fact that intestinal peristalsis occurs at about 3–4 times/minute?

Ans. Auscultation must be done for these movements for a full minute. If done only for a few seconds which is often the practice, a wrong conclusion that the intestinesare "silent" may be arrived at.

Q.7. What is the corollary to the fact that inadequate gluconeogenesis may occur in children who have had severe diarrhea?

Ans. It is important to ensure adequate caloric intake in children when they have diarrhea.

Q.8. Whatever the cause of diarrhea/dysentery, what will the major clinico-physiologic implications?

Ans. Loss of fluid electrolytes, anemia, and hypo-proteinemia (the latter two particularly because of IBD). In severe cases, hypoglycemia, dehydration, hypotension and hypo-volemic shock may be the consequence, particularly in children and the elderly.

Q.9. From the above, derive the cardinal clinical features of intestinal obstruction?

Ans. Colicky pain abdomen, vomiting, absolute constipation and abdominal distension. If dehydration is marked, symptoms and signs of hypovolemic shock will be seen. Appearance of tenderness along with the pain will suggest gut infarction.

Q.10. Why does malabsorption and malnutrition occur in patients of chronic pancreatitis?

Ans. The digestive enzymes from the pancreas are deficient. This prevents proper breakdown of the carbohydrates, fats and proteins, and hence their absorption.

Chapter 8: Pathophysiology of Obesity and Weight Loss

Q.1. Explain the polycythemia in these patients.

Ans. This is secondary to the hypoxia which in turn induces erythropoietin secretion by the kidneys and stimulation of the bone marrow.

Q.2. Explain why hypoproteinemia causes pitting edema. Why are patients of hypoproteinemia susceptible to infection?)

Ans. Hypo proteinemia causes a reduction in colloid osmotic pressure. Therefore water can not be easily retrieved at the venous end of the capillaries. Low protein means low levels of globulins hence the infections.

Q.3. Why is it that a patient with Protein Energy Malnutrition (PEM) may not demonstrate low Hb level even when other clinical manifestations of anemia may be present?

Ans. There is an accompanying hypovolemia in starvation. This masks the relative decrease in Hb because of hemoconcentration. In the early phases of recovery, the Hb deficiency becomes obvious.

Q.4. What simple investigation would you do to find out whether a person is using fats to sustain energy needs?

Ans. Measure ketones in blood.

Chapter 9: Pathophysiology of Musculo-skeletal Disorders

Q.1. What is a motor unit?

Ans. A number of muscle fibres innervated by a single alpha motor neuron constitutes a motor unit. The number of muscles fibres innervated may vary. In muscles used for fine and precise functions, up to 10 fibres may be innervated by a single α motor neuron, while in large muscle masses there may hundreds of fibres which constitute a unit. Each such unit follows its own "All or none" law. More the number of motor units recruited, more is the strength of the contraction, and greater the power generated.

Q.2.. What type of fibres would most suit i. a sprinter; ii. a soccer player; and iii. a long distance runner?

Ans. A sprinter should have Type II fibres while a soccer player should have a mixture of both type I and type II fibres as such a person is required to have both fast movements as well as sustained movements. A long distance runner on the other hand will do well with plenty of type I red, slow fibres.

Q.3. What do you think will be the cellular changes in a hypertrophied muscle?

Ans. There is an increase in the number of myofibrils. More the load, greater the increase. The mitochondria increase by as much as 100–120% with a 60–80% increase in ATP and phosphocreatine. The glycogen storage capacity must increase, as also the triglyceride store. The aim is to enhance both aerobic and anaerobic capabilities of the muscle.

Q.4. How are the various chemical substances named responsible for the symptoms and signs of muscular injury?

Ans. These produces vasodilatation and increased capillary permeability which present as swelling and edema at the site of injury. The prostaglandins and bradykinin may stimulate the free nerve endings in the region to produce the pain.

Q.5. How do you explain the observation that repetitive stimulation will improve the muscle contraction in LEMS?

Ans. This will make more Ca^{++} available at the presynaptic membrane, and in turn help to release more Ach molecules

Q.6. Explain the physiological basis of impairment of neuromuscular transmission if the number of receptors diminish.

Ans. Even if the amount of Ach produced is adequate, it can trigger off only an EPSP because of the pathology, and not a full fledged action potential.

Q.7. How does repeated nerve stimulation help confirm diagnosis of MG?

Ans. Electrical stimulation of a motor nerve is done at a frequency of 1–2 /sec. In a patient of MG, the amplitude of the recorded EMG will reduce with successive impulses while this will not happen in a normal person at this frequency.

Q.8. Why is weight training not feasible during space flight?

Ans. In the zero gravity environment, the weights will have not exert any weight!

Chapter 10: Circadian Rhythms and the Pathophysiology of Stress

Q1. Why is it that the circadian disturbances were not seen in the pre-air travel era?

Ans. The changes in time zones occurred very slowly while travelling across continents by ship. Because of this, the rhythms were able to keep pace with the changing time zones.

Q.2. Why does Diabetes mellitus react adversely to a stressful event?

Ans. The stress induced hypercortisolism further aggravates Insulin resistance, and affects immune responses adversely.

Q.3. Why is the finding that the high metabolic rate is "Confounding" in patients who have chronic stress induced insomnia?

Ans. One might expect that the increased metabolism will be responsible for a loss in weight in such patients. However the hypercortisolism causes insulin resistant diabetes mellitus and obesity

Chapter 11: Pathophysiology of Common Blood Disorders

Q.1. How do you explain the functional murmur heard in a patient of severe anemia?

Ans. Anemia lowers viscosity of blood thereby increasing velocity of blood flow which then results in a functional murmur

Q.2. Reference to Fig. 2: Why is it that red blood cells in stages 4 and 5 during development are unable to synthesize DNA nor undergo mitosis?

Ans. The nucleus by this time is either absorbed or extruded out of the cell.

Q.3. If a hand is immersed in iced water, it may turn blue. Please explain?

Ans. One reason for the cyanosis is the slowing down of circulation because of a significant and rapid fall in the temperature. The sluggishness of circulation is induced by the change in viscosity. Also contributing to this is the sympathetic vasoconstriction.

Q.4. Patients of chronic renal failure suffer from fairly severe anemia. Explain?

Ans. This happens because Erythropoietin formation reduces because of CRF

Q.5. What is the difference between hypoxia of anemic origin and hypoxic hypoxia?

Ans. In the former, the tissue hypoxia occurs because there is deficiency of Hb in the blood, and hence the carrying capacity of Hb for O_2 is less. The arterial PO_2 in patients of anemia is normal The latter is a result of low oxygen tension in blood because of lung or heart disease, or physiologically, exposure to high altitude where PO_2 in the atmospheric oxygen is low despite normal O_2 concentration (20.93%).

Q.6. Why does jaundice occur at all in hemolytic anemia, even when the liver may be quite normal, and why is the bilirubin mostly the indirect type?

Ans. The load presented to the liver is beyond its capacity of metabolizing the bilirubin generated. The bilirubin is of the indirect type as it has not been conjugated by the liver because of the excess burden.

Q.7. How do you explain the appearance of gall stones in patients with repeated bouts of hemolytic anemia?

Ans. The circulating bilirubin in these patients is high and thus when a lot more than normal is metabolized by the liver and secreted in to the bile, it gets precipitated in the gall bladder as gall stones.

Q.8. Why does HbF not release its oxygen easily?

Ans. Because its affinity towards 2,3 biphosphoglycerate which helps to release O_2 from Hb is poor. Ability of the HbA to combine with this compound is dependent upon theon β chains which have been replaced in HbFby γ chains. This is also the reason as to why tissue hypoxia gets accentuated in β Thalessemia where HbF predominates.

Q.9. What do you think will be the bleeding and clotting time in such patients (eg. those with Vitamin C deficiency) and why?

Ans. They should be within normal limits. This is because the defect is in the vessel wall, and not in the platelets or other coagulation factors.

Q.10. Why is aspirin contraindicated in patients with dengue fever?

Ans. Dengue suppresses platelet function and numbers. Because of its depressive action on platelet adhesiveness and activation, aspirin will aggravate the problem.

Chapter 12: Pathophysiology of Some Aspects of the Central Nervous System

Q.1. Why is there no retrograde transmission across a synapse?

Ans. There are no receptors for the concerned neurotransmitter on the presynaptic membrane

Q.2. What will happen to the CS Fpressure in the sitting posture?

Ans. It will increase as the hydrostatic pressure is directly dependent upon the height of the fluid column. When measured the CSF pressure is about 40–60 mm Water higher while sitting.

Q.3. What will happen if a Valsalva manouevre is done in a patient with raised ICP and why?

Ans.: The patient will develop a headache because the Valsalva's manouevre blocks venous drainage and increases intra cranial volume suddenly, enhancing the stretch on the dura.

Q.4. In whichposition should you place a road side accident victim who may have had a cervical spinal injury, and what supportive equipment will you use to transport him?

Ans. Lying in a supine position so that the spine may not flex. It is best to use a cervical collar to stabilize the cervical spine.

Q.5. What immediate treatment should be instituted in a patient who has developed autonomic hyper reflexia and what preventive measures should be taken by the nursing staff?

Ans. Immediately remove the offending sensory stimulus eg. Bladder and rectal distension etc. The nursing staff must ensure proper bladder and bowel hygiene in such patients and must avoid vigorous scrubbing etc of the skin while nursing below the level of the transection.

Q.6. Why should a person who has had a syncope be allowed to remain supine?

Ans. The hydrostatic pressure in the supine posture is equal through out the cardiovascular system. This allows the venous return to increase, there by allowing the stroke out to increase helping restoration of the cardiac out put and hence perfusion to the brain. This aids return to consciousness.

Q.7. What side effects can the use of typical anti psychotics have?

Ans. Excessive activation of the extra pyramidal system (dyskinesias) as dopamine availability may be curtailed to very low levels by these D2 receptor blockers.

Chapter 13: Pathophysiology of Common Thermoregulatory Disorders

Q.1. W hatcould be the disadvantage to the patient who drinks a large volume of plain water?

Ans. It is likely to induce hyponatremia. This in turn may cause cerebral edema followed by CNS symptoms.

Q.2. Why is anti-pyretic medication not helpful in heat induced hyperpyrexia or heat stroke?

Ans. Anti-pyretics act by reducing release of PGE_2 which is secreted by the hypothalamus in response to pyrogens. The latter are not involved in hyperpyrexia induced by hot environment exposure.

Q.3. Why is it important to exercise care while treating children with diarrhoea and vomiting with anti-cholinergic medication in the summer months?

Ans. These medications can reduce /block sweating and thus inhibit evaporative cooling which is necessary under the circumstances, especially if the GIT infection is accompanied by fever.

Q.4. Can you explain the possibility that those who do severe physical exertion in sub zero temperatures could develop heat hyperpyrexia/heat stroke?

Ans. In sub zero conditions prevailing in the arctic and high altitude Himalayan regions, the clothing that is worn has very high insulation. The sweat that is produced may not evaporate to cool off the body which is accumulating high degree of metabolic heat that is being generated. Dehydration can occur rapidly.

Q.5. Which common blood parameters will help you to monitor the clinical status of a patient of heat stroke?

Ans. Electrolytes, urea and creatinine; liver function tests; bleeding and clotting time.

Q.6. Increased blood concentration of which hormone explains the low volume and high coloured urine in such people?

Ans. ADH

Q.7.Why does the GFR reduce in severe hypothermia after the initial rise?

Ans. Possible reasons: 1. Reduced perfusion of renal arteries because of poor cardiac output. 2. Aggravation of the former by increased blood viscosity which reduces blood flow velocity.

Q.8. Can you work out why the main pH disturbance in severe hypothermia is a metabolic acidosis?

Ans. This is because tissue hypoxia results in lactic acid accumulation. Added to this in more severe hypothermia is the failure of the renal tubules to secrete H^+.

Index

A

ABGs (102, 104)

ACE (186)

ACE inhibitors (58)

ACI (53)

ACTH (163)

ADH (12)

AIDS (143)

ARDS (78, 101) ARF (22)

ATP (5, 53)

AVP (168)

acid- base balance (17)

acetylcholine (151,153)

acidosis (18)

acceleration (210)

acute renal failure (24, 25)

acute HF (66)

adaptation to stress (175)

adenosine (45)

adrenal medulla (173)

adrenaline

adventitious sounds (94)

age (166)

aging (7)

airflow obstruction (93)

airway resistance (95, 96, 97)

alarm (173)

alcoholism (206)

aldosterone (75)

alkalosis (18)

alpha1 anti trypsin (98)

alveoli (90)

Alzheimer's disease (166)

amplitude (162)

amygdala (215)

anaerobic metabolism (6)

anemia (105, 183)

angina (62)

angiotensin II (49)

anion gap (18)

anorexia (69)

anorexia nervosa (135)

antipyretics (232)

antipyrin (237)

anxiety (200, 216)

aplasia (3)

apoptosis (5, 209)

arterial baroreceptors (64, 75)

arterial hypoxia (102)

arrhythmias (61)

astrocyte (205)

atonic bladder (38)

atrial natriuretic peptide (62)

atrial receptors (12)

atrophy (3)

aural temperature (228)

automatic bladder (37)

autonomic nervous system (212)

auto regulation (45)

B

BER (111)

Bernard Claude (1)

BMI (135)

baro reflex sensitivity (59)

behavioural abnormalities (214)

bilirubin (205)

biphosphoglycerate (185)

blood brain barrier (200, 205, 206)

blood- functions of (178)

blood pressure (54, 55)

blood properties (179)

blood rheology (179)

blood viscosity (179)

body cell mass (143)

body temperature (221, 228)

body water (9)

bone marrow (182)

bone morphogenic proteins (155)

bone remodeling (156)

bronchial asthma (96, 100)

bronchial tree (90)

bronchiectasis (95, 98)

bronchoconstriction (53)

bronchospasm (139)

bulimia (135)

burns (14, 80)

burns classification (80)

burns cardiovascular response (83)

burns immunosuppression (85)

burns metabolic problems (85)

burn zones (82)

C

CAD (53)

CAR (163, 173)

CHF (62, 64)

CKMB (49)

CO2 (102)

COPD (66, 89, 98, 100, 101)

CRH (136, 168)

CRF (22)

CSF (200, 215)

CTZ (112, 113)

CVA (207)

cachexia (68, 143)

cancer (141)

cannabinoid receptors (144)

cardiac index (213)

cardiac output (42, 56)

cardiopulmonary receptors (74)

callus (158)

cellular changes in shock (76) cell death (5)

cell injury (4)

chemoreceptor (65, 105)

chest pain (52, 94)

Cheyne Stokes breathing (67)

choline esterase (154)

chromatin (5)

chronic renal failure (29)

chronic pancreatitis (130)

chronobiotics (167)

chronotherapy (167)

circadian rhythms (161)

circumventricular organs (205)

classification of shock (74)

clock genes (163)

clubbing (94)

coagulation (179, 195)

collateral vessels (46)

collision (210)

coma (78)

compliance (101)

control of respiration (105)

core temperature (221)

coronary blood flow (44)

cor-pulmonale (99)

coronary steal (47)

corrected reticulocyte index (183)

cough (92)

cough reflex (93)

critical closing pressure (78)

cyanosis (52, 65, 70, 94, 98)

cytokines (128, 169, 194)

D

DIC (196)

DLCO (105)

DNa (6)

DOMS (152)

deceleration (210)

defensins (19)

dehydration (227)

delayed sleep phase syndrome 9166)

depression (200, 214)

deuterated water (237)

diaphoresis (53)

diarrhea (114, 122)

diastole (44)

diastolic dysfunction (65)

Diabetes - Insipidus (238), Mellitus (34, 207), diffusion (89, 100, 103)

disseminated intravascular coagulopathy (178)

dizziness (53)

dopamine (218)

dyslipidemia (138)

dysphagia (109)

dyspnea (65, 67)

dysplasia (4)

dysreflexia (212)

E

ECF (9)

ECG (52)

EDHF (45)

EDRF (45)

EMG (152)

ET1 (45)

echocardiography (52)

emphysema (101)

encephalitis (206)

endocannabinoid system (111)

end organ damage (60)

end stage renal disease (29)

endothelial dysfunction (45, 53)

epigastric pain (131)

epilepsy (107, 166)

epithelium (91)

erythropoiesis (181)

esophageal sensations (118)

essential HBP (57)

external environment (11)

F

FEF (100)

FEV1 (97, 99)

FEV1/FVC ratio (99)

FFAs (140)

FRC (94)

FVC (97, 100)

Fahraeus- Lindqvist (181)

fasciculations (153)

fever (221, 229, 231)

fibrin degeneration products (197)

fibroblasts (152)

fibrosis (95, 101)

food intake (136)

force transmission (210)

Frank Starling law (62)

G

G cells (127)

GERD (109, 117)

GFR (22, 24, 29, 235)

GnRh (170)

Gz (211)

gall bladder (138)

gastric motility (120)

gastroparesis (109, 119)

general adaptation syndrome (173)

GIT bleeding (114)

GIT function (110)

GIT hormones (111)

GIT reflexes (112)

GIT symptoms (113)

glomerulonephritis (34)

glucose (11)

glucose transporters (187)

glycogen depletion (151)

growth (171)

growth factors (3)

growth hormone (150)

H

HBP (55, 56, 139)

HPA axis (161, 163)

head up tilt (213)

hemorrhage (14, 74)

heart failure (43, 47, 139)

heat- acclimatization, exhaustion, illnesses stroke, hyperpyrexia (221)

heat exhaustion (225)

heat illnesses (225)

Helicobacter pylori (127)

hematoma (158,201)

hemoglobinopathies (189, 191)

hemophelia (195)

hemolytic anemia (185, 189)

hemostasis (195)

Hb-oxygen saturation (101)

hepatomegaly (68)

hippocampus (174)

histamine (152)

hot cockpit (228)

hydrolytic enzymes (129)

hydrogen peroxide (5)

hydrostatic pressure (14)

hydroxyapatite (155)

hyperbaric oxygen therapy (105)

hyperglycaemia (35, 139)

hyperkalemia (239)

hyperlipidaemia (36)

hypernatremia (15)

hypertensive heart disease (61)

hyperthyroidism (152)

hypertrophy (3, 43)

hyperplasia (3)

hyperpyrexia (227)

hypertonic (11)

hypervolemia (9, 27)

hyperkalaemia (16, 27)

hypertension (34)

hypoalbumenia (36)

hypokalaemia (16, 240)

hyponatremia (15, 16, 27)

hypopituitarism (201)

hypoplasia (3)

hyporeflexia (212)

hypothalamus (174, 222, 232)

hypothermia (221, 233)

hypotonia (212)

hypovolemia (9)

hypotonic (11)

hypotensive haemorrhage (75)

hypoventilation (99)

hypoxia (6, 65, 105)

I

IBD (123)

ICF (9)

IDA (115, 188)

IHD (46, 53, 139)

inflammation (5, 141)

inflammatory mediators in shock (77)

inotropic support (79)

insomnia (172)

Insulin (150)

integrins (192)

interleukin (3, 100)

interstitial fibrosis (105)

intestinal motility (122)

intestinal obstruction (124)

intra cranial haemorrhage (207)

ion pumps (209)

isotonicity (11)

ischemia (6, 53)

J

jet lag (164)

J receptors (53)

JVP (52, 68)

K

kernicterus (206)

Kussmaul's breathing

L

LEMS (153)

LES (116)

LVEDV (65)

Lactate dehydrogenase (50)

lactic acid (75)

lactic acidosis (187)

Lamin A (7)

left ventricular hypertrophy (60)

leptin (135)

leucocytosis (178)

leukotrienes (100)

leukopenia (194)

limbic (170) lipiduria (36)

locus ceruleus (168)

low birth weight and HBP (57)

lungs (90)

lung chest wall mechanics (95)

lymphocytosis (194)

M

MAO (215)

MMCs (111, 119)

MODS (73, 77, 78, 83, 86)

macrophages (90)

malabsorption (131)

Mania (200, 214)

margination (192)

mast cell (100)

melatonin (166)

melena (114)

Meissner's plexus (110)

metabolic syndrome (138)

metaplasia (4)

micro

trauma (152)

motility (109)

multi organ dysfunction (73)

myasthenia gravis (148, 154)

myenteric plexus (110)

myocardial stunning (51)

dysfunction (73)

murmurs (69)

muscle fatigue (151)

muscle fibre types (149)

muscle fibrillation (152)

muscle injury (152)

muscle strain (152)

myocardium (47)

myoglobin (152)

N

Na+/ K+ ATPase (235, 239)

NANC (92)

NMDA (218)

necrosis (5)

nephrotic syndrome (34)

nerve gases (148)

Newtonian fluids (180)

NM junction (148, 153)

nitric oxide (5)

non hypotensive haemorrhage (74)

norms for blood pressure (54)

neurogenic vasodilatation (74)

neuropeptides (173)

O

ORS (124)

obesity (134)

obstruction to airflow (95)

oliguria (68, 78, 227)

organelle (5)

orexins A and B (135)

orthostatic hypotension (214)

orthostatic stress (213)

osmolality (9)

osteoblasts (155)

osteoclasts (155)

osteoporosis (156)

osteopenia (156)

osteoprotegrin (156)

P

PCV (181)

PTH (156)

PaCO2 (97, 102)

PCOS (141)

pH (9, 97, 99, 126)

pain (175)

Paintal AS (53)

pain relief (54)

pancreatitis (109, 129)

PaO2 (97, 9, 102)

parietal cells (126)

Parkinsonism (166)

pathogenesis of heat induced illness (2275)

penumbra (209)

peptic ulcer disease (109)

performance (174)

perfusion (89)

period (162)

pernicious anemia (187)

phototherapy (167)

pineal gland (163)

plasma (179)

plasma renin activity (57)

pleural effusion (95)

pneumocytes (90, 101)

pneumothorax (101)

Polio myelitis (151)

post concussion syndrome (200, 201)

post traumatic stress disorder (216)

potassium (16)

post renal failure (29, 37)

pressure volume relationship (62)

procallus (158)

Progeria (7)

prostaglandins (56, 126)

proteases (6)

protein caloric malnutrition

protein energy malnutrition (142)

proteoglycans (155)

pulmonary circulation (43)

pulmonary edema (68)

pyloric obstruction (109, 121)

pyrogens (232)

Q

Quetelet's index (134)

R

RAA system (57)

RARs (65)

RANKL (155)

ROS (5)

RBC maturation (182)

RBC morphology (181)

red blood cells (178)

regurgitation (69)

renal osteodystrophy (32)

reperfusion injuries (6)

respiratory failure (107)

respiratory muscles (102)

respiratory zones (91)

reticulocytes (183)

rhabdomyolysis (229)

rhythm (162)

rigor mortis (148, 153)

S

SCN (163)

SIADH (202)

STEMI (49)

sarcopenia (134, 141, 150)

Schizophrenia (200, 217)

selectins (192)

set point (227, 232)

shear stress (45, 47)

shock (73,74)

sibling (237)

sickle cell disease (187)

silent abdomen (126)

silent zone (90)

skin temperature (228)

sleep (166)

small airway (90)

somatostatin (171)

space flight (164)

spinal cord injury (200, 209)

spirometry (100, 102)

steatorrhea (129)

stenosis (69)

stress (59, 161, 173)

stress, adaptation to (175)

stress incontinence (38)

stress manifestations of (175)

stress metabolic effects (171)

stroke classification (207)

stroke volume (63)

subarachnoid hemorrhage (207)

surfactant (101)

systemic circulation (43)

swallowing reflex (116)

sweating (14, 227)

T

TBW (237)

TLC (100)

TNFα (63, 100, 140)

TRH (136)

telomeres (7)

telomerase (7)

temperature- body, core (221)

Thalassemia (190)

thermoregulatory disorders (221)

thirst centre (12)

thorax (89)

thyroid function and stress (170)

tissue perfusion(79)

tissue resistance (101)

thromboxane A2 (45)

total body water (9)

transfusion (74)

transient ischemic attack (207)

triggering mechanisms in DIC (194)

tritiated water (237)

troponin I (50)

troponin T (50)

trypsin inhibitor (129)

U

urea (11)

uremia (30)

urinary bladder dysfunction (36)

V

VC (100)

V/Q (104, 107)

valvular heart disease (612)

vasoconstrictors (45) vascular resistance (56)

vasodilators (45,79)

ventilation (89, 101, 105)

ventricular fibrillation (61)

vomiting (14, 112)

W

WBC (178, 192)

W/H ratio (135)

water deficit (13)

Wallace's rule of 9 (82)

work shifts (164)

whiplash injury

X

X-ray (52)

Z

Zollinger Ellison syndrome (127)